MAPS WITH THE NEWS

MAPS WITH THE NEWS

The Development of American Journalistic Cartography

Mark Monmonier

THE UNIVERSITY OF CHICAGO PRESS
Chicago and London

MARK MONMONIER is professor of geography at Syracuse University. He is the author of *Maps, Distortion, and Meaning; Computer-assisted Cartography: Principles and Prospects;* and *Technological Transition in Cartography.*

The University of Chicago Press, Chicago 60637
The University of Chicago Press, Ltd., London

Printed in the United States of America

98 97 96 95 94 93 92 91 90 89   54321

LIBRARY OF CONGRESS CATALOGING-IN-PUBLICATION DATA

Monmonier, Mark S.
Maps with the news : the development of the American journalistic cartography / Mark Monmonier.
p. cm
Bibliography : p.
Includes index.
1. Maps in journalism—United States—History. I. Title.
PN4888.M37M66 1989
070.4'449912—dc19   88-23829
ISBN 0-226-53411-1   CIP

The paper used in this publication meets the minimum requirements of the American National Standard for Information Sciences—Permanence of Paper for Printed Library Materials, ANSI Z39.48-1984.

For Richard Edes Harrison,
whose "unplanned career" in journalistic cartography
enhanced public awareness of the potential of news maps

Contents

Preface and Acknowledgments

This is a book about maps in newspapers, news magazines, television news programs, and electronic news reporting systems. Its title, *Maps with the News*, has a double meaning: news maps not only present geographic patterns useful in understanding news reports, they are also a part of the packaging of news. Appreciation of the news map's dual role as both information and decoration requires examination of the maps themselves and of the technological developments that have shaped them. But it is also imperative to examine journalism as both an industry and a profession. Indeed, the rise of journalistic cartography is related at least as much to development of organized sections, precision journalism, pleasing page layout, and an increased appreciation of color, as it is to machines that can generate maps rapidly and distribute them economically.

The book cuts across several disciplines. In examining the evolution of the form and content of news maps, it contributes to the history of cartography in an area largely ignored by historians of cartography. In documenting the rise of cartographic illustration of news stories and in delving into the changing attitudes held by editors, publishers, and art directors toward maps, it adds to the history of journalism. It contributes as well to the history of technology by studying the effects upon journalistic cartography, information graphics, and news presentation of the technological transition from eighteenth-century wood-block engraving to late-twentieth century microcomputers and telecommunications. In exploring the institutional settings in which news maps originate and the conflicting attitudes among journalists toward news illustration, it is also a study of certain facets of contemporary American journalism. In investigating how timeliness, technology, and artistic trends affect the visual quality of news maps and in analyzing how the news business—largely unaware of its important role—functions as an influential cartographic institution, it is an essay on contemporary cartography. In describing the roles in journalistic cartography of wire services, news syndicates, television networks, newspaper groups, design consultants, and local newspapers, it offers insights useful to those concerned with geographic literacy and the geography of mass communications.

The book evolved from a plan to study more broadly the effects of electronic publishing on cartography. My decision to focus on the news business, and to ignore for the time being both government mapping agencies and commercial map and atlas publishers, reflects a personal curiosity about journalism, as well as an appreciation of both the wide audience for news maps and the important role of electronic publishing in journalistic cartography. Moreover, the mid-1980s have been an exciting time for studying news maps. The print and electronic media have become highly enthusiastic about graphics and adventuresome with computers and telecommunications. Requirements that news maps be up-to-date and sometimes dramatic have given news publishers a mandate to explore uses of mapping software and computer graphics, and to incorporate cartographic displays with electronic page make-up systems, television graphics systems, and electronic information retrieval systems. Although the reality of the mid-1980s has tempered the hyperbole of the decade's beginning, the print and electronic news media are very much aware of both the short-term potential of electronic publishing and the cloudy but inevitable long-term effects of electronic technology upon the collection and dissemination of news.

Maps with the News views the current enthusiasm for computers, electronic displays, and communications satellites from the perspective of past successes and failures. Always secondary to words and pictures of people, news maps have been exploited when technology permitted and when editors and publishers desired. This historical perspective provides a useful glimpse into the future: because maps currently illustrate but a fraction of news reports with geographic themes, the potential for further growth in journalistic cartography is substantial. By providing access to a much wider range of information than now available to newspaper readers and television viewers, electronic technologies such as videotex promise a still-richer array of news maps.

Methodologically the book is varied. Bibliographic research provided most of the information on the historical development of news publishing and cartographic production and dissemination. Much of the writing is descriptive and interpretative, and most of the analysis is graphical. Informed study of the rise of journalistic cartography required the systematic sampling and coding of microfilmed newspapers. I also coded a number of current newspapers and news magazines, and examined many others less formally in order to cultivate a sense of the variety of news maps and how they related to the publication's content and design. Visits to over 25 news organizations, large and small, and unstructured interviews with numerous editors, artists, and art directors provided valuable insights, as did auditing three graduate courses in Syracuse University's Newhouse School of Public Communications. My need to learn more about the news business and its operations led to three short articles on the geography of the newspaper industry and a coauthored study of the design of newspaper weather reports.

The book's principal limitation is its North American focus. As a few examples in the introductory chapter attest, journalistic cartography is by no means an exclusively American phenomenon. An unsystematic survey of foreign newspapers and magazines revealed a substantial use of maps in *Le Monde*, *La Stampa*, and a few other prestigious non-American newspapers. Map use in Third World publications might be described as "occasional to rare." Although the section on newspaper history acknowledges the evolutionary links between the British and the American press, and the section on videotex notes pioneering developments overseas and in Canada, the only news publications outside the United States examined systematically were the *Times* of London, the *Economist*, Toronto's *Globe and Mail*, and *Macleans*. I trust the reader will appreciate the logistical constraints of an even partially comprehensive survey of the foreign press. Yet, through reading and admittedly unsystematic observation, I have assured myself that, aside from videotex systems, principal recent developments are almost wholly American. In a decade or so, I should think, a thorough study of the transfer of electronic publishing to the Third World press should prove most illuminating in its revelation of an increased global use of news maps.

Another limitation is that the illustrations of news maps are not fully representative. To control the cost of publication I have avoided color illustrations; black-and-white halftones provide adequate but less effective reproductions of several maps originally in color, including all of the examples from television and videotex. In addition, to promote reproduction quality, I have omitted illustrations with fine-texture area symbols, which many newspapers use for water or background; even original artwork for large news maps with fine-dot patterns would not withstand reduction from newspaper to book format.

Research for this book was carried out during a year of study made possible by a grant from the John Simon Guggenheim Memorial Foundation. Syracuse University provided additional support, including grants from the Maxwell School's Appleby-Mosher Fund for research travel to London and Canada.

David Woodward, at the University of Wisconsin, encouraged me to combine my interests in the history of cartography and in computer-assisted cartography in an examination of the effects of electronic publishing.

At the Newhouse School of Public Communications, Professors William Babcock, Mario Garcia, and Henry Schulte graciously shared insights on the history of communications, on newspaper design, and on the American newspaper, respectively.

Several newspapers, news magazines, news syndicates, electronic media firms, and equipment manufacturers have generously provided illustrations. In addition to the individual acknowledgments in the figure captions, I would like to express particular appreciation to Don DeMaio of the Associated Press for cheerfully responding to my frequent requests.

Numerous informants in the news business talked freely about their work, their firms, their attitudes, and journalism in general. Respondents interviewed formally are listed in the bibliography. I am especially grateful to Ron Couture, David Driver, Mario Garcia, and Robert Lockwood for extended discussions of trends in newspaper design, and to Richard Edes Harrison for a delightful afternoon discussing his work during the 1930s, 1940s, and 1950s for *Fortune*, *Life*, and *Time*.

In carrying out my research, I used numerous libraries, large and small. I am particularly indebted to the staffs of the Geography and Map Division of the Library of Congress, the St. Bride Printing Library (London), and the Onondaga Historical Association (Syracuse), and to the interlibrary loan staff of Syracuse University's Bird Library.

At the Syracuse University Cartographic Laboratory, Michael Kirchoff and Marcia Harrington were most helpful in photographing the facsimile examples and preparing the pen-and-ink line drawings. At the Production Center of the Syracuse University Advanced Graphics Research Laboratory, Steven Segal executed most of the statistical diagrams, and Kristina Ferris drafted most of the explanatory line drawings.

Finally, for their encouragement and tolerance, I would like to thank my wife Margaret and our daughter Jo Kerry.

MAPS WITH THE NEWS

Nothing tells WHERE better than a map. In the case of storms, fires, murders, floods—people like to orient themselves. They like to know WHERE the action took place.

John H. Sorrells, *The Working Press: Memos from the Editor about the Front and Other Pages*

Where? The place in which an event occurs is necessary, but often minor, information.

Leon Whipple, *Current Events: A Guide to an Appraisal of the News*

1 Introduction

News maps accompany and support words—the written words of the newspaper and news magazine and the spoken words of the television newscast. Words structure information linearly, in one dimension, whereas maps structure information graphically, in two dimensions. Readers and viewers seem more at home with words than with maps, and journalism uses many more paragraphs than maps to convey its facts and opinions, which are largely nonspatial. But "Where?" is a question of journalistic concern because some news has an important geographic component, most nonlocal news at least has a dateline, and many persons have an only rudimentary knowledge of their own city. When the reader's or viewer's mental map is sketchy, a map is an efficient means for showing location and describing geographic relationships.

Although news maps are a small part of the news business, journalism is one of a few industries that provide the public most of its information about places and geography. The other major suppliers of geographic information are commercial and government cartography, censuses and statistical surveys, educational systems, and the travel industry. Of these, journalism is distinguished by the wide availability of its products and its event-driven focus on timely occurrences, often of ephemeral significance. The cartography industry is far more systematic in its coverage, with a bias toward topographic description, but its products often are incomplete or out-of-date. Censuses and surveys are periodic as well as narrowly focused on demographic and economic phenomena. Geographic education is largely concentrated in the preadult years; it can be a highly concentrated, richly rewarding scholastic endeavor but often suffers the neglect of being little more than a boring principal-exports-and-place-names adjunct to "social studies." Travel can yield superb insights but is inherently individual and unsystematic, with significant coverage of foreign areas available only to the deeply committed and the economically elite. Hence most adults receive most of their geographic information indirectly, selected and packaged by journalists. News maps can be an important part of the packaging.

Roles of the News Map

Maps have several uses in journalism. Some news maps are principally informative, others are largely decorative, and many are both. The most common news map is the simple locator map showing a single site and whatever more familiar neighboring features might be needed to serve the reader as a geographic frame of reference. Large-scale locator maps, such as Figure 1.1, often provide a detailed view of the local street grid to show precisely the site of an accident, crime, or proposed real estate develop-

Figure 1.1. A typical large-scale locator map with text block and pointer. Inset maps at a smaller scale are sometimes included to show location of area of detail. [Source: Baltimore *Evening Sun*, 11 February 1985, 1.]

Figure 1.2. Globe inset on this Washington Post foreign-area locator map provides a dramatic, visually interesting accent as well as a useful portrayal of China's proximity to India and the Soviet Union. [Source: *Washington Post*, 13 February 1985, A29.]

ment. An inset map, as at the upper left of Figure 1.1, might be required to portray the location of this area of detail within the county or metropolitan area. Small-scale locator maps frequently refer to unfamiliar foreign areas. As in Figure 1.2, several places might be equally prominent, and an inset map can reveal a region's or small country's geographic situation within the continent. But these examples demonstrate that few locator maps are purely informational: the text block in Figure 1.1 supplements the story's headline, and the eye-catching globe inset in Figure 1.2 signals the story's focus on eastern China.

More cartographically complex and geographically rich news maps may portray routes or the distributions of area or point phenomena. Istanbul, Turkey's *Cumhuriyet* used Figure 1.3 to show its readers the

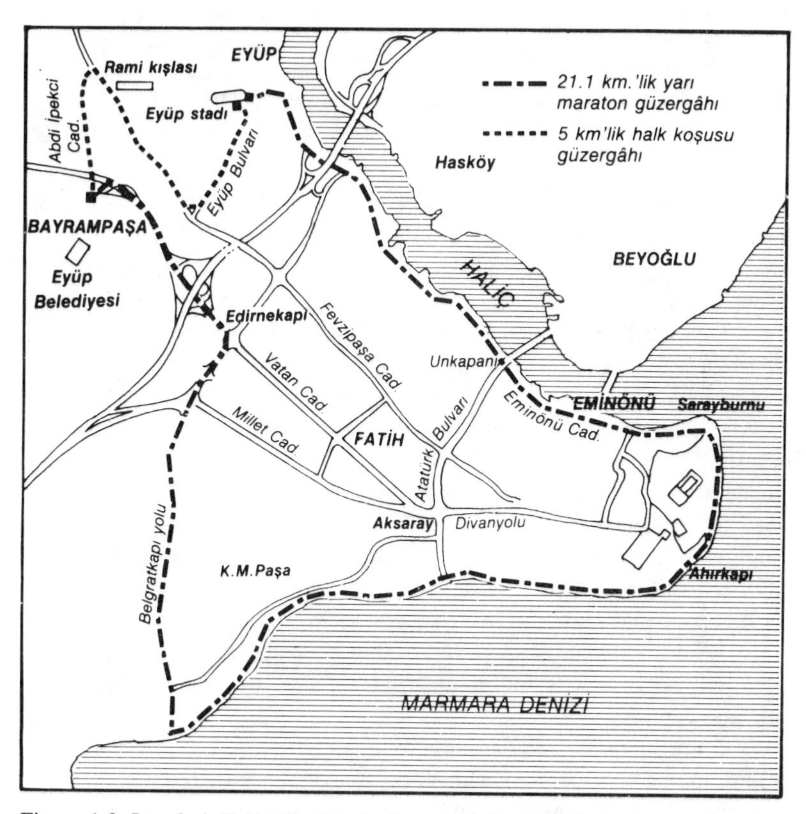

Figure 1.3. Istanbul, Turkey's *Cumhuriyet* used this map on its front page to show where a 21-km (34-mi) marathon around the Golden Horn and a 5-km (8-mi) noncompetitive race would interrupt traffic in Istanbul. [Source: *Cumhuriyet*, 2 May 1987, 1.]

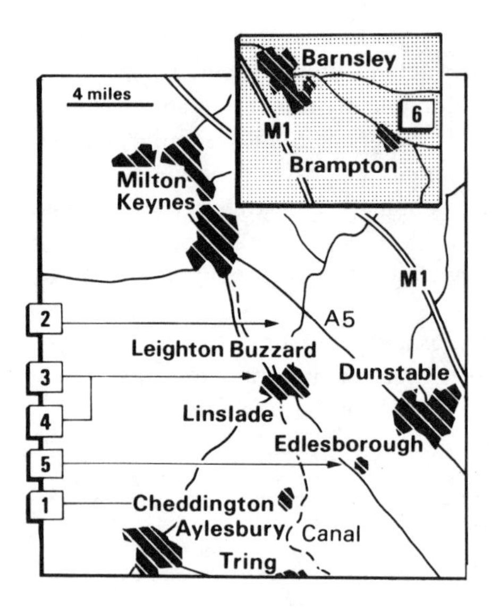

Figure 1.4. The *Guardian*, an elite British national newspaper printed in London and Manchester, used this number-key distribution map to show six sites of sexual assaults perpetrated by a recently captured burglar-rapist. [Source: *Guardian*, 27 February 1985, 3.]

routes of two foot races that would interrupt local traffic the following day. Britain's *Guardian* keyed the numbers on Figure 1.4 to its discussion of six brutal assaults attributed to a burglary suspect. Newspapers and television stations might use maps to summarize election results or to show the distributions of acid rain or radon contamination.

Occasionally a news map is crucial to a story's explanation of a battle, a bombing, a geopolitical strategy, or an environmental threat. Words alone are both more awkward and less dramatic than maps for describing spatial relationships. The *Christian Science Monitor*, for instance, used the near-polar perspective in Figure 1.5 to show Canada's need for northern defenses against the Soviet Union. *Time* magazine employed Figure 1.6 to restage the sequence of events in Indian Prime Minister Indira Ghandi's assassination in 1984 by two Sikh guards. Explanatory maps address two questions: "Where?" and "Why?" Occasionally an explanatory map addressing a comparatively minor news event will serve as a "stand-alone graphic," self-contained except for its title and caption.

Some maps are regular parts of the newspaper. The *Houston Post*, for instance, uses a daily "Road Work" map (Figure 1.7) to warn readers of traffic delays because of highway construction. Other newspapers include fishing maps or maps showing parks and recreation areas as daily or weekly cartographic features. Retailers in large metropolitan areas often supplement their display ads with store-location maps, such as

Figure 1.8. Some newspapers group their classified real estate advertising according to zones described on a standing map. For the reader interested in a particular region, the map's two-dimensional format is usually easier to search than a one-dimensional, alphabetically organized table.

In the newspaper or on television, the most common, regularly appearing news map is the weather map. For those who understand the elements of meteorology, the weather map is an explanatory graphic. The forecast weather map affords a glimpse of how tomorrow's weather will affect comfort or discomfort at home or on a trip. Weather maps offer visual variety as well as spatially organized information. TV weather

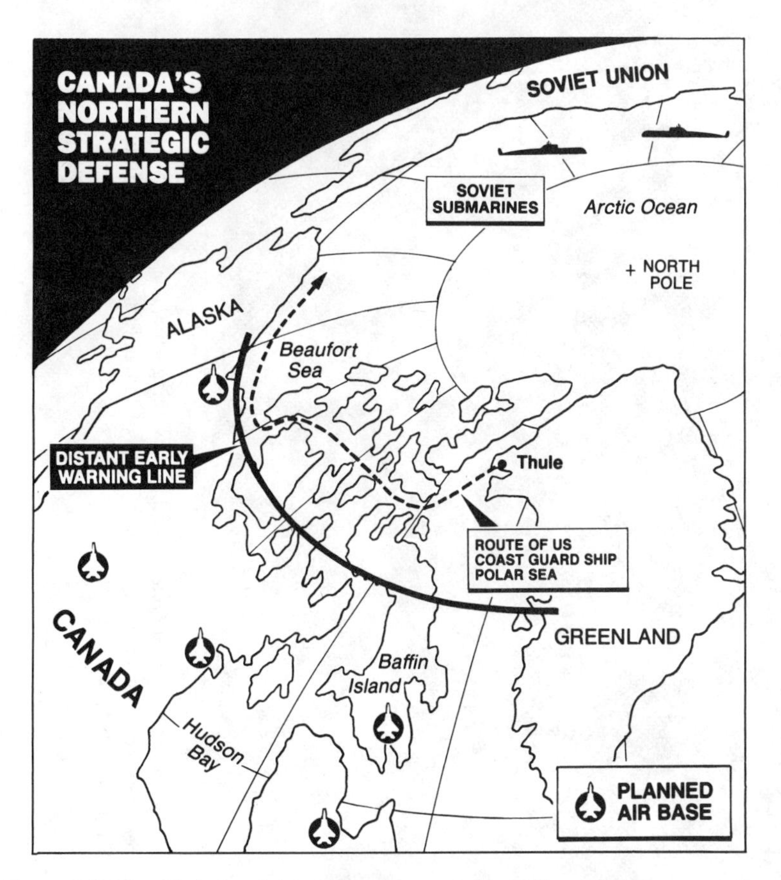

Figure 1.5. The *Christian Science Monitor* used this oblique globular projection to illustrate an article on Canada's plans to enhance its defense against a possible Soviet attack from the north. [Source: *Christian Science Monitor,* 23 March 1987, 9. Forbes in The *Christian Science Monitor* ©1987 TCSPS.]

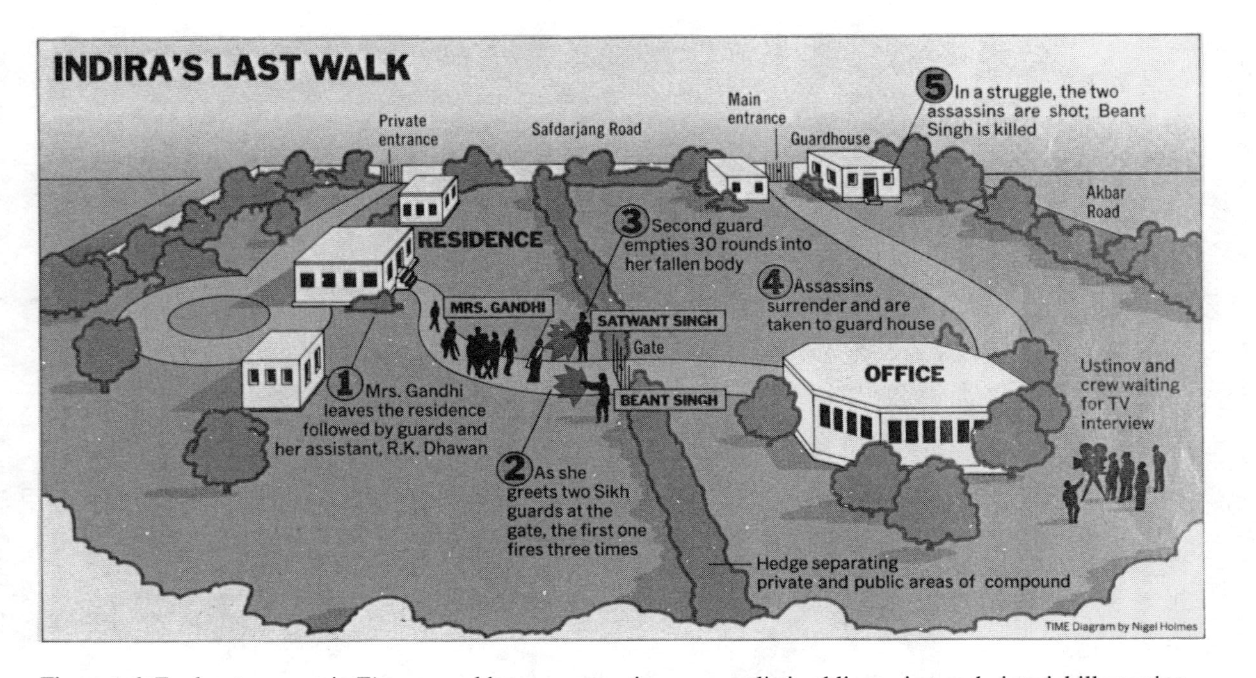

Figure 1.6. Explanatory map in *Time*, a weekly news magazine, uses realistic oblique view and pictorial illustration to describe sequence of events in the assassination of Indian Prime Minister Indira Ghandi. [Source: *Time* 124, no. 20 (12 November 1984), 45.]

graphics are colorful, numerous, dynamic, and spectacular. *USA Today*'s large, colorful weather map is one of the newspaper's hallmarks as well as a service to its readers who travel. Regional weather maps cater to local travelers and, like the *Detroit Free Press*'s Michigan weather map in Figure 1.9, promote the identification of the newspaper with a state or region.

Maps are especially useful in travel sections, where they not only show distances and relative locations but also add "flavor" to an article by allowing the reader to imagine a walk through a colonial city's historic

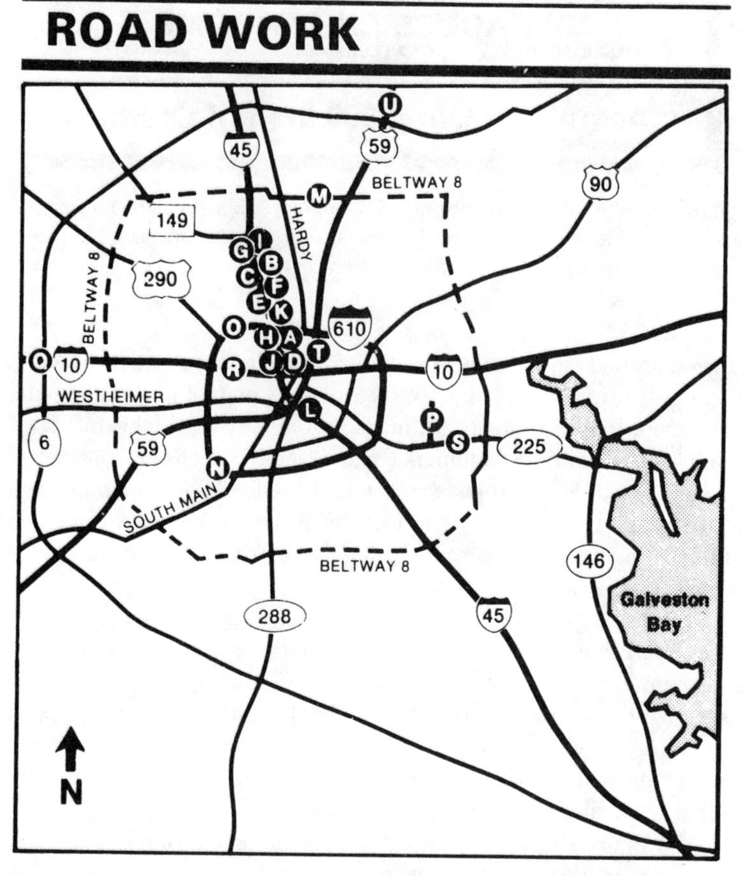

Figure 1.7. The *Houston Post*'s daily "Road Work" map uses letter symbols keyed to a more complete description below the map (not shown here) to identify places and times where construction is likely to delay traffic. [Source: *Houston Post*, 18 July 1985, 4B. Reprinted by permission of The Houston Post.]

Figure 1.8. Silo Inc. uses both simple locator maps (above) and a noncartographic address block (below) at the bottom of full-page newspaper advertisements for its appliance stores. [Source: Courtesy of Silo Inc.]

district, around the grounds of a medieval castle, or among the monoliths of a prehistoric monument. Like *La Stampa*'s map of plazas and palaces in Viterbo, Italy (Figure 1.10), maps accompanying travel stories commonly show multiple landmarks and resemble maps in tourist guide books. A well-designed page in a travel section uses a map as well as headlines, text, and photographs to draw the attention of affluent readers, who in turn attract advertisers. Perhaps more than weather maps, travel maps decorate as well as inform.

Maps add interest and balance to a page layout, especially for the carefully planned front page of the Sunday travel section. Figure 1.11 has as its central visual element a map listing 75 tourist attractions. Yet this large map fits well with the section nameplate, headlines, bodytype, and three small and one medium-size photos that share the page. Page designers find maps useful both as complements to photographs and as substitutes when a suitable photo is not available.

At times a map or maplike artwork is almost purely decorative.[1] Sometimes an old map, compass rose, or cartouche adds an antique flavor to illustrations concerned with buried treasure or the area's early settlers. More common is a geographic shape likely to be recognized by a majority of readers or viewers. Decorative illustrations often contain subtle signals, and a maplike element is ideal for advertising the accompanying story's geographic focus. In Figure 1.12, for instance, artist Jean-Francois Allaux blended Italy's familiar boot-shape with the image of a smoldering ruin to provide a cartographic icon for an article in the *Atlantic* on Italian response to radiation from the nuclear plant explosion at Chernobyl, in the Soviet Union. The result not only breaks up the monotony of the type but also offers a clever juxtaposition of icons.

Less impressionistic cartographic icons provide numerous useful place-related signals.[2] In television newscasts, title frames with national

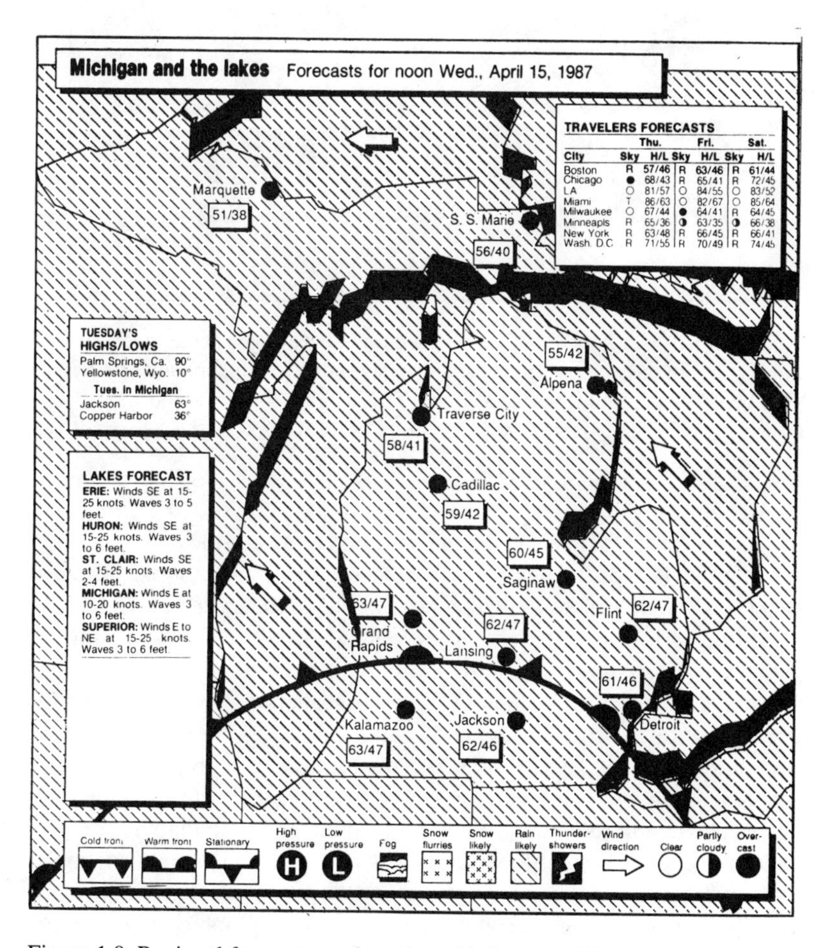

Figure 1.9. Regional forecast weather map with focus on Michigan reinforces the *Detroit Free Press*'s image as a regional newspaper. Dashed parallel-line area symbol for "Rain likely" provides a visually effective, albeit somewhat dismal, portrayal of the forecast. [Source: *Detroit Free Press*, 15 April 1987, 2A.]

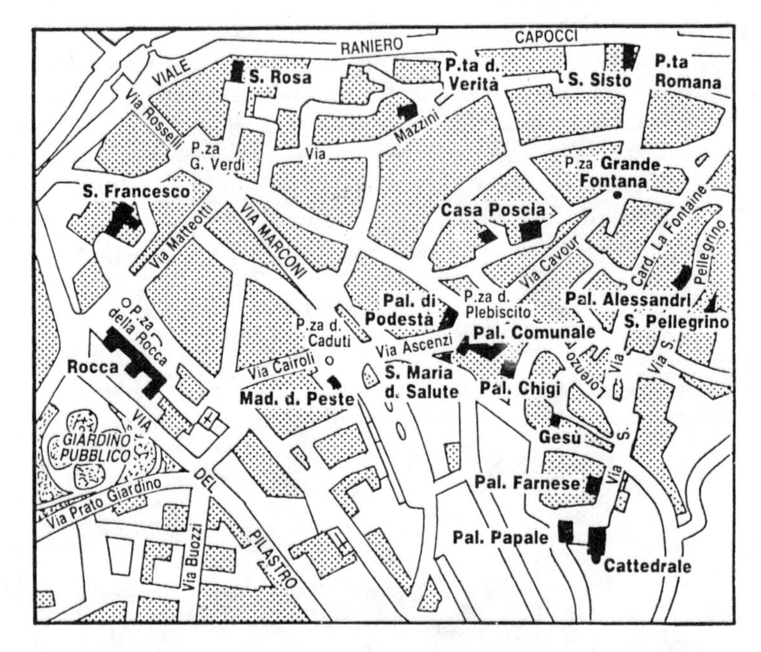

Figure 1.10. Turin, Italy's *La Stampa* used this cartographic centerpiece for a travel story on the plazas, churches, and palaces of Viterbo, a popular tourist site about 80 km (50 mi) north of Rome. Etruscan in origin, Viterbo was a refuge of the Popes from the twelfth century onward. [Source: *La Stampa*, 9 April 1987, 2.]

or state outlines herald a change in locale and reinforce the news reader's fleeting mention of the news item's place of occurrence. Some newspapers use a simple map outline like the national boundary in Figure 1.13 to identify a series of geographically grouped news briefs—short, single-paragraph news stories. These cartographic cues contribute to the newspaper's structure by helping regular readers locate quickly short items of world, national, state, and local news.

Locator maps, in a sense, play a dual role as cartographic icons, by supplementing the headline and sending the reader a strong signal of place. The iconic role would seem paramount when the map has only a vague geographic focus, as in Figure 1.14, a very general map of Afghanistan dominated by the name of the country. *Pravda* used this map principally to signal the accompanying story's geographic focus, rather than to locate Kabul or show Afghanistan's neighbors. But cartographic icons can be far smaller, and a bit more subtle, as the *Times* of London demonstrated with the half-column-wide representation of the European Economic Community, or Common Market, in Figure 1.15.

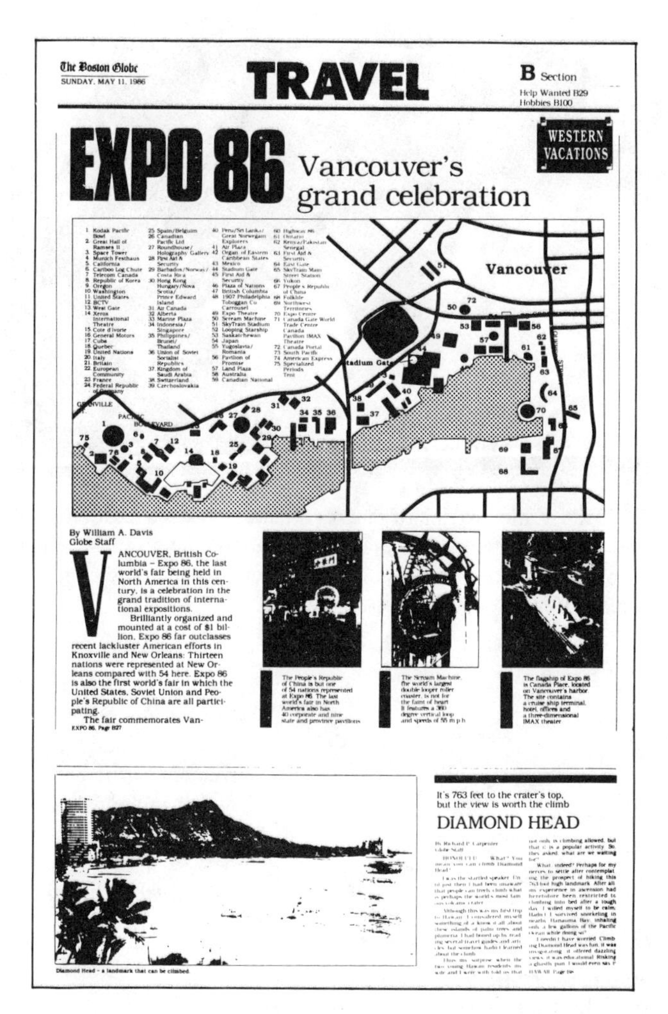

The Boston Globe
SUNDAY, MAY 11, 1986

TRAVEL

B Section
Help Wanted B29
Hobbies B100

WESTERN VACATIONS

EXPO 86 Vancouver's grand celebration

By William A. Davis
Globe Staff

VANCOUVER, British Columbia – Expo 86, the last world's fair being held in North America in this century, is a celebration in the grand tradition of international expositions.

Brilliantly organized and mounted at a cost of $1 billion, Expo 86 far outclasses recent lackluster American efforts in Knoxville and New Orleans: Thirteen nations were represented at New Orleans compared with 54 here. Expo 86 is also the first world's fair in which the United States, Soviet Union and People's Republic of China are all participating.

The fair commemorates Van-

EXPO 86, Page B27

The People's Republic of China is but one of 54 nations represented at Expo 86. The last world's fair in North America also has 40 corporate and nine state and province pavilions.

The Scream Machine, the world's largest double looper roller coaster, is not for the faint of heart. It features a 360 degree vertical loop and speeds of 55 m.p.h.

The flagship of Expo 86 is Canada Place, located on Vancouver's harbor. The site contains a cruise ship terminal, hotel, offices and a three-dimensional IMAX theater.

Diamond Head – a landmark that can be climbed

It's 763 feet to the crater's top, but the view is worth the climb

DIAMOND HEAD

Figure 1.11. Map by Deborah Perugi, showing locations of 75 tourist attractions in Vancouver, serves as the visual centerpiece for a section front. [Source: *Boston Globe*, 11 May 1986, B1. Reprinted courtesy of *The Boston Globe*.]

Figure 1.12. Illustration by Jean-Francois Allaux invokes a cartographic flavor appropriate to a correspondent's report from Italy on the official and popular response to radiation from the Chernobyl accident. [Source: *The Atlantic* 259, no. 1 (January 1987), 30.]

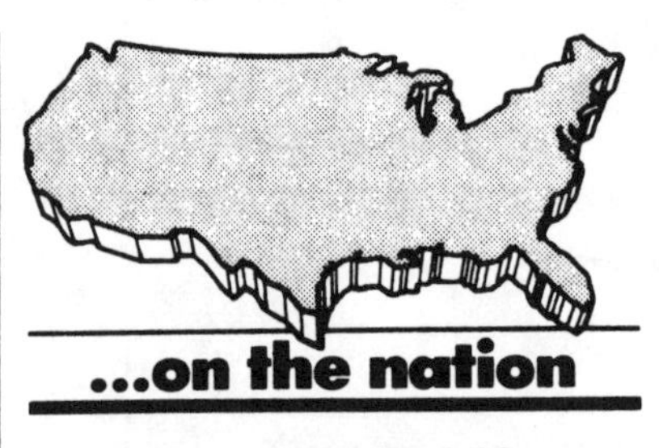

...on the nation

school bus carrying 60 children slid off a highway into a ditch and overturned, injuring eight passengers.

Doctors bail out: **Faced with a growing number of**

Figure 1.13. A simple outline map of the United States is a convenient cartographic icon for identifying national news briefs—short, one-paragraph stories. [Source: Baltimore *Evening Sun*, 21 February 1985, A2.]

Figure 1.14. The Soviet newspaper *Pravda* used this map of Afghanistan and its neighbors to illustrate a story discussing alleged attempts by China, Pakistan, and the United States to undermine the "rightful government" in remote parts of this Soviet-occupied country. [Source: *Pravda*, 28 January 1980, 6.]

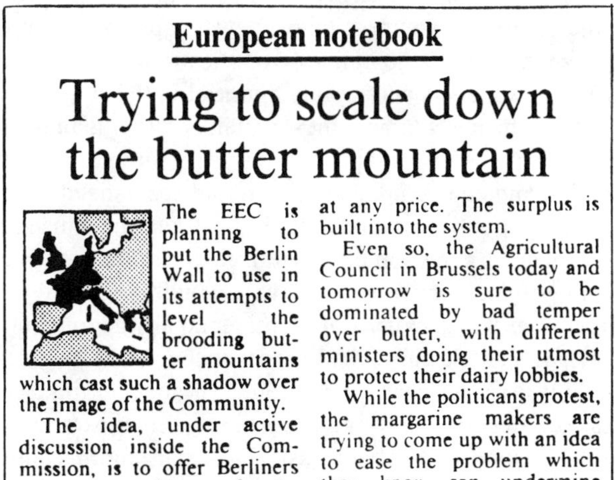

European notebook

Trying to scale down the butter mountain

The EEC is planning to put the Berlin Wall to use in its attempts to level the brooding butter mountains which cast such a shadow over the image of the Community.

The idea, under active discussion inside the Commission, is to offer Berliners two packets of butter for the [illegible] for a six-week trial [illegible]

at any price. The surplus is built into the system.

Even so, the Agricultural Council in Brussels today and tomorrow is sure to be dominated by bad temper over butter, with different ministers doing their utmost to protect their dairy lobbies.

While the politicans protest, the margarine makers are trying to come up with an idea to ease the problem which they [illegible] can undermine their [illegible]

Figure 1.15. A small cartographic icon identifies the European Economic Community, or the Common Market, as its story's geographic focus. [Source: *Times* (London), 25 February 1985, 7.]

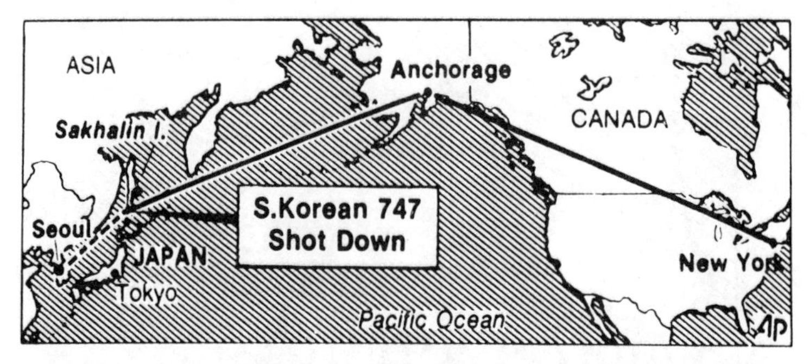

Figure 1.16. Associated Press Laserphoto map showing bend in route of ill-fated passenger airliner KAL 007, shot down while off-course over the Soviet Union in September 1983, reflects an unwise choice of a standard, equatorially centered cylindrical base map. A carefully curved line might have depicted the plane's route more accurately on this projection, whereas a gnomonic projection would have allowed a straight line, or one nearly so, to provide an accurate representation. [Source: AP map published in the Syracuse *Herald American*, 14 July 1985, Stars magazine section, 11.]

News Maps As a Cartographic Genre

In design and content, maps used in the news media constitute a distinctive cartographic genre. Unlike most maps, news maps are a comparatively minor element in a larger, more diverse publication that is largely textual in the case of newspapers and largely pictorial in the case of television. Because their audience is large—most certainly larger than for any other cartographic genre—news maps generally are simple in content and symbolization, and can be understood without specialized training. Unfortunately, news publishers tend to hire artists untrained in cartographic principles, and news maps sometimes reflect an ignorance of map projections or cartographic conventions. As an example, Figure 1.16 portrays with straight lines a near-polar transcontinental airline route more accurately shown as a gentle curve; had the artist chosen the projection more knowledgeably, a single, nearly direct route, without the kink, would have been possible. Because they must be timely in illustrating breaking stories, news maps often reflect haste in design and execution. Wire services and graphics syndicates prepare many news maps and distribute them electronically to client newspapers, whose editors often have a large array of maps, photographs, features, and news stories from which to choose a few to fill the paper's comparatively small "news hole." Publication of a news map reflects not only a belief in the news value of the story it accompanies but often the absence of a suitable photograph.

News maps in print differ substantially from those on television, reflecting differences between these two media in format and image creation. In newspapers and weekly news magazines, the basic planning unit is the page, which calls for a balanced layout, and which imposes the structure of a grid usually four, five, or six columns wide. A news map might illustrate a particular story, but it also is a part of the page layout—which might account for the presence of the map as well as govern its size. News maps also reflect printing quality, which varies from the blotchy, over-inked graytones of obsolete letterpress printing, on the coarse newsprint of some ailing big-city daily newspapers, to perfectly registered full-color offset lithography on the smooth, expensive paper of mass-circulation weekly news magazines. Advances in offset printing have made color graphics more common in newspapers, and the map provides an opportunity to add controlled amounts of color to the page.

Newspapers, which have used computers for more than two decades for typesetting, are now finding electronic workstations efficient for page layout, picture enhancement, and graphic illustration. Enthusiastic adoption of laser printers, microcomputers, and simple graphics software has led to numerous news maps like Figure 1.17, which bears the imprint of the MacDraw software's "drawing tools" and "menus" in the same way that, say, a painting reflects the artist's brushes, paints, and canvas. The easy movement of electronic maps and charts over telephone lines, from graphics syndicates and wire services such as the Associated Press, as well as among chain newspapers in distant cities, further encourages the use of computer-produced news maps.

Television maps also reflect their parent medium. All viewers of a news broadcast receive the same programmed sequence of images. On the screen for less than a minute, TV news maps inherently need to be simpler than maps in the print media. Screen size, shape, and resolution constrain format, detail, and use of labels. Like all TV graphics, news maps are broadcast in color, but many viewers see them in black and white because of color blindness or lack of a color receiver. Poorly tuned or deteriorating receivers and various sources of electronic interference add distortion. Ranging in complexity from simple cartographic icons used in title frames to elaborate, information-rich weather graphics, TV maps also include wall-size outline maps decorating studio sets used for news telecasts, as Figure 1.18 illustrates.

Electronic display makes possible map symbols that blink and move, and computer graphics systems can generate animated maps. Video display technology and modern telecommunications offer the potential for linking highly dynamic cartographic display software with large geographic databases. Videotex, a type of interactive information retrieval and display system, can generate a data map in a carefully programmed sequence, focusing the viewer's attention, in succession, on the title, the map base, the classification, the high-value areas, the low-value areas, and the average areas. Should electronic news databases ever emerge as a mass-communications medium, dynamic news maps would become commonplace.

As a cartographic genre, news maps reflect the objectives and values of journalistic institutions. They are timely and narrowly focused, and present highly selective views of whatever portions of the globe

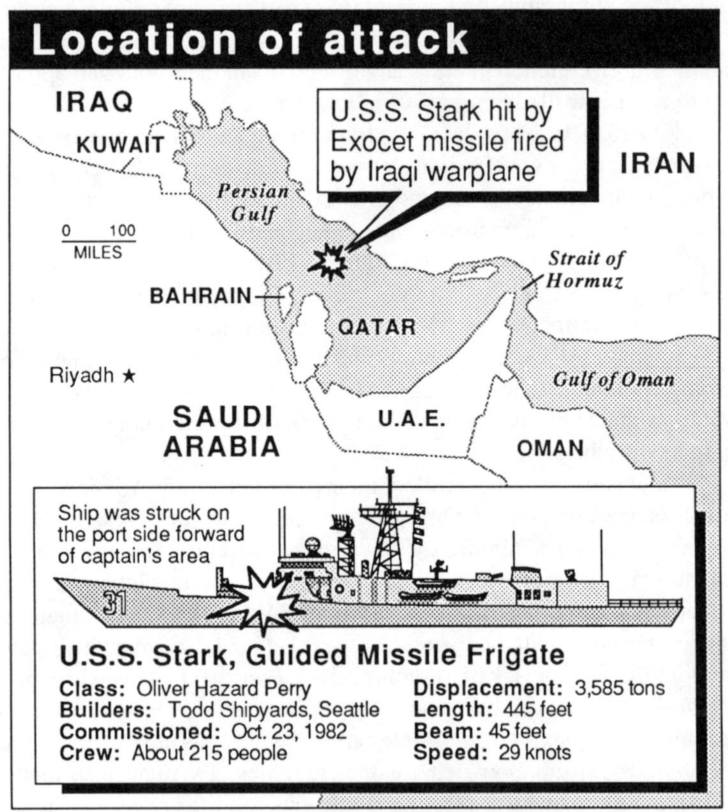

Figure 1.17. The Knight-Ridder Graphics Network used an Apple Macintosh microcomputer to produce this intricate information graphic describing the U.S.S. Stark, attacked by an Iraqi fighter plane over the Persian Gulf. [Source: Syracuse *Post-Standard*, 19 May 1987, 1. Reprinted courtesy of Knight-Ridder Newspapers and the Knight-Ridder Graphics Network.]

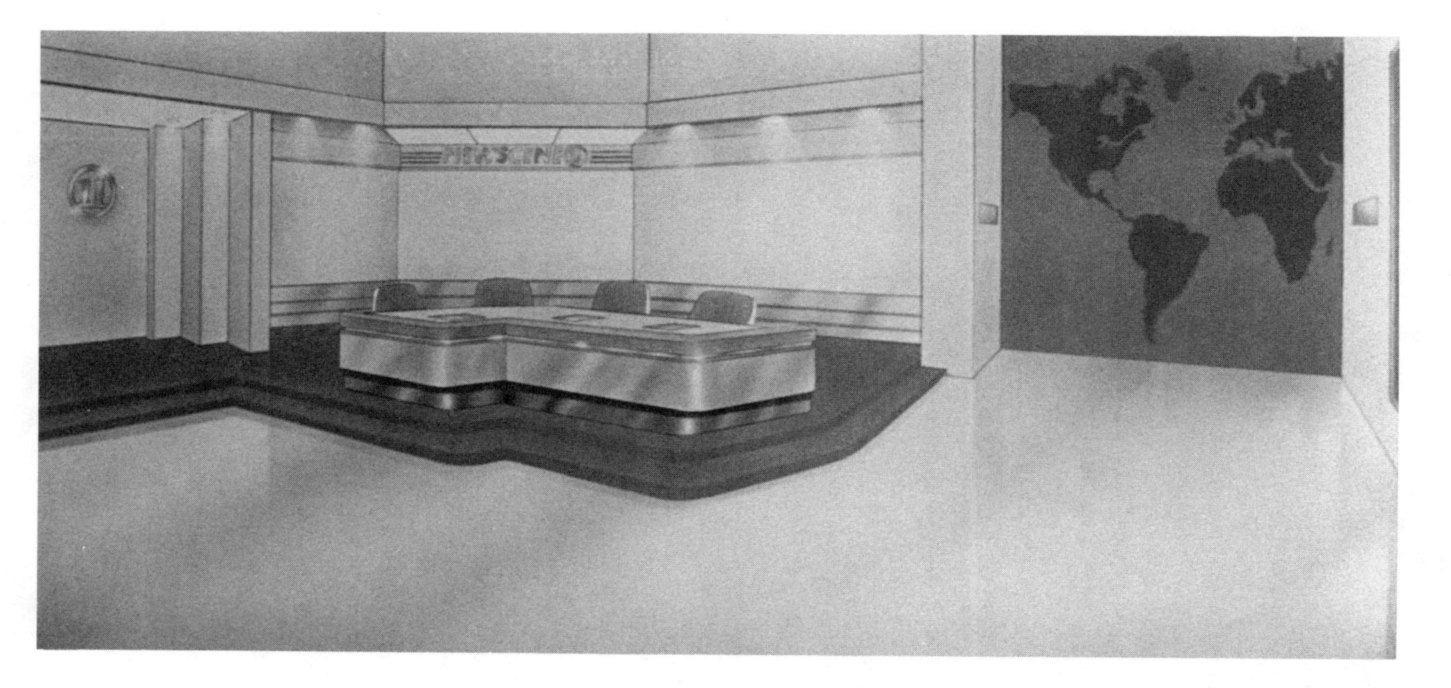

Figure 1.18. Wall-size map decorating television newsroom promotes a creditable image for a local TV news program. [Source: Courtesy of G & G Designs/Communications.]

editors deem noteworthy. Indeed, the press accounts for far more expository maps than the writings of professional geographers. Since "newsworthy" often means "sensational," though, news maps, more than any other cartographic product, tend to focus on crime, war, accidents, and natural disasters. Yet in a profession with an inherently greater respect for words than for pictures, the news map has been somewhat of an outcast—lumped with photos, illustrations, cartoons, and stock charts, and drawn in the "art department" by nonjournalists. Despite the efforts of a minority of journalists who frequently have seen the wisdom of using maps to locate places and explain spatial relationships, the greatest impetus for an increased use of news maps has been the newspaper industry's concern with the appearance and packaging of its product. News publishing is much more than the profession of journalism—it is a business, and information graphics has become an arena in which newspapers compete with each other as well as with the electronic media.

As a cartographic type, news maps are unique in their vast audience. In the United States, for instance, over half the adult population regularly watches evening television news programs, and 60 percent are regular newspaper readers.[3] No other cartographic genre has as large—or as captive—a following, although road maps and other way-finding guides also have a large constituency.[4] Regular use of news maps on local newscasts, almost all of which feature a variety of weather maps, firmly establishes the genre's ascendancy. Moreover, the variety and number of news maps, as well as their timely presentation to the reader or viewer, balances their relative simplicity, in establishing the importance of journalistic cartography.

That cartographic scholars have only recently begun to examine news maps might seem surprising. Yet the five-volume *Bibliography of Cartography*, compiled at the Geography and Map Division of the Library of Congress and covering the period 1875 through about 1972, lists a single study: Walter Ristow's 1957 essay on maps of World War II and the cold war in daily newspapers and weekly news magazines.[5] Moreover, a two-volume supplement covering 1972 through 1979 includes not a single reference to "news maps" or "journalism."[6] This neglect might result at least in part from a professional disdain for overly simple, seemingly trivial maps.[7] Academic cartographers, who direct their research toward visually complex data maps, aesthetically impressive atlases and thematic maps, and extensive series of topographic and image maps, easily ignore the cumulative importance to society of individually unimpressive news maps.

The Study of News Maps

Journalistic cartography warrants scholarly attention for a variety of reasons. As a focus for humanistic study, news maps reflect the influence of technological change on both map publishing and journalism. As a focus for linguistic studies, journalistic cartography has much to reveal about the iconography of the print media and the need for the two-dimensional language of graphics to supplement the one-dimensional language of words, sentences, and paragraphs. As as focus for social science, news maps are important as a source of geographic information for the majority of citizens—and hence as an important influence on the development of perceptions of foreign, domestic, and local problems. As an institution, the news media are society's most significant cartographic gatekeeper and its most influential geographic educator.

The effect of technological change on the map image and its information content is an important focus in the history of cartography.[8] Journalistic cartography reflects a range of technical developments at least as wide as that for any other cartographic genre, including topographic maps and land-cover maps.[9] News maps reflect not only significant advances in printing and engraving technology but also the adoption of telegraphy for the rapid acquisition of geographic data and, later, for the distribution of pictorial images. In the 1980s, electronic publishing has had a particularly significant effect on the number, variety, and appearance of news maps. A phase of centrally prepared news maps, linked to the initiation of the Associated Press Wirephoto network in 1935, is being modified by electronic distribution systems, no longer captive to the needs of halftone photographs, and by microcomputer graphics workstations, with which newspapers conveniently can customize network graphics. Television news broadcasts, cable news and weather programs, and two-way electronic information retrieval systems all employ timely news maps with a form and content unique to these media. Whatever journalistic media evolve in the next century to expedite the delivery of news to homes—and to replace the mass-circulation, centrally manufactured newspaper—surely will have further substantive effects on the appearance and utility of news maps.

News maps also reflect massive institutional changes in news publishing. The vertically structured, monotonously gray appearance of the first front page of the New *York Times* (Figure 1.19) lacked not only news maps but other types of illustration as well. Both technology and professional mores conditioned its editors to describe events in words alone, and geography was evident only in story datelines. The front page

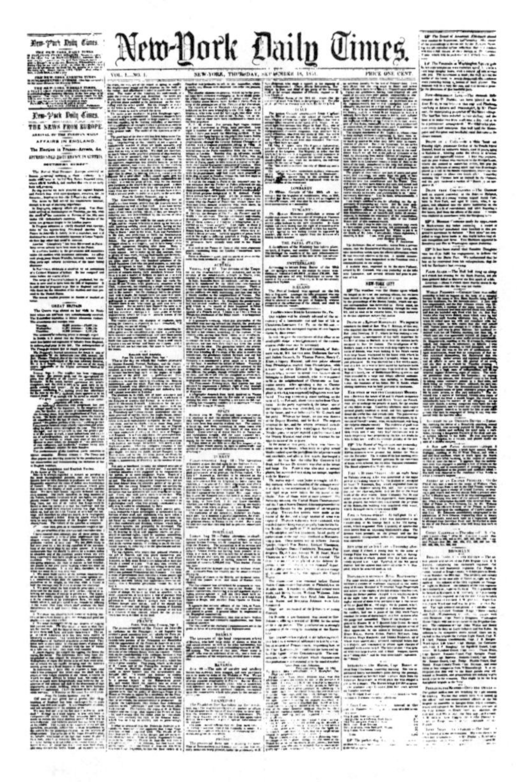
New-York Daily Times.

Figure 1.19. Front page of the *New York Times*'s first issue, for September 18, 1851, reflects the gray, vertical appearance of most nineteenth-century newspapers. [Source: Augustus Maverick, *Henry J. Raymond and the New York Press for Thirty Years* (Hartford, Conn.: A. S. Hals and Co., 1870), foldout opp. p. 88.]

of a recent issue of the New York *Daily News* (Figure 1.20) is the graphic antithesis of its mid-nineteenth century counterpart. Although few other present-day front pages are as flamboyant, headlines have become longer and bigger, pictures are commonplace, and news maps are moderately frequent in most newspapers and very common in some. These changes reflect both a gradual evolution in newspaper design and a more recent revolution, starting in the early 1970s and spurred by competition from television. Newspapers have expanded their art departments, invited designers to editorial meetings, hired artists specifically to draw maps and charts, and installed graphics editors to coordinate the creation of news illustrations with the needs of reporters and editors. The revolution in newspaper design has been a threefold revolution involving systems, editorial marketing, and attitudes. As a revolution in systems, it has displaced typesetters, engravers, and other skilled technicians, and has given editors, writers, and managers a fuller control of the newspaper and its content. As a revolution in editorial marketing, it has broadened the scope of editorial content with a wider use of so-called lifestyle features and has led to a greater concern for the organization, packaging, and appearance of the newspaper as a consumer product. As a revolution in attitudes, it has occasioned an increased appreciation of graphics, especially maps and other information graphics, among senior editors and publishers as well as among many writers. Indeed, substantial investments in art staff and computer graphics workstations would have been

GM/HUGHES IN MEGAMERGE

Story on Page 51

Manhattan
★ ★ ★ ★
Racing
Final

DAILY NEWS

30¢ Thursday, June 6, 1985 NEW YORK'S PICTURE NEWSPAPER® Morning clouds. High 75. Details p. 2

HONOR STUDENT SLAIN IN W'CHESTER

The half-nude, bound and gagged body of high school honor student Mana Azevedo, 17 (photo right), was found in a Westchester County motel room, police said. She had been strangled. Page 3

In downpour, Koch orders water curbs

Page 3

Worldwide heroin ring smashed here

Page 5

Claus begged ex-mistress not to testify

Page 2

Figure 1.20. The New York *Daily News* used a pair of maps to locate the scene of a brutal murder. Victim's photograph is the visual focus of this front page, which is nicely balanced with maps at the left, promo boxes at the right, and headlines and nameplate above. [Source: New York *Daily News*, 6 June 1985, 1. © 1985 New York News Inc. Reprinted with permission.]

unlikely without a respect for information graphics among professional journalists at all levels. An examination of the rise of journalistic cartography thus must study institutional change as well as technological change.

The occasionally enthusiastic but often slow and sporadic adoption of news maps holds a perplexing revelation about how well the media treat geographic aspects of news. Are maps necessary in news reporting, and if so, for what types of story? Is a low level of map use in the past an implicit indictment of previous standards in journalism? Is the higher current level of map use evidence for substantive improvement in news reporting, or is it merely meretricious cartographic decoration of newspapers and television broadcasts? Although the present study makes no attempt to measure the effectiveness of news maps in instilling insight and molding public opinion, it will reveal considerable variation in map use among the newspapers studied.[10] This variation suggests a minimal consensus in the journalistic community about the importance of geographic information to an understanding of news, particularly foreign news. It might also reflect a preoccupation with personalities and an insider's overestimation of the reader's grasp of location, place, and basic geographic relationships. Yet particularly noteworthy is the frequent and prominent use of maps by such elite newspapers as the *Christian Science Monitor* and the *New York Times*, which attempt to offer their readers not only the immediate facts of the news but also an understanding of causes and possibilities. Also noteworthy is the fuller cartographic coverage of metropolitan news by a growing number of daily newspapers, including the prestigious *Washington Post*. Exemplary map use by these newspapers is a convincing normative argument that electronic technology finally has enabled the media to add the maps their stories often require. Moreover, if the print media are to differentiate their product, beyond the headlines, simple images, and glib quotes delivered more efficiently by television, news maps will become a significant part of the "precision journalism" many media critics believe essential to the newspaper's vitality.[11]

Alarming studies of geographic ignorance point to a substantial vacuum that the news media might address, and thereby demonstrate the fallacy of assuming that most persons have a mental map suitable for understanding the significant geographic components of many important news stories. Recent surveys have focused upon high school and college students, many of whom know surprisingly little about nearby areas.[12] For example, 36 percent of Boston high school seniors could not list the six New England states, and a quarter of seniors in Dallas could not identify Mexico as their nearest foreign neighbor. Among North Carolina college students, 65 percent could not associate the Seine River with France, and more than 92 percent did not know that Madras was in India. On a map test administered to a sample of college freshmen in Indiana in the mid-1980s, 95 percent failed to identify Vietnam. Gallup polls in 1947 and 1955 revealed substantial geographic ignorance among American adults, among whom 28 percent could not identify Italy two years after World War II and more than half could not correctly locate New Jersey and Massachusetts in the mid-1950s.[13] No doubt the neglect of geographic education in the nation's elementary and secondary schools underlies much of this ignorance.[14]

With their large, regular audiences of readers and viewers, newspapers and television news programs play a fundamental role in adult geographic continuing education. They are the prime gatekeepers in communicating facts of location and place, and they are also the principal cartographic gatekeepers, potentially able to deliver when appropriate the maps needed to understand or interpret important world, national, and local events. Atlases and other cartographic resources abound in libraries and some bookstores, but even in homes with a good collection of cartographic references, the effort required to consult them frequently is a significant deterrent. As Herbert Simon points out, ours is an information-rich world, and the truly scarce resource is public and individual attention.[15] The media have that attention, and how they use it should be important to geographic educators and concerned citizens at all levels. The National Geographic Society recognized the significance of the press when, as part of an outreach effort, it formed a news features service providing newspapers with free articles illustrated with such rich, carefully crafted maps as Figure 1.21.[16]

Media maps also educate the public in the advantages of maps in general. Readers frequently exposed to media maps—and encouraged to examine these maps by the writer's directive, the attractiveness of the map's symbolization, or their own strong interest in the story—might tend not only to appreciate maps as spatially organized information but to expect them, particularly since newspapers have demonstrated that maps, like photographs, are now easy to provide. Thus newspapers, news magazines, and television news stimulate a need for formal and informal education in "graphicacy," the graphic equivalent of skills in reading, writing, computing, and speaking.[17] Although the inherent simplicity of most news maps might present a poor or misleading image for maps in general, journalistic cartography fosters a focused approach whereby the map becomes either a largely independent graphic essay or an important graphic extension of a verbal essay, rather than a graphic encyclopedia or database. In this sense, news maps are an impetus for a fuller appreciation of the map as an instrument of communication and as a medium for creative enlightenment.

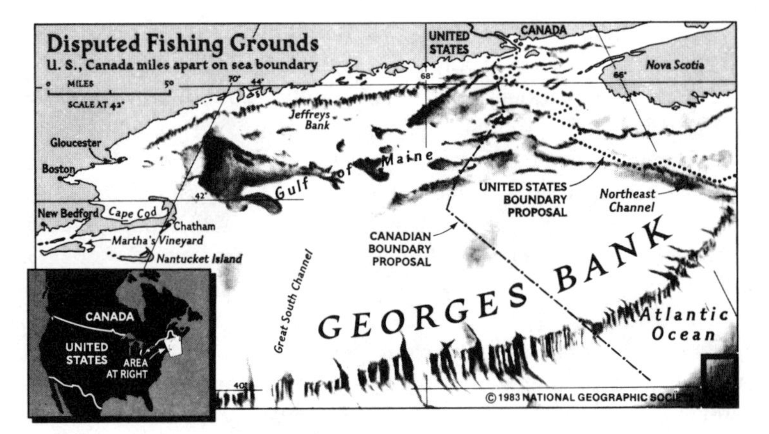

Figure 1.21. Map provided with feature story distributed free to newspapers in 1983 by the National Geographic Society's News Service Division employs shaded relief symbols and oblique perspective in a striking cartographic portrayal of Canadian and United States claims to Georges Bank, argued before the World Court in the early 1980s. [Source: Courtesy National Geographic Society.]

Organization of the Book

News maps are subordinate to the newspapers, the news magazines, and the television newscasts in which they appear, and an understanding of their evolution and importance is impossible without an understanding of the media in which they are embedded. The next chapter examines the rise of journalistic cartography within the context of press history and the history of printing technology, with a focus on elite newspapers and weekly news magazines. Chapter 3 explores the role of syndicates and wire services in making news maps and weather maps widely available to smaller daily papers with limited resources. Chapter 4 addresses a number of important institutional issues, including the role and organization of editorial art departments, journalists' and artists' attitudes toward cartographic art, and the growing acceptance of microcomputer graphics workstations for drafting news maps and customizing wire-service graphics. Chapter 5 studies the unique graphic limitations and strengths of the electronic media—the mass communications of broadcast television and the demassified information delivery of videotex. Chapter 6 concludes with an appraisal of the likely effects of future technological and institutional change on newspapers and news maps.

2 Engraving Technology and the Rise of Journalistic Cartography

Technology for printing and engraving controls the timeliness and visual appearance of news publications. Reproduction technology is particularly important for news maps because how an image is reproduced constrains the method of composition, and hence its information content and visual effectiveness. Improved printing technology has lowered per-copy cost and made possible large circulations, thereby increasing the number of readers who might benefit from a particular map. Production time, skill level, and cost of preparing the map have fallen as well, and news graphics have become more common. By promoting the integration of maps and charts with bodytype and headlines, modern printing technology has created a fuller, more important role for the map in news reporting and news analysis.

This chapter examines the development of reprographic technology and its effects on journalistic cartography, and traces the evolution of the newspaper from the small, sporadic news pamphlets of the early seventeenth century to the contemporary print media, with their colorful, often lavish graphics. Particularly significant advances in the technology of image transfer and multiplication include the steam-driven rotary printing press, which made *mass* communications possible, and photoengraving, which replaced the painstaking artistry of the wood-block engraver with the expedient delineation of the line-block illustrator. Although introduction of speedier, less costly means for making news maps did not lead automatically to their widespread adoption, technological innovation was a prerequisite to modern journalistic cartography. Only after improvements in pictorial printing made it possible for journalists to seriously consider using news maps with their stories did political and economic events lead to an increased use of news maps.

Printing and the Origin of the News Business

In printing, an image typically is transferred from an inked surface, which is impressed upon a sheet of paper. *Block printing*, developed in China in the sixth century, uses a raised image surface that is linked with a roller or cloth.[1] When pressed against a sheet of paper, the inked block defines a reversed image on the receiving surface, and the nonimage areas remain uninked. Block printing was first used in Europe about 1420 for illustrations in hand-copied religious books.[2] The invention of movable type in the mid-fifteenth century by Johann Gutenberg, a German goldsmith, was a major innovation because word messages could be composed for printing by selecting the appropriate sequence of type characters and setting them, letter-by-letter and line-by-line, in a frame for inking and impression.[3] Individual letters were cast in a wrong-reading form and by pouring molten lead into right-reading brass molds. Called *relief printing* because of its raised image, or *letterpress* because of its association with type, this strategy for image transfer dominated news printing through the middle of the twentieth century.

Invention of movable type did not lead immediately to newspapers. In Gutenberg's day, only a privileged minority could read, and monarchs and church authorities recognized that printing could disseminate not only the holy word of God but the subversive opinion of Man. For almost two centuries after Gutenberg's invention, most nonlocal news still was circulated in hand-written newsletters. In England provincial businessmen and the wealthy elite in country manors paid correspondents in London and major cities on the continent to write weekly letters with gossip about the court and foreign affairs. The first surviving English-language newspapers are double-sided, two-column news sheets, called *corantos*, printed in 1620-1621 in the Netherlands.[4] The oldest surviving periodical news publications printed in England are seven corantos published in 1621 by London bookseller Nicholas Bourne.[5] News-books, or news pamphlets, printed in book form with title pages, and containing letters from several correspondents in distant cities as well as material from other publications, appeared in 1622. Dated, numbered serially, and often published weekly, under license by the Star Chamber, these news pamphlets provided little domestic news.[6] In 1632 government efforts to control news publishing led to a Star Chamber decree that withdrew all licences and prohibited "Stationers, Printers, and Book Sellers" from printing and selling "the ordinary Gazetts and Pamphletts of news from forraigne partes."[7]

After the Star Chamber was abolished in 1641, news pamphlets were again prominent. In 1655, an official publication, the *Oxford Gazette*, revived the coranto format to become the first English newspaper. After printing 23 issues at Oxford, the paper moved to London. When Parliament dropped the licensing act in 1695, many new papers formed, including provincial weeklies and hand-written newsletters that converted to printing. Britain's first daily, the *Daily Courant*, formed in 1702, soon expanded from two columns on a single side of a sheet to four full pages. Advertising became important in the early 1700s, and expansion

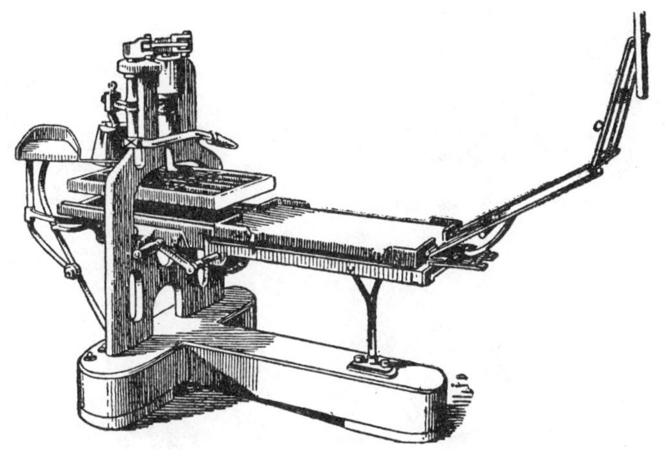

Figure 2.1. Stanhope's cast iron platen press, ca. 1800. The type frame faced upward toward the heavy platen, which was lowered by pulling the bar and raised with the aid of a large counterweight, shown just to the right of the pressman's seat. [Source: Encyclopaedia Britannica, 11th ed., vol. 21 (New York: *Encyclopaedia Britannica*, 1910-11), 351.]

of the postal service in the mid-1700s allowed several London papers to extend their reach.[8]

Early newspapers in the American colonies derived their equipment and appearance from Britain. The first colonial paper publishing more than a single issue was the *Boston News-Letter*, established in 1704 by the city's postmaster, and soon renamed the *Massachusetts Gazette*. As a loyalist organ, the *Gazette* closed when the Revolutionary Army took Boston from the British in 1776. Before the Revolution, in 1775, the colonies had 33 newspapers, mostly weeklies and no dailies.[9] Several cities, such as Boston and Philadelphia, had more than one paper, and many newspaper publishers were also job printers and pamphleteers. Before, during, and for decades after the Revolution, the American press was highly opinionated, independent, and partisan. Although papers supporting Britain ceased with the Revolution, by 1810 the United States and its territories were home to 359 separate newspapers—with no less than 71 in one state, Pennsylvania.[10]

During the first half of the nineteenth century the creative ingenuity of the Industrial Revolution radically altered the character and economics of the news business. The iron printing press, invented by Lord Stanhope in 1800, provided a stability and pressing force impossible with the wooden presses of the seventeenth and eighteenth centuries. The full-sheet platen and lever-and-screw pressing mechanism shown in Figure 2.1 enabled the printer to take a clear impression of the entire type surface with a single "pull." After the upward-facing type in the frame was inked,

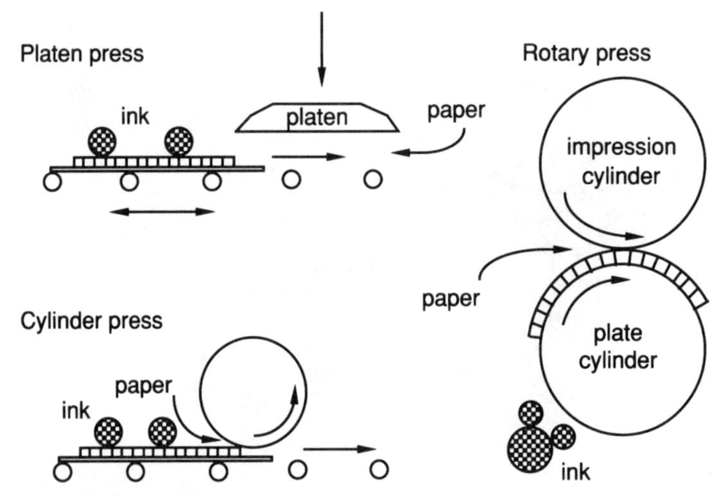

Figure 2.2. Diagrams illustrating the operating principles of the platen press (upper left), the cylinder flat-bed press (lower left), and the rotary press (right). After receiving ink from the ink rollers, the plate on a platen press moves under the platen; the printer inserts a sheet of paper and the platen moves downward to make the impression. A cylinder press inks the plate in a similar manner, but the plate moves beneath the impression cylinder as paper is inserted. Ink rollers on a rotary press apply ink to a curved plate, and paper is inserted between the plate cylinder and the impression cylinder. Printing is slowed by the back-and-forth, stop-and-start motion of the plate on the platen and flat-bed cylinder presses, whereas there is no reverse movement of the plate on the rotary press.

the printer carefully placed a sheet of paper on the top of the type, lowered the platen against the type to make the impression (Figure 2.2-upper left), reversed the lever to raise the platen, and removed the printed sheet.

Efforts to substitute steam power for manual effort led to a radically different press design in which an impression cylinder replaced the impression platen. The type frame was mounted face-upward in a track, along which it was pushed first under a set of inking rollers and then under the impression cylinder, with a sheet of paper attached to receive the image (Figure 2.2-lower left). In 1814, the *Times* of London attained 1,100 impressions per hour on a steam-driven press with two impression cylinders, designed by Frederick Koenig to take fuller advantage of the motion of the type frame.[11]

Further efficiency awaited the type-revolving rotary press, with the type for several pages locked into curved frames mounted on a cylinder so that the type image could be moved from ink rollers to impression cylinder without stopping to reverse direction (Figure 2.2-right). As Figure 2.3 shows, in a single revolution the rotary press could take sheets

Figure 2.3. Eight-cylinder type-revolving printing press, with the type form revolving on a single large type cylinder as eight workmen attend individual smaller feeding-impression cylinders. Richard Hoe initiated this design in 1847 to increase the number of impressions without increasing the speed of the type cylinder, which would fling type about the pressroom if made to revolve rapidly. Some Hoe presses with this design had as many as twelve feeding cylinders, which collectively could make 400 impressions a minute. [Source: Albert S. Bolles, *Industrial History of the United States* (Norwich, Conn.: Henry Bill Publishing Co., 1879), 294.]

of paper from four or more feeders, each with its own impression cylinder registered to the appropriate type frame on the cylinder, and make impressions for several pages at once. The first successful steam-driven rotary press, built in 1847 by Richard Hoe for the Philadelphia *Public Ledger*, used a horizontal type cylinder 1.7 m (5-1/2 ft) in diameter to print up to 4,000 copies per hour of a four-page newspaper.[12] A further advance was the "perfecting" press, which printed both sides of the paper and assembled a complete, folded newspaper.

Greater printing speeds awaited the adaptation to rotary printing of two earlier innovations: the *web* of paper manufactured in a continuous roll, and curved *stereotype* plates, with the type for each page cast as a single metal plate to withstand the huge centrifugal forces of a rapidly turning type cylinder. Stereotyping was practiced as early as 1725 by William Ged, a Glasgow goldsmith who took metal castings from plaster molds made from a frame of type. Eighteenth-century book printers used stereotyping to reduce wear on their type and to make a limited supply of type available for further composition once a page was set and proofed. Paper-mache replaced plaster in the 1830s. In 1854, Charles Craske cast curved full-page stereoplates for the New York *Herald*.[13] Louis Robert invented a machine for manufacturing paper in continuous rolls, and the brothers Henry and Sealy Fourdrinier bought the patents in 1804, made extensive improvements, and reduced the time required to make paper from three months to one day.[14] But Britain's newspaper tax, which until 1861 required a tax stamp on every sheet of newsprint, delayed adoption there of the web press for newspaper printing.[15] In 1865, for a press at the Philadelphia *Inquirer* inventor William Bullock used stereotyping and paper in rolls to overcome problems with mechanical paper "grabbers" and high-speed printing cylinders, which sometimes would fling type across the pressroom.[16] Whereas Bullock's press cut the roll into sheets before making impressions, the Walter rotary web press installed at the *Times* of London in 1868 printed on the web before cutting. Named after John Walter III, whose grandfather had founded the paper in 1785, the Walter press printed on both sides of the paper and reduced printing time substantially.[17]

Inexpensive paper to feed these giant presses was essential to the large-circulation daily newspaper. Around 1850 paper still was manufactured mostly from rags, cut into scraps and reduced by machine to a pulp, which was spread on wire frames to dry into sheets, or pressed, dried, and wound onto a roll. Demand outpaced supply, though, and as the price of rags rose, paper manufacturers turned occasionally to less expensive substitutes, such as straw and corn husks. Although wood pulp prepared with caustic soda or ground mechanically had been used earlier for making paper, wider use of this substitute awaited the invention by Hugh

Burgess and Charles Watt, in 1851, of a process for boiling wood chips in caustic alkali to separate the fibers.[18] By 1880, most paper for printing newspapers contained 40 to 50 percent wood pulp.[19] Although not as durable as rag paper, newsprint made from wood chips promoted substantial increases in newspaper size and circulation during the 1870s.

Mechanical typesetting was yet another technological boon for news publishers. For most of the nineteenth century a printer holding a *composing stick* sat in front of a type case, from which he selected the appropriate sequence of type and spaces for a line of text. Each line of type was both set and justified in the composing stick, and then added to the type already in the frame. Like Gutenberg, the nineteenth-century printer set his type at about one line a minute, until the Linotype, invented in Baltimore by Otto Merganthaler in 1886, increased the rate to four or five lines a minute.[20]

Enormous increases in newspaper circulations and in the number of newspapers paced these technological advances in newspaper manufacturing. In the United States the number of daily newspapers rose from 27 in 1810, to 141 by 1840, and to over 2,200 by the end of the century.[21] Weekly papers, always more numerous than dailies, increased from 289 to nearly 13,000 during the same period. As shown in Figure 2.4, most of the increase in daily newspapers came after 1880, when European immigration intensified and many new cities and towns were formed. After reaching a peak of 2,600 in 1909, the number of daily newspapers has declined steadily, as mergers and closings have eliminated competition in many cities that once supported several dailies.

The rise of the "Penny Press" in the early 1830s is perhaps the single most significant event in the evolution of the daily newspaper. The six-cent price of most newspapers in 1833 was prohibitive to workers earning less than a dollar a day.[22] As some publishers recognized, though, a profit could be made by lowering the price and increasing substantially the number of copies sold. The first penny paper was the New York *Morning Post*, which survived only a few weeks after its initial issue on January 1, 1833. Later that year Benjamin Day, a journeyman printer with a sense of marketing, launched the highly successful New York *Daily Sun*. Day's strategy was twofold: he offered readers a daily account of the city's crime and gossip in easily comprehended short sentences, and he enlisted as partners a platoon of small boys to whom he sold 100 copies for 67 cents. When Day sold the paper four years later, in 1837, he had a daily circulation of 30,000.[23]

Prominent among Day's contemporaries were James Gordon Bennett, Horace Greeley, and Henry J. Raymond. In 1835, Bennett founded the somewhat more literary and varied New York *Herald*, and by 1837 he had built its circulation to 20,000. After doubling the price to two cents, he continued to attract readers, and by 1860 had attained a circulation of 60,000.[24] Greeley, a philosopher, idealist, and reformer, in 1841 founded the *New York Tribune*, which by 1860 had attained a circulation of 300,000, larger than any other newspaper in the world.[25] Raymond founded the more sedate *New York Times* in 1851, with only $70,000 in capital from his partners, but by 1861 he had developed a daily circulation of 75,000, at two cents a copy.[26] New York and many other cities supported several daily papers with divergent styles, and many persons read two or more newspapers.

Newspaper circulation in the United States increased throughout

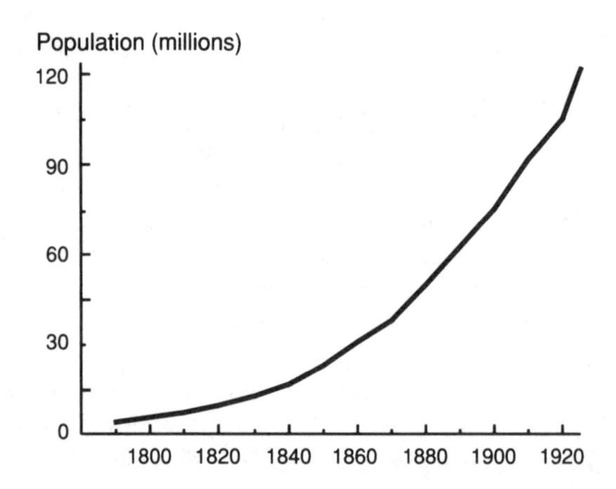

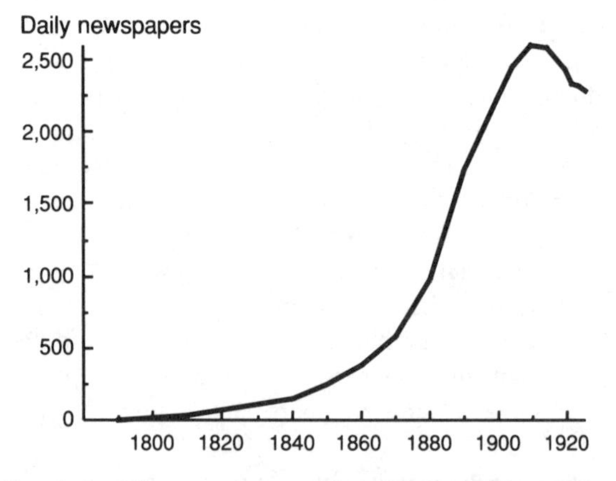

Figure 2.4. Trends in daily newspapers and population, United States, 1790-1925. [Data from William A. Dill, *Growth of Newspapers in the United States* (Lawrence, Kans.: Department of Journalism, University of Kansas, 1928), 11-12 and 28.]

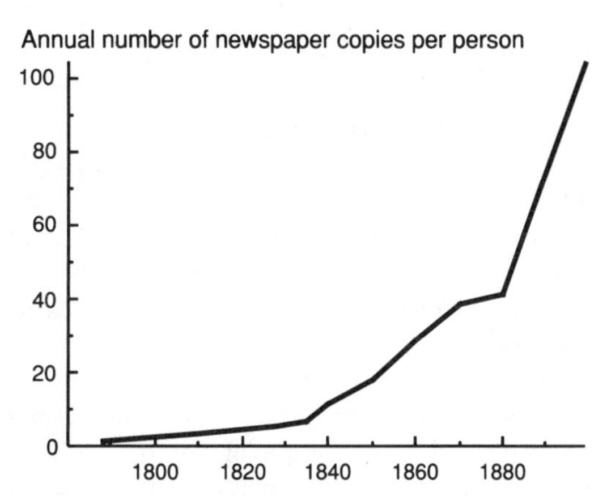

Figure 2.5. Trend in annual number of newspaper copies distributed per person, United States, 1788-1899. [Data from William A. Dill, *Growth of Newspapers in the United States* (Lawrence, Kans.: Department of Journalism, University of Kansas, 1928), 22.]

most of the nineteenth century, as shown in Figure 2.5. The annual number of copies printed per inhabitant rose from one in 1788 to over 104 in 1899, with the rate doubling between 1835 and 1840, and again between 1880 and 1890. Most of these increases represented the growth of the daily newspaper, which accounted for 55 and 63 percent of all copies sold during 1850 and 1899, respectively.[27] Increased sales reflected enhanced prosperity, particularly during the 1830s and the years immediately after the Civil War. Real wages for nonfarm workers rose 10 to 20 percent during the 1830s, and as much as 10 percent during the 1850s.[28] Gains during the latter part of the century were more modest, on the average, as employers took advantage of the abundance of semiskilled or unskilled new immigrants. Foreign-language papers, which helped these non-English speakers settle and assimilate, also contributed to the totals of newspapers published and copies sold. By 1890, the average number of subscriptions passed one per person.[29]

Heightened prosperity and a more specialized, less self-sufficient labor force increased the reader's need to consult newspaper ads and enhanced the publisher's revenue from advertising. Retail trade, which increased in value-added national income from $500 million for 1869 to $1.3 billion for 1899, was attracted by the large circulations and extended "reach" of newspapers.[30] Between 1880 and 1899 the percent of newspaper revenue derived from advertising rose from 44 to 55 percent.[31] Newspapers had become not only widespread and frequently read but also an essential element in the market economy.

Engraving, Cartoons, and News Illustration

Early American newspapers were gray, graphically bland presentations of words, and sometimes numbers. Often the only artwork was in the nameplate, engraved in type metal and changed only when worn or broken.[32] When used at all, other illustrations were printed from wood-block engravings. Unlike the typical woodcut print, with the image cut below the surface to yield white lines on a black background, a wood engraving is formed by carving away the wood in nonimage areas. Commonly the image was copied onto the smoothed and polished end-grain sections of a piece of medium-grained wood, such as apple or beach. The engraver then pushed a burin or graver over the surface to lower the nonprinting sections (Figure 2.6). Intricate details required a knife or stylus. Mounted in the type frame with the uncut parts of the original surface "type-high," the engraved wood block printed the desired black-line image. Used for book illustrations since the middle of the fifteenth century, and in China six centuries earlier, small wood engravings were employed occasionally in the news-books and corantos of the sixteenth century. From the seventeenth through the nineteenth centuries wood-block engraving was used for maps and other small illustrations in text books, dictionaries, and encyclopedias.[33]

Some of the more famous American wood-engraved news maps are better known as political cartoons. To persuade the colonies to unite in the war against France, Benjamin Franklin designed the "snake device"

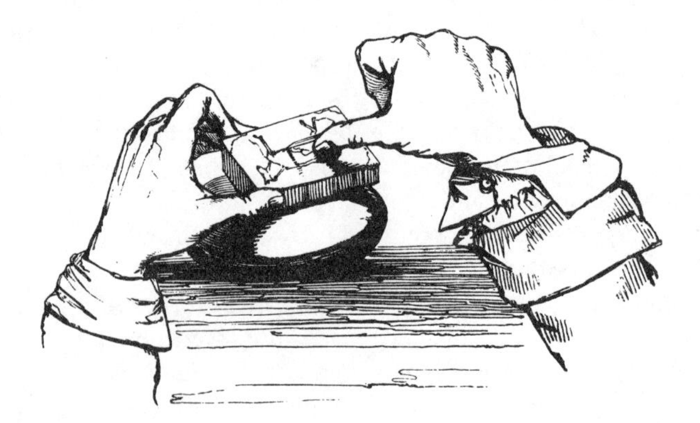

Figure 2.6. Holding the graver and block to make small, precise cuts required a variety of special, carefully sharpened engraving tools, as well as several years of apprenticeship to develop the painstaking skill of the wood engraver. [Source: John Jackson, with W. A. Chatto, *A Treatise on Wood Engraving, Historical and Practical*, 2nd ed. (London: Henry G. Bohn, 1861), 580.]

Figure 2.7. Benjamin Franklin's snake device, printed in the *Pennsylvania Gazette* for May 9, 1754. [Source: Albert Matthews, "The Snake Devices, 1754-1776, and the Constitutional Courant, 1765," *Publications of the Colonial Society of Massachusetts* 11 (1906-7), 409-53; plate 1, opposite p. 416.]

shown in Figure 2.7 and printed it in his *Pennsylvania Gazette* for May 9, 1754. Based on an old wives' tale that a snake cut apart will live if joined together by sundown, Franklin's design within the month appeared with minor variations in at least four other colonial newspapers.[34] Revived in 1765 for protests against the Stamp Act and used again during the American Revolution, this political slogan portrays the sequence along the East Coast of the larger colonies, with those in New England aggregated for simplicity under the regional initials "N.E." and Delaware and Georgia omitted altogether. Historians of graphic humor consider Franklin's wood-engraved snake the first American political cartoon.[35] It also might well be the country's first newspaper map.

Early American newspapers occasionally used maps to show legislative districts, as demonstrated in Figure 2.8, a map of Essex County, Massachusetts, printed in the Boston *Weekly Messenger* for March 6, 1812. The parallel dashed lines show how the Democrats controlling the state legislature proposed to divide the county to the advantage of their own candidates. Governor Elbridge Gerry won political immortality by not vetoing the plan. A Federalist editor, stimulated by a colleague's observation that the contorted upper Essex district resembled a salamander, is said to have exclaimed, "Salamander! Call it a Gerrymander."[36] Drawn by artist Elkanah Tisdale and engraved in metal, the Gerrymander cartoon inspired by this outburst (Figure 2.9) appeared first in the Boston *Gazette* for March 26, 1812, and later in other eastern Massachusetts papers, using the same engraving.[37] The intricate detail of the Gerrymander cartoon provides a marked contrast to the comparatively crude details of its cartographic precursor.

The explanatory power of news illustrations called for improvements to wood engraving. In the eighteenth and early nineteenth centuries engraving techniques were crude, particularly in lettering place names on a wood block. Examination of Franklin's snake map (Figure 2.7) and the

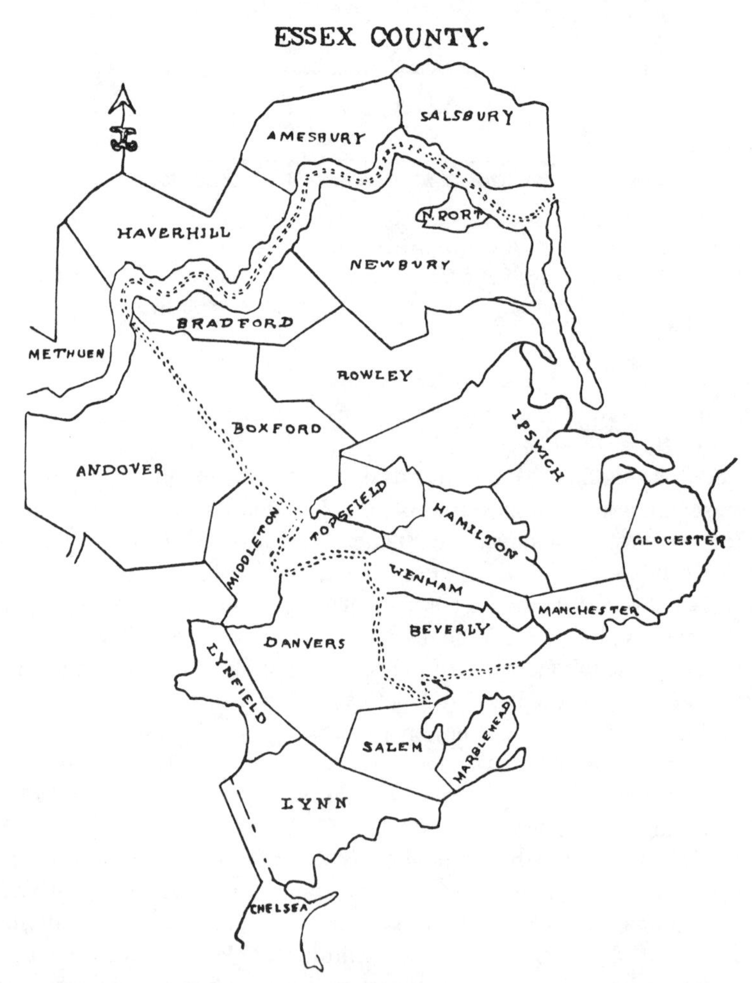

Figure 2.8. Map of proposed Senatorial District boundary through Essex County, Massachusetts, printed in the Boston *Weekly Messenger* for March 6, 1812. [Source: Elmer C. Griffith, *The Rise and Development of the Gerrymander* (Chicago: Scott, Foresman and Company, 1907), 69. The original map measures 19.7 cm high by 15.2 cm wide, or 7-3/4 in. by 6 in.

Figure 2.9. The Gerrymander cartoon-map printed in the Boston *Gazette* for March 26, 1812. [Source: James Parton, *Caricature and Other Comic Art* (New York: Harper and Brothers, 1877), 316. The original print measures 16.5 cm. high by 15.2 cm wide, or 6-1/2 in. by 6 in.]

Essex County map (Figure 2.8) reveals inconsistencies in lettering not found for cast metal type. Curved letters in particular appear unnaturally squarish where the engraver was forced by crowded details to make a limited number of cuts with a squarish blade. To produce neater labels and presumably save time, some engravers would cut a slot into the block and insert metal type, letter by letter, or a small stereotype plate with the complete label.[38] Other improvements were introduced in the 1790s by Thomas Bewick, who advocated use of end-grain sections of boxwood, a close-grained wood able to withstand long press runs and hold relatively intricate details without being excessively difficult to cut.[39] Larger, more easily worked pieces of pear, cherry and other medium-grained woods, cut parallel to the grain, had provided comparatively larger engraving surfaces for earlier wood engravings. In contrast, boxwood was a smaller

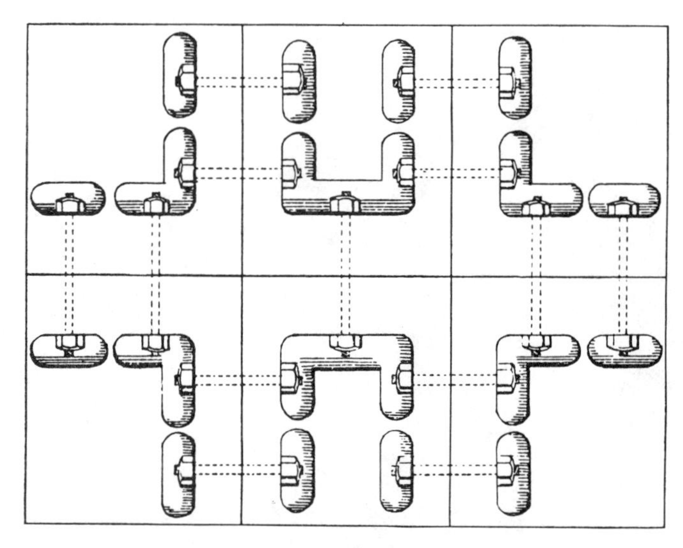

Figure 2.10. Wood-engraved block, constructed of seasoned pieces of boxwood fastened together at the back by brass nuts and bolts. Individual sections engraved by different craftsmen could be joined to print a single large illustration. [Source: Mason Jackson, *The Pictorial Press: Its Origin and Progress* (London: Hurst and Blackett, 1885), 316.]

tree, and when cut perpendicular to the grain, allowed for relatively small illustrations. But, as shown in Figure 2.10, individual blocks of boxwood, usually smaller than 10 cm (4 in.) square, could be joined together by nuts and bolts inserted through recesses and holes cut into the blocks.[40] Locked together for drawing and printing, the blocks were easily separated for engraving. Particularly useful for meeting news deadlines, the system allowed several engravers to work simultaneously on the same picture. Many early nineteenth-century wood-block illustrations reveal this approach in a grid of fine white lines between sections loosened slightly by the stress of printing.

Emergence of wood engraving as a specialized craft, with instruction manuals and apprenticeships, fostered the growth in the nineteenth century of several prominent illustrated weekly, fortnightly, and monthly magazines. Founded in 1832, *Penny Magazine*, a monthly printed in London, was the forerunner of such acclaimed picture magazines as *L'Illustration*, started in Paris in 1842, and *Frank Leslie's Illustrated Weekly* and *Harper's Weekly*, established in the United States in 1855 and 1857, respectively.[41] But the most successful, most enduring pictorial magazine was the weekly *Illustrated London News*, founded in 1842 and still published.[42] The sale of 26,000 copies at sixpence each for its 16-page first issue demonstrated the demand for news graphics. An innovator as the first illustrated weekly newspaper, the *Illustrated London News* was responsible for a number of important firsts, including the use in 1860 of a technique for transferring photographic images onto a sensitized wood block.[43] Metal printing plates made with an electroplating process avoided having to stop the presses to tighten the bolts holding together a multipiece wood-block engraving. A wax impression was taken of the wood block, and the waxed image was covered with a thin coat of blacklead and dipped into a solution of copper sulfate. A strong electric current deposited copper on the conductive lead coating to form a tough duplicate of the engraved block. When filled with a lead backing and mounted on wood, the resulting *electroplate* could easily withstand the pressures and vibrations of long runs on powerful iron presses.[44] Electrotyping was both more durable and more costly than stereotyping, and was more commonly used in book publishing, which required higher quality presswork than the newspaper.[45]

Wars in Europe and colonial areas heightened the public's appetite for news and challenged the ingenuity of the pictorial press. The *Illustrated London News* demonstrated exceptional foresight by sending an artist to Balaclava in 1854, to sketch towns and terrain where British troops later clashed with the Russians in the tragic Charge of the Light Brigade. These sketches, as well as battlefield drawings by field artists, contributed to the weekly's coverage of the Crimean War, during which circulation rose to an unprecedented 150,000.[46] In the United States the Civil War of the early 1860s provided another opportunity for illustrated journalism. Allotted only a minor role in news reports of the Crimean War, maps were particularly prominent in the coverage of the war between the states. Several daily newspapers occasionally published a full-page map or a special supplement with numerous maps showing a variety of battle sites and fortification plans. For example, on July 2, 1862, the front page of the Philadelphia *Inquirer* included a large map (Figure 2.11) showing sites of battles on three successive days near Richmond, Virginia, only a week or so earlier. Civil War newspaper maps were largely wood-block engravings, but an electrotype was sometimes made for use on a rotary press. Engraving and press set-up required two weeks.[47]

Photographic Engraving

Photographic line-engraving, developed late in the nineteenth century, changed radically the appearance and frequency of news maps and illustrations in general. Whereas wood engraving involved not only drawing the original sketch but also transferring a wrong-reading image

R 55,000. PHILADELPHIA, WEDNESDAY, JULY 2, 1862.

FIELD OF McCLELLAN'S OPERATIONS.

Scene of the Recent Battles in Front of Richmond—And the Late Important Military Operations in Eastern Virginia.

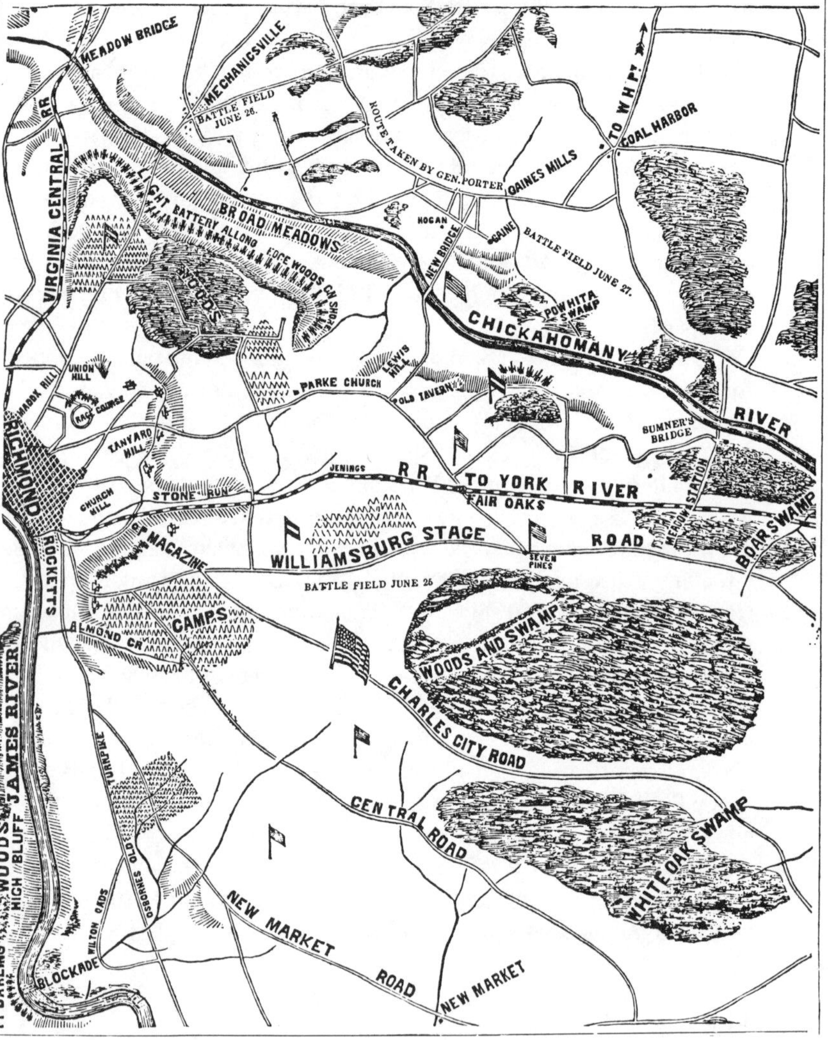

Figure 2.11. "Field of McClellan's Operations," from the front page of the *Philadelphia Inquirer* for July 2, 1862. [Source: Library of Congress, Civil War map no. 650.6.]

of the sketch to the wood block and removing carefully the nonprinting parts of the surface, photoengraving required only a sharp image of solid black lines on white paper or cardboard. All details of any high-contrast drawing, no matter how crude, could be reproduced exactly—or nearly so, given the variations in image quality of letterpress on newsprint. For maps the shift to photoengraving made lettering easier and promoted the use of mechanically ruled parallel-line and cross-hatched area symbols and point symbols such as circles, which the wood engraver's knife and burin made unnaturally squarish.[48] Compare, for example, the crude lettering and angular details of the Philadelphia *Inquirer*'s 1862 map of McClellan's exploits in eastern Virginia (Figure 2.11) with the smooth circumference and intersecting symbols of the Syracuse *Sunday Herald*'s 1899 county road map (Figure 2.12). Not a work of skilled craftsmanship, the Syracuse map is highly detailed and potentially more informative than

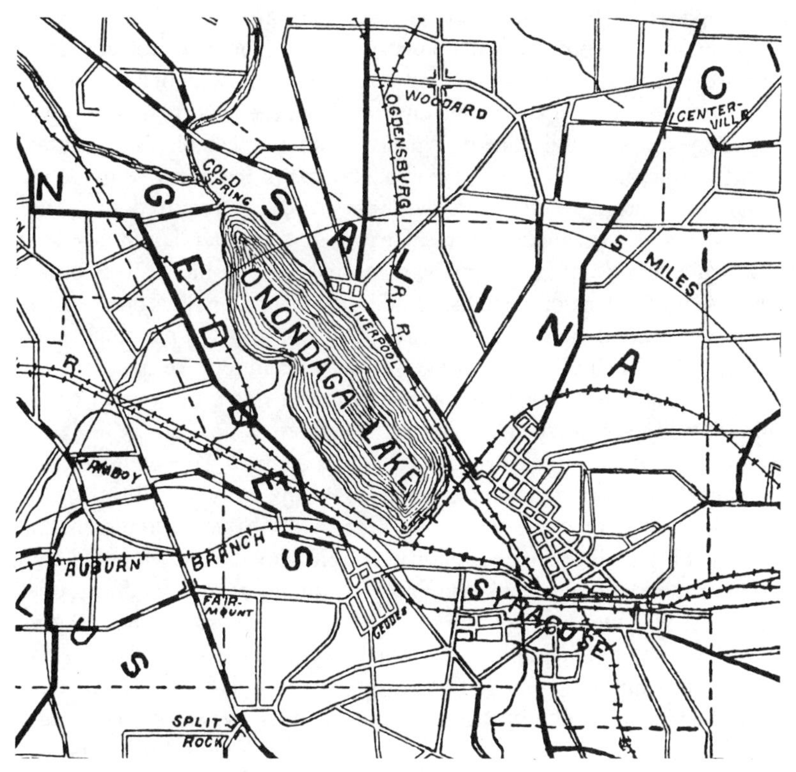

Figure 2.12. Portion of "The Herald's Road Map of Onondaga Co.," published in the Syracuse *Sunday Herald* on May 7, 1899. [Courtesy of the Onondaga Historical Association, Syracuse, New York.]

a wood-block map of the same size. Because of photoengraving, maps became both less costly and more timely. Whereas a wood-block engraving might have required either several days of carving by a single craftsman, or perhaps a day and night of simultaneous effort by several engravers, a large or medium-size paper with its own engraving plant could convert a line drawing to a relief printing plate in an hour or less.[49]

As practiced by newspapers at the end of the nineteenth century, the photoengraving of a line drawing was a straightforward manufacturing process.[50] Figure 2.13 describes a few of the essential steps. A strong electric light illuminated the pen-and-ink artwork, fastened to a copyboard in front of the camera (A) used to reduce the "up-size" drawing to a printing-size negative. In an era before modern plastics, the photographic "film" consisted of a thin, photosensitive emulsion on one side of a flat glass plate. In the camera the emulsion side faced the drawing to record a wrong-reading image. After developing the negative using a series of chemical baths to remove the emulsion from the image areas and fix an opaque coating of silver in the non-image areas, the photoengraver made a contact exposure of the negative onto a zinc plate coated with a photosensitive solution of albumen and bichromate of potash. For this exposure, the film negative was removed from the glass plate, reversed, and deposited on another sheet of glass, which was then placed in contact with the zinc plate in a contact frame. Reversing the negative made the image transferred to the plate wrong-reading, as required for relief printing. Negative and plate were exposed for several minutes to a powerful electric light (B), which penetrated the translucent lines on the image parts of the negative to harden the corresponding lines on the plate. After exposure, the engraver rolled a thick ink over the plate, and then soaked it in a bath of cold water (C) to soften and remove the emulsion from the unexposed nonimage areas. When dusted with a resin called "dragon's blood" and then heated (E), the hardened, now-sticky linework would resist the diluted nitric acid used for etching. This etch bath (D) attacked the unprotected, nonimage areas, "biting away" and lowering the plate surface except in the image areas, which were protected and remained high—like the uncut portions of a wood block. Mounted on a wood or metal block to make the image "type high," the resulting relief "line cut" was then ready to be placed in a type frame for stereotyping with the bodytype, headlines, and other engravings for the page.

Photoengraving can be traced to the experiments of Joseph Nicéphore Niepce, a French lithographer intrigued by the sensitivity to light of Guaicum gum. Niepce, who experimented with photolithography as early as 1813, demonstrated in 1824 that a solution of asphaltum in oil applied to a metal plate and dried provided a photosensitive emulsion useful for engraving and printing. When exposed to light, the asphaltum

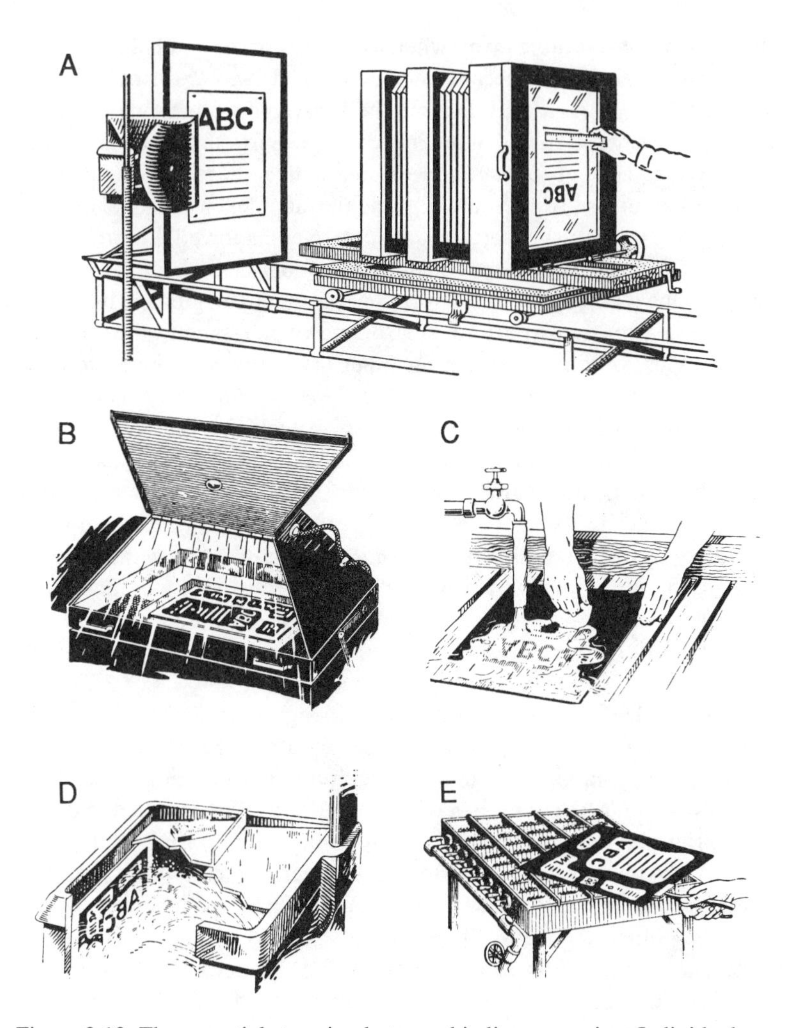

Figure 2.13. The essential steps in photographic line engraving. Individual pieces of "line art" are photographed (A) and the negatives arranged carefully on a glass plate, which is then placed above a polished zinc plate coated with a bichromate solution and exposed in a plate-making frame to a strong light (B). The exposed plate is then coated with an ink or dye, and bathed under cold water (C) to remove the plate coating in areas not exposed to light. Next the plate is etched in an acid bath (D) to lower the nonexposed, nonimage areas. A powdered resin is brushed onto the plate, which is then heated (E) to make the raised, exposed "line image" areas resistant to the next etching stage, or the "second bite." A third bite was common, and possibly a fourth, with powdering and heating preceding each. [Source: Ben Dalgin, *Advertising Production: A Manual on the Mechanics of Newspaper Printing Techniques* (New York: McGraw-Hill, 1946), 10-16.]

coating became insoluble in oil, whereas the part of the emulsion shielded from light remained soluble and could be washed away to expose the bare metal. Etching with an acid lowered these bare portions to yield a plate for *intaglio* printing, in which the inked image is formed in the low areas of the plate. Use of a camera was impractical because of the long exposures required in direct sunlight, and the later development of camera photography by Niepce's partner Daguerre overshadowed this promising discovery for many years. But around 1850 Niepce de St. Victor, a cousin, improved the method after drawing upon the equally neglected discovery of the high photosensitivity of bichromate of potassium, published in 1839 by Mongo Ponton. Numerous other inventors contributed to photographic engraving, including Fox Talbot, who around 1850 made prints with bichromated albumen, and Charles Gillot, who in 1872 used an albumen-bichromate emulsion directly on the metal plate.[51] Experimentation continued, and a variety of variations of zinc line-engraving arose throughout the 1870s and 1880s.[52] By 1900, though, line-engraving technology was well established, including the process camera, whirlers for applying to the plate a uniform coating of emulsion, the thick-glass printing frame and electric arc lamps for exposing the negative to the plate, and efficient methods for developing the negative and etching the plate.[53]

In the late 1880s and 1890s, photoengraving technology made possible a publishing revolution marked by abundant graphics, sensational reporting, and highly competitive circulation-building. Named for the Yellow Kid, a character in a widely read comic strip printed in color, the "yellow press" of the late nineteenth century had its origin in the New York *World*, bought in 1883 by Joseph Pulitzer.[54] With a brisk style and an emotional appeal to lower- and middle-class readers, the *World* vigorously supported labor unions, income and luxury taxes, civil service reform, and safeguards against monopolies. Sociologist Robert Park described Pulitzer's formula as "love and romance for the women, [and] sports and politics for the men."[55] Pulitzer used abundant line illustrations to add interest and realism, as in the amusing and catchy portraits of several prominent Wall Street personalities sketched with large, highly accurate heads on small bodies.[56] From near bankruptcy and a daily circulation of only 10,000 in 1883, when he bought it, Pulitzer increased the paper's circulation to 100,000 within a year and to nearly 375,000 by 1892. The *World*'s success was a major incentive for other dailies to adopt a heavy use of line-engraved illustrations.

The New York *Daily Graphic*, started in 1873, was another New York newspaper prominent for its pictures. The earliest heavily illustrated daily paper in the world, the *Daily Graphic* was quite different from other newspapers—one side of its rag-paper pages was printed on lithographic

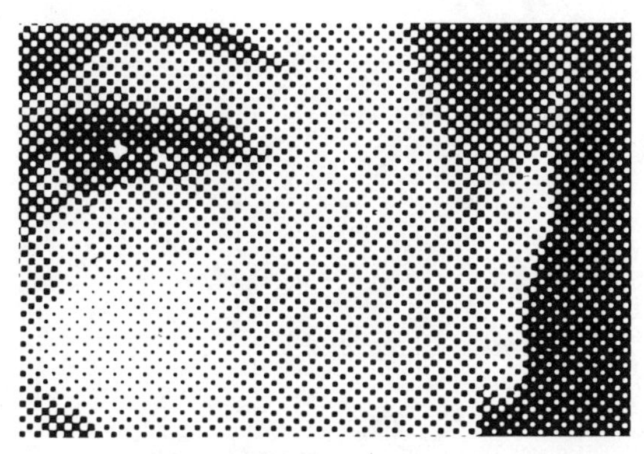

Figure 2.14. Portion of a halftone engraving, enlarged to show the pattern of perpendicular rows of closely spaced dots. In the late nineteenth century, in particular, halftones were also based on a single pattern of closely spaced parallel lines. When viewed from a normal reading distance, the texture of the dots or lines was far less noticeable than the variation in graytones resulting from larger black or small white dots (in relatively dark areas) on a dot halftone and thicker black or thin white lines (in darker areas) on a line halftone. [Source: Erwin Jaffe, *Halftone Photography for Offset Lithography* (New York: Lithographic Technical Foundation, 1960), 2. Courtesy of the Graphic Arts Technical Foundation.]

stone, the opposite side was printed on a high-speed Hoe "Lightning" letterpress, and a copy of the eight-page newspaper sold for five cents, well above the price of other New York dailies.[57] Perhaps its greatest contribution to news publishing was the successful demonstration on March 4, 1880, of halftone photoengraving, in which a metal plate was made directly from a continuous-tone photographic negative. Widely used today for news and sports photos, the halftone process represents the continuous-tone image with a pattern of tiny dots, arrayed in lines, 55 or more to the linear inch, as shown in Figure 2.14. Stephen Horgan, manager of the paper's photomechanical department, founded his own photoengraving firm after the *Daily Graphic* closed in 1884 and wrote widely on the technique.[58] Several other photographic experimenters, notably Frederick Ives and the brothers Max and Louis Levy, worked to perfect the process and the cross-line glass screens used to break an image into a pattern of dots. By the late 1890s halftone photographs were common in daily newspapers.[59]

Surprisingly perhaps, newspapers did not shift from sketching and line-engraving to photography and halftone-engraving until six years after 1891, when the Ives process and Levy screens became available. By that time a properly adjusted web press could print halftone photographs

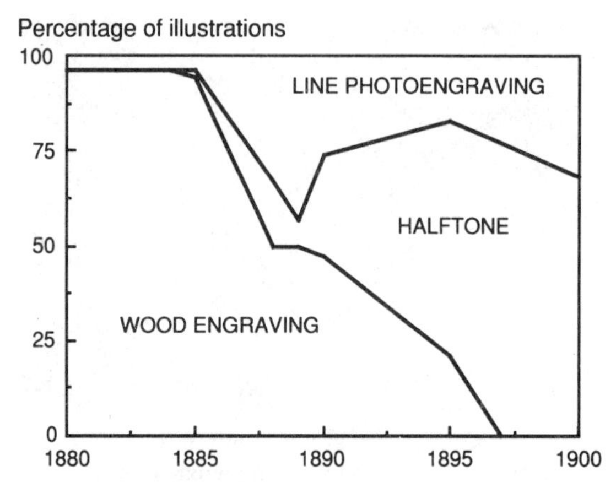

Figure 2.15. Changes between 1880 and 1900 in the method for reproducing illustrations in *Harper's Magazine*. [Source: David Woodward, "The Decline of Commercial Wood-Engraving in Nineteenth-century America," *Journal of the Printing Historical Society* no. 10 (1974-75), 56-83; adapted from Figure 11 on p. 76.]

without blurring or overinking, and photojournalism had become not only more rapid but less expensive. According to photographic historian Smith Schuneman, the delayed adoption of halftones was due largely to pressure from the more than 1,000 newspaper artists and wood engravers who each week produced 10,000 drawings.[60] Through a propaganda campaign, these craftsmen convinced printers and editors alike that photographs were not art and would not be acceptable to readers. But efficiency ultimately triumphed. Although the Boston *Transcript* resisted halftone photographs until 1907, at most large-city newspapers the transition was rapid and essentially complete by the turn of the century. Cartographic historian David Woodward, in a systematic examination of illustrations in *Harper's Magazine*, portrayed graphically (Figure 2.15) the demise of wood engraving between 1885 and 1897, the rise of halftone engraving after 1889, and a rapid acceptance of photographic line-engraving after 1885.[61] By 1901, according to the trade journal *Newspaper Maker*, New York had only a handful of wood engravers, struggling not as news illustrators, but as artisans.[62]

Although the demise of any craft in which people have invested years of training is an occasion for sadness, the shift from wood engraving to photoengraving was a major advance for journalistic cartography. As with other innovations, though, progress sometimes compromised quality. Cartographic historian Arthur Robinson has noted that photography removed from cartographic reproduction the "design filter" provided for centuries by the skilled engraver or printer, who as the artistic partner of the cartographer was often the sole champion of aesthetic standards.[63] In contrast, the cost-effective photoengraver could reproduce "as is" whatever a cartographic draftsman ignorant of graphic design chose to produce—and the many sloppy maps of the early twentieth century attest to the importance of the printer as an artistic mentor of the map maker. Yet for news maps, which had seldom attained elegance at the hands of deadline-driven wood engravers, the improved timeliness possible with photoengraving more than offset the loss of whatever traditions and standards the wood engraver might have contributed. Moreover, news maps were now more easily labeled, and hence more informative. And when reduced photographically from an up-size drawing, lines and other symbols were sharper than before. Because of photoengraving, cartographic illustration could keep pace with advances in photojournalism, a trend toward specialized reporting, and a growing need for locational information, local as well as foreign.

Photocomposition, Offset Lithography, and Color Graphics

Further advances in printing and engraving during the twentieth century have enhanced the newspaper's ability to reproduce pictures. But many graphic improvements were little more than serendipitous side-benefits of publishers' quests for solvency and profit through cost control and increased circulation and advertising sales. The growth of television in the 1950s was a serious threat, which cost newspapers much of their readers' attention and contributed to many closings and mergers. Between 1945 and 1960, for instance, the number of television stations operating in the United States rose from six to over 500, while the number of daily newspapers dropped from 1,890 to 1,750. Cities that once supported four or more dailies now supported four or more television broadcasters and a single newspaper, perhaps with separate morning and afternoon editions.[64] Younger people, raised on television, did not inherit their parents' appreciation of large, colorless sheets of printed news, which were often notorious for causing dirty hands and soiled clothing because of ink rub-off. Communications researcher John P. Robinson estimated that between 1966 and 1976 the time devoted by the average adult to newspaper-reading dropped by 25 percent, from 28 to 21 minutes a day.[65] That most twentieth-century innovations in newspaper publishing occurred after 1950 is not surprising, given the strong competition from television as well as the opportunities presented by advances in electronics technology.

Perhaps the most fundamental change in newspaper production was the replacement of manually-operated line-casting machines with computer-generated "cold type." *Photocomposition*, as it is also called, produces on photosensitive film or paper an image of correctly sized, justified, and hyphenated type, set at a rate of several hundred lines per minute from either optically projected negative images or electronically stored fonts.[66] Its origins lie in the 1930s, when perforated paper tape was used to drive Linotypes; because there was no need to pause at the end of each line, a typist producing a paper tape could work more rapidly and with fewer errors than a compositor at a Linotype keyboard. Additional savings accrued from the use of paper tape generated by wire-service teletypewriters—at the expense, though, of typographers' jobs and of a more thorough editing of wire-service copy. In the 1960s computers provided automatic hyphenation and line justification, and the ability to alter, insert, and delete words. By 1980 most newspapers had computerized typesetting systems with which reporters could store both notes and stories and editors could receive and edit copy before releasing it for typesetting.

For maps and other artwork the principal benefit of photocomposition was a shift from page make-up with Linotype relief type and photoengraved blocks, to stick-up *mechanicals*, produced on layout boards with knife, paste, and positive paper-copy images of computer-generated type and process-camera artwork. Although editors still design the layout of pages on a rough sketch or "dummy," the shift to cold type promoted last-minute changes to accommodate late news, as well as greater flexibility in column width, story shape, and integration of text and art. Particularly noteworthy is the abandonment in the late 1960s of a tendency toward vertical layout, in which the reader might have to follow a story to the bottom of the page and then jump to the top of the next column (Figure 2.16-left). With the more easily read horizontal layout of the 1970s, a long story was divided into several relatively short columns and extended across, not down, the page (Figure 2.16-right). The layout editor could more freely add rules or borders, perhaps to organize as a module a story's headlines, bodytype, and illustrations.[67] The made-up page was then photographed, and the resulting negative used to "burn" press plates. Although halftone negatives for photographs had to be "stripped in" on the page negative just before making the plate, line illustrations, including most maps, were laid out on the mechanical with the type.

Concurrent with the widespread shift to photocomposition in the 1970s, further changes in newspaper production eliminated the need for hot metal of any variety, even for stereotyping. Most small and medium-size newspapers switched to offset lithography, and even those larger papers retaining their relief presses adopted plastic letterflex plates or an equivalent process for the photographic transfer of a coldtype image onto the page negative used to make inexpensive pressplates.[68] The many papers changing from letterpress to offset printing not only saved in labor, equipment, materials, and plant space, but were able now to reproduce more clearly and cleanly both news pictures and display advertising.

Literally meaning "stone printing," *lithography* in its modern form is derived from the early lithographic processes of the 1790s, which used smooth, flat slabs of stone, instead of the metal plates that now permit

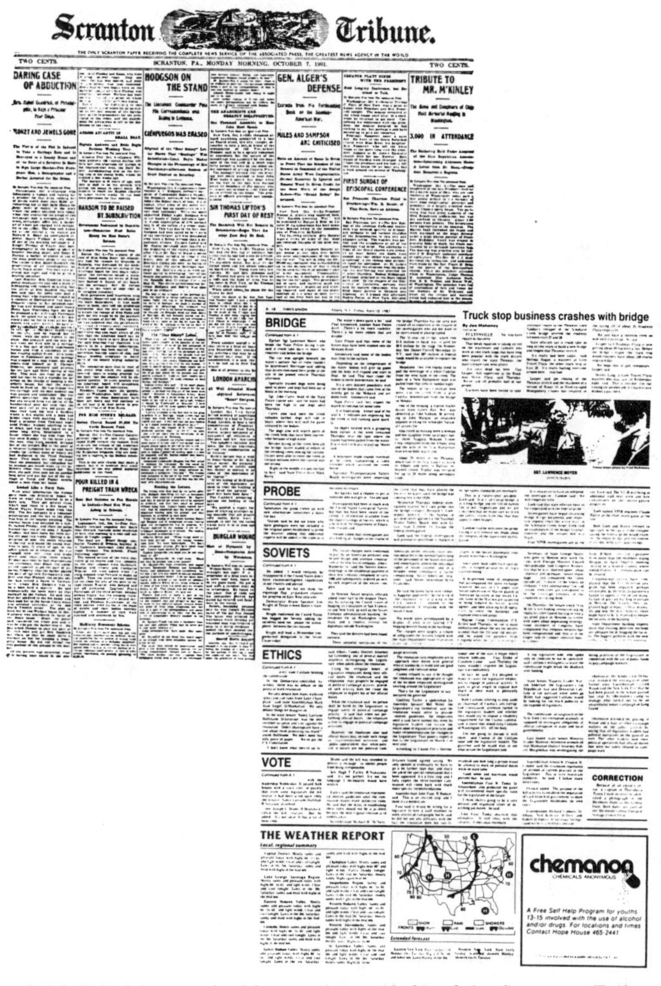

Scranton Tribune.

TWO CENTS. SCRANTON, PA., MONDAY MORNING, OCTOBER 7, 1901. TWO CENTS.

DARING CASE OF ABDUCTION

HODGSON ON THE STAND

GEN. ALGER'S DEFENSE

TRIBUTE TO MR. M'KINLEY

MILES AND SAMPSON ARE CRITICISED

3,000 IN ATTENDANCE

FIRST SUNDAY OF EPISCOPAL CONFERENCE

SIR THOMAS LIPTON'S FIRST DAY OF REST

FOUR KILLED IN A FREIGHT TRAIN WRECK

LONDON ANARCHISTS

BRIDGE

Truck stop business crashes with bridge

PROBE

SOVIETS

ETHICS

VOTE

CORRECTION

THE WEATHER REPORT

Local, regional summary

Extended forecast

chemanon

CHEMICALS ANONYMOUS

A Free Self Help Program for youths 13-15 involved with the use of alcohol and/or drugs. For locations and times Contact Hope House 465-2441

Figure 2.16. Primitive vertical layout (upper-left) of the *Scranton Tribune*'s front page for October 7, 1901, contrasts with the strongly horizontal layout (lower-right) of the "jump," or continuation, page of the Albany, New York, *Times Union* for April 10, 1987. [Source of upper-left example: O. F. Byxbee, *Establishing a Newspaper* (Chicago: Inland Printer Co., 1901), 29.]

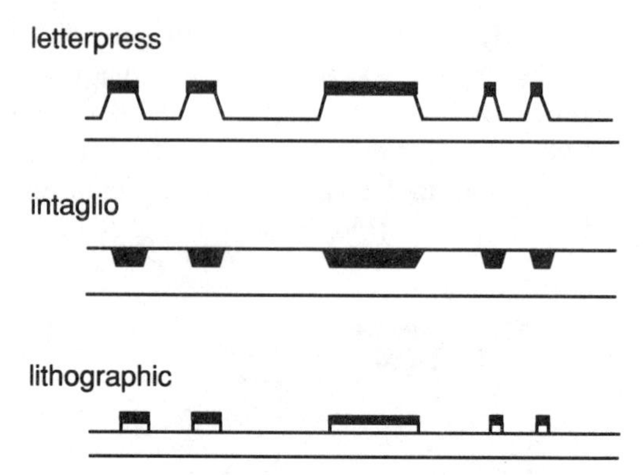

Figure 2.17. Differences in printing plate among letterpress, intaglio (gravure), and lithographic printing. For letterpress, or relief printing, ink is deposited on image areas rasied above the non-inked areas on the plate surface, whereas for intaglio printing, ink is squeezed into image areas engraved or etched below the plate surface. In lithographic printing, also called planar printing, ink adheres only to portions of the plate with a thin waxy coating. Areas with this waxy coating on a lithographic plate are much lower than the raised, inked "type-high" areas on a letterpress plate.

high-speed rotary "offset-litho" presses to make hundreds of impressions a minute. Also called *planographic printing* to contrast it with the raised image of relief (letterpress) printing and the incised image of intaglio printing, lithography records the printing image with a thin waxy substance on the flat surface of the printing plate (Figure 2.17). Early lithographers drew a wrong-reading image with crayon, but in modern offset lithography the image is transferred photographically from the page negative to the photosensitive surface of a thin aluminum plate. Mounted on a cylinder, the plate rotates past a set of rollers that apply water and then past another set of rollers that apply ink (Figure 2.18). Its waxy image areas do not pick up the water and are thus free to receive the oily ink, whereas the nonimage areas accept a thin layer of water that rejects the ink. The right-reading image now used on lithographic plates is first offset in reversed form onto a rubber-coated blanket cylinder, from which it is transferred again, in right-reading form onto the paper. Employed for cartographic printing since about 1830, lithography was never widely used for printing newspapers until the 1960s, when the high-speed web-offset press became an attractive alternative for smaller papers.

Another graphic-arts innovation particularly attractive to smaller papers was the electronic scanner engraver, which sensed tonal differences on a photograph mounted on one spinning drum and generated a corresponding halftone image on a sheet of plastic mounted on a second drum rotating in synchronization with the first. Although developed before World War II, electronic scan-engraving was not widely used until the early 1950s, when the Fairchild Camera and Instrument Corporation offered an improved but inexpensive model called the Scan-A-Graver.[69] Newspapers unable to afford a photoengraving department—a considerable expense for a small paper in the hot-type era—no longer had to rely upon syndicates and outside engravers for halftone cuts. With an electronic scan-engraver, small papers could now easily include local news and sports photos as well as halftone artwork for local display ads. Although the images often lacked the clarity of photoengraved line cuts, the Scan-A-Graver could also produce plastic relief cuts for maps and other line drawings.

In the 1970s and 1980s offset printing and scanners of another sort increased the use of color in large and medium-size daily newspapers. Color printing requires different plates and cylinders for each ink color, with each "separation" printed in register on the same page. Colored ink

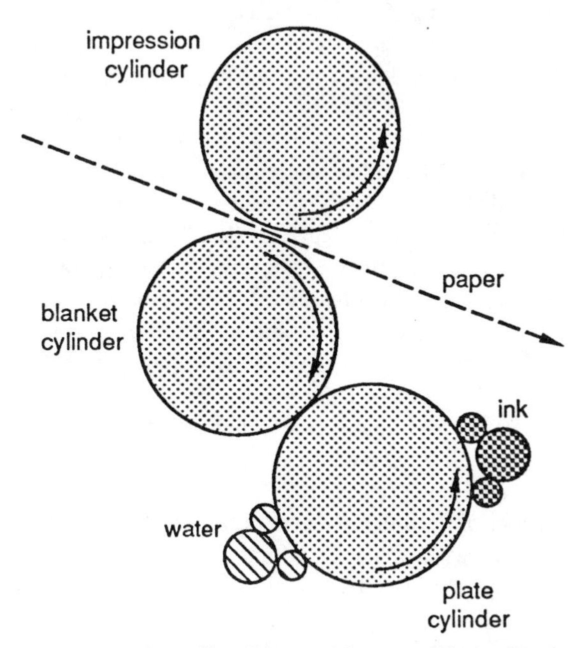

Figure 2.18. Elements of an offset lithographic press. Water adhering to noncoated portions of the lithographic plate repels ink, whereas waxy-coated portions of the plate accept ink. Inked image is transferred to the rubber-coated blanket cylinder and then "offset" onto the paper, which is pressed between the blanket and impression cylinders.

can be applied as a *flat tint* to fill a designated area with pigment, or as a *screened tint*, with pigment distributed in a pattern of dots, arrayed in a grid as on a halftone photo. Screened colors are lighter than solid colors for the same ink, and different screened tints can be overprinted to produce new colors, as when screened blue and yellow are combined to print green. Inks for the *process colors*—black plus the subtractive primaries of yellow, magenta (a purplish red), and cyan (a bright blue)—can be combined in various screened proportions to yield a wide range of colors, and virtually any hue. Electronic color-separation scanners, available since the mid-1970s, automatically examine a color photograph and produce the halftone negatives needed for each of the four process colors. They also allow the operator to adjust contrast and retouch the image with an "electronic airbrush."[70] For maps and other line drawings, a black-and-white image can be scanned, and colors selected from a palette on the screen can be painted with a light pen onto lines and areas designated on the displayed drawing. Because of more consistent inking, offset lithography is more suitable than letterpress for printing color-separated artwork.

In the late 1970s technical advances in color separation and competition with color television encouraged daily newspapers to increase their use of color, especially for front-page and feature-page photographs.[71] Papers without a scanner either produced separations with a photo-optical system using filters or had them made at an outside color laboratory. Wire services and syndicates also supplied color separations. And for maps and charts a staff artist could prepare the needed separations mechanically, drawing or cutting one "flap" for each color. With an increased number of advertisers supplying color separations for their displays, newspapers could conveniently "piggy back" editorial (that is, nonadvertising) color artwork onto another page printed from the same cylinder. Thus a full-color ad on the back page of the first section subsidized the color photo on the front page. In 1983, a new national newspaper, *USA Today*, convincingly demonstrated the possibility of consistent, high-quality presswork, and encouraged many newspapers to order scanners and new presses.[72]

The appeal of graphics to readers is perhaps best revealed in the history of the Sunday newspaper, which for almost a hundred years has been the most highly illustrated paper offered American readers. In the competitive news business, the inherent receptiveness and expectations of consumers enjoying their traditional day of rest were strong incentives for innovation.[73] Perhaps the strongest believer in the strength of the Sunday edition was Adolph Ochs, who bought the faltering *New York Times* in 1896. Three weeks after he took over, Ochs started a Sunday supplement with halftone photographs.[74] In 1914, he started the first

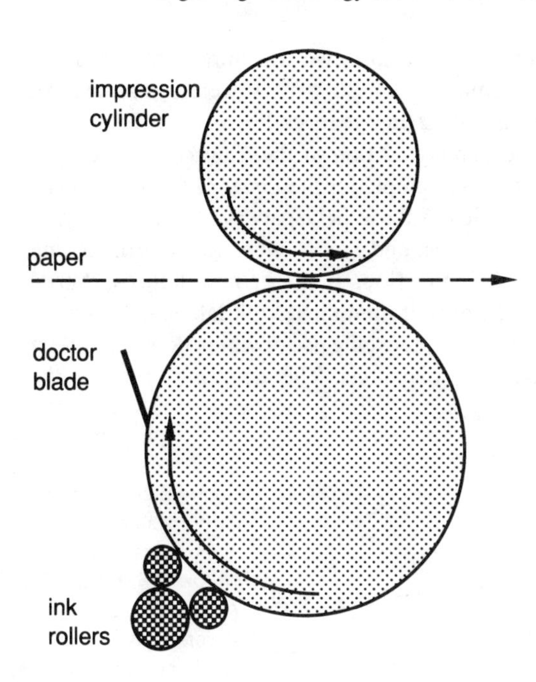

Figure 2.19. Elements of a rotogravure press. Rollers spread ink into etched areas below the plate surface, and doctor blade wipes surface ink from nonimage areas. Paper receives image when pressed between the plate and impression cylinders.

rotogravure Sunday supplement in the country. Developed commercially in England late in the nineteenth century, *rotogravure* is the adaptation to a high-speed rotary press of intaglio printing, in which ink-bearing parts of the plate are etched below the nonimage surface, kept clean of ink by a large "doctor" blade located between the ink rollers and the impression cylinder (Figure 2.19).[75] The present-day Sunday newspaper is noted for its color comics, its color magazine (which for smaller papers is usually provided by a syndicate and printed weeks ahead), and a variety of special sections. Cartographic artwork has become particularly important on Sundays, for enhancing "backgrounder" articles on national and foreign political and economic affairs, as well as for providing directions or "flavor" for stories on travel or recreation.

Trends in Map Use by Elite Newspapers

A sampling of map use by five elite newspapers provides an empirical base for exploring the rise of journalistic cartography. The five papers examined are the *Times* of London, the Toronto *Globe and Mail*, the *New York Times*, the *Christian Science Monitor*, and the *Wall Street Journal*.[76] The first three are the newspapers of record for Britain, Canada, and the

United States, respectively, and the last two are important American dailies with a national circulation. These papers were examined for January and July, for 1985 and all earlier years ending in zero. To provide a British counterpart to the Sunday edition of the *New York Times*, the *Sunday Times* was included in the sample for the *Times* of London.[77] Data collected from microfilm copies of these newspapers are based only on maps for news and feature articles, and ignore maps in advertising and decorative artwork, as well as the floor plans of new houses. Aerial photographs were considered only if annotated with labels or symbols. This analysis also ignores weather maps, the use of which does not reflect individual, day-by-day editorial decisions. The sampling unit was the

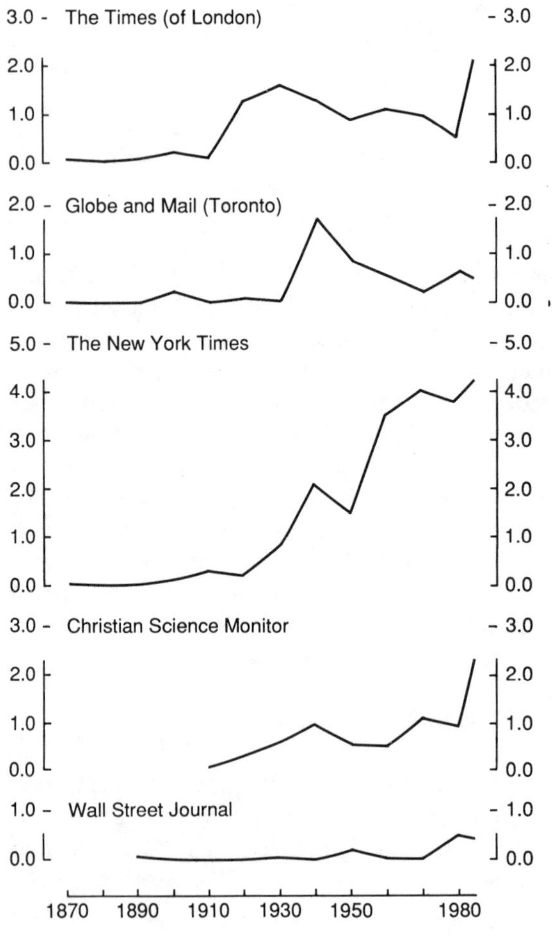

Figure 2.20. Average daily mean number of mapped articles, 1870-1985, for the *Times* of London, the Toronto *Globe and Mail*, the *New York Times*, the *Christian Science Monitor*, and the *Wall Street Journal*. Means are based on all issues published during January and July of the years sampled.

article, not each individual map, and for articles with two or more maps, map characteristics were only coded for the larger or more important map. For each newspaper a mean rate of map use for each year was first computed for every day of the week, by dividing the number of mapped articles appearing on, say, a Tuesday by the number of Tuesdays in January and July of that year. Summing these adjusted daily means and dividing by the number of days of the week the newspaper was published yielded the average daily means used to construct the graphs and tables for this section.

These newspapers varied widely in their use of maps, as shown in Figure 2.20 for 1870 through 1985. No maps were detected in the sample until 1870, and news maps were rare indeed until early in the twentieth century. The *Times* did not attain an average rate of one nonweather map per day until the 1920 sample, and the *New York Times* and the Toronto *Globe and Mail* did not reach this frequency until 1940. The *Christian Science Monitor*'s use of maps reached the map-per-day average in 1970, but the *Wall Street Journal* was still averaging one mapped article every other day in 1985. The *Journal*'s most notable increase occurred between 1970 and 1980, in contrast to a much earlier surge at the *Times*, between 1910 and 1920. The *New York Times* registered two significant increases, between 1930 and the outset of World War II in 1940, and between 1950 and 1960, reflecting an expansion of feature stories and an improved coverage of national and local events.[78] The importance of maps for wartime reporting underlies a peak for the *Globe and Mail* in 1940, and a marked decline thereafter. Map use by the *Monitor* also showed a temporary rise for 1940, but more significant is the increase from slightly less than a map per day for 1980 to well over two maps per day for 1985. Between 1980 and 1985 the *Times* increased its average daily frequency from less than 0.6 to over 2.1, reversing a general decline since 1930, and its New York namesake advanced to an overall high, among all five papers for all years sampled, of 4.3 maps in the average daily issue. In general, the trends in Figure 2.20 suggest an adoption of journalistic cartography delayed well past the 1890s, when photographic line-engraving had become common and efficient. World War II was a major but generally short-lived incentive for more news maps. An increased use of maps in the 1970s and 1980s by all five newspapers reflected a design revolution in which graphics in general were accorded a fuller role in news presentation.

Map use by these five newspapers also has reflected a variety of unique circumstances and persons. The *Journal,* for instance, has long eschewed halftone news photographs and, as a financial newspaper, has focused on stories with particular relevance to businessmen and investors.[79] The ascendancy in Canada of the *Globe and Mail* reflects the shift of that nation's business center from Montreal to Toronto in the 1960s and 1970s, when Quebec restricted the use of languages other than French.[80] Founded by the Christian Science church but largely secular in outlook, the *Monitor* almost from its inception has addressed a national audience, most of whom receive their paper a day late, through the mail. An emphasis on thoroughness and monitoring important national and foreign events is central to both its philosophy and its survival.[81] The *Monitor*'s reputation for excellence in journalistic cartography began with artist Russell Lenz, who drew thousands of hand-lettered maps for the paper between 1928 and 1970.[82] Lenz trained his assistant, Joan Forbes, to draw maps, and a *Monitor* redesigned as a tabloid in 1974 maintains this tradition with several artists, cold-type labels, and a more contemporary graphic design. With an editorial art department that includes a ten-person "map group," the *New York Times* was a pioneer in the "sectional revolution" of the 1970s.[83] Maps have been an important part of the artwork for the cover and inside pages of its numerous weekday and Sunday feature sections. In the mid-1950s, its publisher, Arthur Hayes Sulzberger, who liked maps, decided to use smaller maps but more of them.[84]

In London the *Times* had fallen on hard times in recent decades, and lost money for a number of different owners, who absorbed the losses to preserve a prestigious British institution. Cost containment, a traditional preference among journalists for words rather than pictures, and a decline in Britain's influence abroad would seem to underlie the post-1930s erosion of journalistic cartography at the *Times*. In 1978-1979 the *Times* and the *Sunday Times* suspended publication for 11 months in a labor dispute. Bought in 1981 by Rupert Murdoch, an Australian publisher who also owns a number of sensational tabloids, the *Times* began to use more maps and other information graphics in the early 1980s after the arrival as editor of Harold Evans, who had edited another Murdoch paper, the *Sunday Times*. This renewed emphasis on information graphics continued after Evans left the *Times* in 1982, following a dispute with Murdoch over editorial policy and management style.[85] Mid-1980s circulation for this flagship of Murdoch's newspaper group was around 340,000.[86]

Average daily rates of map use mask considerable variation throughout the week, as shown in Figure 2.21 for the period 1920 through 1985 for all papers but the cartographically marginal *Wall Street Journal*. Particularly prominent is the high rate of map use on Sunday, evident as early as 1930 for the *New York Times* and by 1970 for London's *Sunday Times*. Sunday news maps, in fact, account for much of the ascendancy of the *New York Times*: for 1985 average daily rates of map use computed just for Monday through Friday are much closer, at 2.7 and 2.3, respectively, for the *New York Times* and the *Christian Science Monitor*.

Moreover, in recent years most of the growth in map use by the New York paper has been in its massive Sunday edition. In a more minor way, maps in the *Sunday Times* in 1970 and 1980 partly offset an otherwise precipitous drop in weekday maps in the *Times*. Without a Sunday edition, Toronto's *Globe and Mail* seems in the 1980s to have offered more maps on Monday and Saturday than on the four intermediate days combined. Weekend editions distributed on Saturday were responsible for cartographic peaks at the *Globe and Mail* in 1940 and the *Christian Science Monitor* in 1930 and 1940. Maps associated with weekend recreational events explain in part the minor peak on Friday for the *Monitor* in 1985 and for the *New York Times* in 1970 and 1980. Like most American metropolitan newspapers, the *New York Times* assembles its Saturday edition with a reduced staff. It has the fewest maps, the smallest number of pages, and the smallest circulation of the week.

High average daily rates of map use in 1940, as shown in Figure 2.20, suggest the importance of war as a cartographic theme, and high daily rates on Sunday and toward the end of the week, as shown in Figure 2.21, suggest travel, sports, and recreation as another broad topic commonly treated by news maps. Table 2.1, a comparison of the samples for 1940 and 1985, verifies these inferences. The percentage of maps devoted to recreation has risen for all papers except the *New York Times*—an anomaly explained by the delay until late 1941 of America's combat role in World War II and the 1940 New York World's Fair, the events of which were covered with a daily map throughout that summer. Although maps covering battles and attacks were less common in 1985, these elite newspapers nonetheless devoted from a fifth to over half of their mapped articles to the broad category of military conflict, defense, geopolitics, and terrorism. Between 1940 and 1985, the *Monitor* actually increased the share of maps covering international affairs, from 49 to 55 percent, and over half of the mapped articles in the *Journal* addressed this subject—52 percent, in contrast to only 29 percent for stories on economic resources, development, and transportation. Articles on science, exploration, and environmental hazards increased in relative importance. Downplaying sensational news, these elite newspapers used comparatively few maps with stories on crime, accidents, and natural disasters. In general, the relative decline in war-related stories was accompanied by enhanced cartographic coverage of public works, planning, and neighborhoods; politics, elections, and legal matters; and a variety of general educational topics, such as history, demography, and news summaries. Without a commitment to local coverage of a metropolitan area, the nationally oriented *Christian Science Monitor* and *Wall Street Journal* differed from the other three dailies in topics such as planning, neighbor-

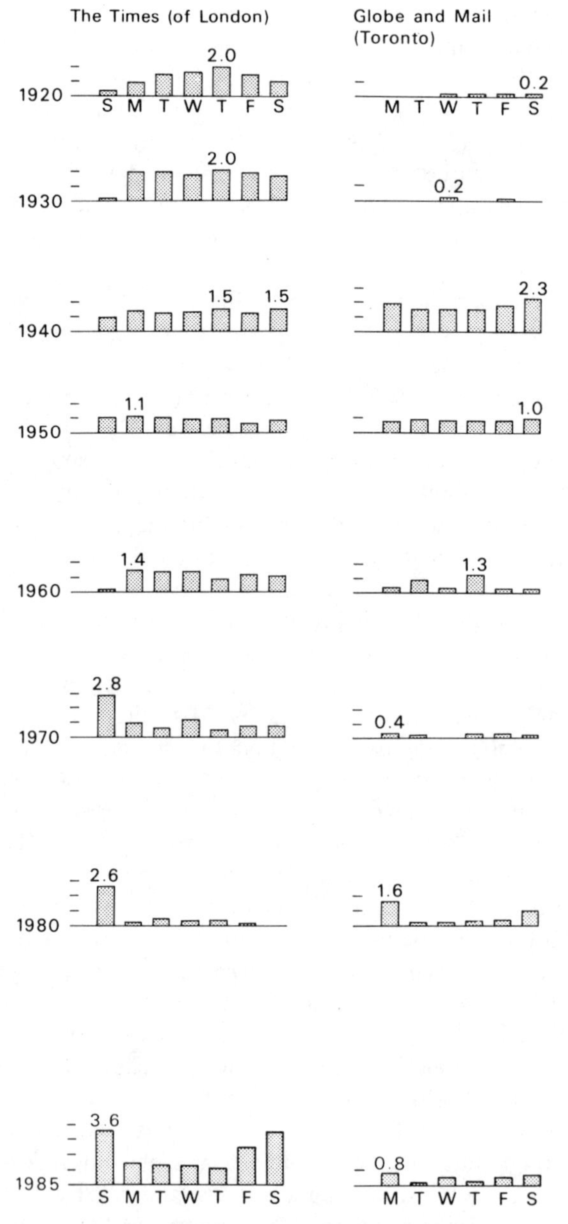

Figure 2.21. Trends in the daily mean number of mapped articles, by day of the week, 1920-1985, for the *Times* of London, the Toronto *Globe and Mail*, the *New York Times*, and the *Christian Science Monitor*. The *Monitor* discontinued its Saturday edition in the 1970s. Means are based on all issues published during January and July of the years sampled.

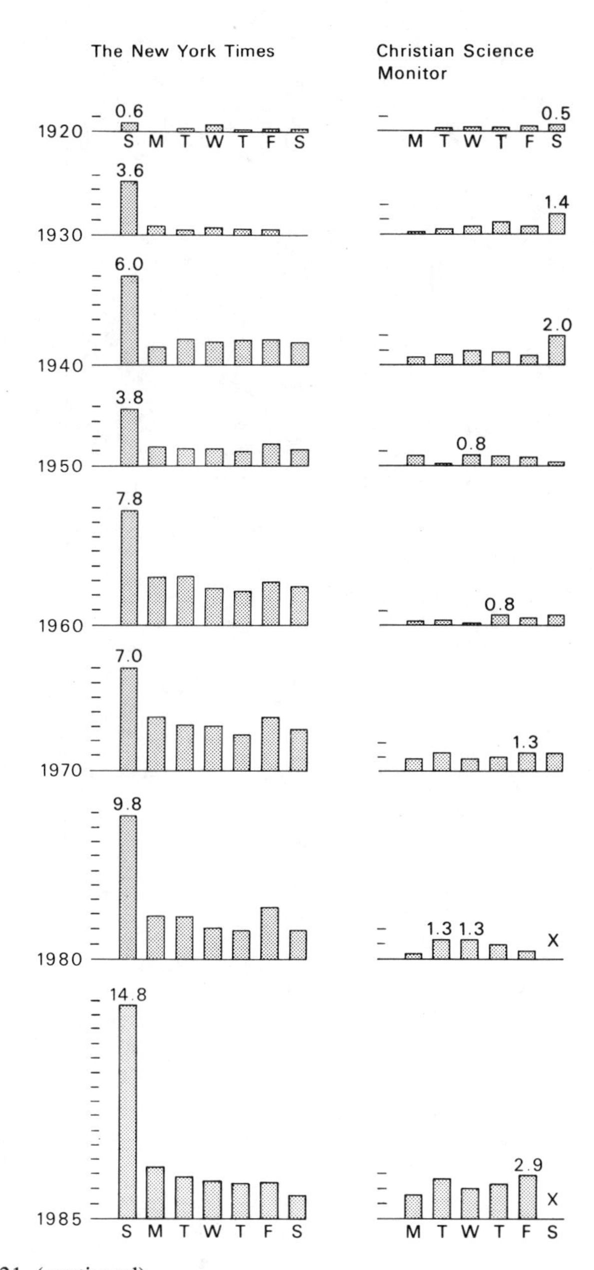

Figure 2.21. (continued)

Table 2.1 General Theme As a Percentage of Nonweather Articles with Maps, 1940 and 1985.

	The Times (London)		*Globe and Mail*		*New York Times*		*Christian Sci. Monitor*		*Wall Street Journal*	
Mapped Articles	1940	1985	1940	1985	1940	1985	1940	1985	1940	1985
Theme of Article:	%	%	%	%	%	%	%	%	%	%
Crime	0	2	1	0	1	2	0	0	—	0
Public works, planning, and neighborhoods	0	10	1	4	5	15	6	3	—	0
Tourism, travel, recreation, and sports	0	22	0	9	32	28	4	9	—	5
Politics, elections, and legal matters	0	4	0	9	1	5	2	8	—	5
Military conflict, defense, geopolitics, threats, and riots	94	37	89	39	57	20	49	55	—	52
Economic resources, development, and transportation	1	4	2	9	2	11	12	6	—	29
Accidents and natural disasters	4	5	6	9	0	5	0	2	—	0
Science, exploration, and environmental hazards	1	7	0	17	2	10	0	11	—	0
Education, history, demography, news summaries, and so forth	0	8	0	4	1	4	27	6	—	10
Number of mapped articles in the sample	78	125	93	23	126	255	49	101	0	21
Average daily mean number of mapped articles	1.3	2.1	1.7	0.4	2.1	4.3	1.0	2.3	0	0.5

Percentages are based on all nonweather mapped articles in the newspaper's sample for the year.
Because of rounding, column percentages might not sum to 100.

hoods, recreation and accidents. More notable than the generally increased use of maps by all five elite newspapers was the greater diversity of themes treated cartographically.

Trends in Map Use by Weekly News Magazines

The survey of map use by six weekly news magazines supplements the case study in the previous section by examining another nationally important news medium. By covering national and world political, economic, and cultural news for a national audience, news magazines function as national newspapers.[87] Unable to cover breaking news with the frequency and speed of daily newspapers, the weekly magazines compensate with comprehensive summaries and thoughtful analyses, often supported by clever and colorful information graphics. Far more people read a weekly news magazine than *USA Today*, the *Wall Street Journal*, or any other daily newspaper with a national circulation, and for many readers the weekly news magazine serves as an influential complement to television news or their local daily newspaper.[88] Now printed in full color on coated paper by offset lithography, weekly news magazines are organized into "departments," or sections, each treating national affairs, foreign affairs, or one of several interest areas such as science and technology, business, and the arts. Much effort is given to monitoring current developments in these areas and to verifying facts that might be used in or as background for news stories. The weeklies have staff correspondents in important foreign news centers as well as in several domestic cities, and receive the same wire-service reports available to daily newspapers. Journalists at Britain's *Economist* even refer to their publication as "the newspaper," and in 1983, the *Washington Post* began distribution of a separate national weekly edition, printed on newsprint as a tabloid, but in a magazine format.[89] Weekly publication seems a workable compromise between the hectic pace of the daily paper and the more reflective and often more literary stance of the monthly magazine.

Like newspapers, magazines owe much of their growth to advances in printing and photoengraving. Magazines existed as early as 1704, when Daniel Defoe started a weekly journal of news and opinion called the *Review*. Defoe's magazine, which lasted only seven years, was followed by a number of other English magazines, including the *Tatler* and the *Spectator*, founded in 1709 and 1711, respectively. Both offered light humor, gossip, and short stories, as well as news and political commentary, and the *Spectator* evolved to become one of modern Britain's leading weekly journals of news and essays. Illustrations were rare in magazines until 1731, when Edward Cave founded the *Gentleman's Magazine*, a monthly that carried engravings and maps. Cave's success is believed to have inspired the first two American monthly magazines, Andrew Bradford's *American Magazine, or a Monthly View of the Political State of the British Colonies* and Benjamin Franklin's the *General Magazine*, both started in 1741.[90] Yet magazine illustration was uncommon in America until the mid-nineteenth century, when *Frank Leslie's Illustrated Newspaper* (1855-1922) and *Harper's Weekly* (1857-1916) offered wood engravings to enhance their news reports and short stories. But the greatest technological advance by picture magazines came in the 1890s, when photographic engraving lowered the cost of preparing illustrations for advertisements as well as for news and feature stories, and magazine publishers lowered prices to attract a mass audience. A halftone photoengraved for less than $20 could replace a page-size wood engraving costing as much as $300. In 1893, *Munsey's Magazine* (1889-1929) lowered its street price from 25 to 10 cents, and a year's subscription from $3 to $1. *Cosmopolitan* increased its circulation from 16,000 in 1889 to 400,000 in 1894, to join a growing number of successful mass-circulation magazines that included *Century*, *Harper's*, *McClure's*, the *Saturday Evening Post*, and *Scribner's*.[91] By 1900, magazine illustration was common, and in varying numbers all large-circulation weekly and monthly magazines used pictures and, occasionally, maps.

Time was the first member of what emerged as a distinct journalistic genre: the American news magazine, devoted to timely reporting and omitting the literary trappings of most earlier magazines that dealt with the news.[92] Founded in 1923 by recent Yale graduates Henry Luce and Briton Hadden, *Time* sought to keep its readers informed by condensing the news into a "curt, clear, complete" narrative for the busy man with neither the time nor the inclination to follow the comparatively long-winded daily press.[93] Unlike *Life*, which Luce founded in 1936 to condense the news in pictures, *Time* consisted largely of text punctuated with generally small, concise photos, charts, and maps. A *Time* article typically was written by several people, using inverted sentences, focusing on personalities, and addressing the origin, importance, and likely future consequences of significant events. Circulation rose from 9,000 in early 1923, to more than a quarter million in 1929, to 4.6 million in the mid-1980s, attesting to widespread public acceptance of "group journalism" and "*Time* style."[94]

Time's success was at least partly responsible for the emergence in 1933 of two additional weekly news magazines, *News-Week* and *U.S. News*. Thomas Martyn, a former foreign editor at *Time*, used $2,250,000 raised from 120 investors to found *News-Week*, which four years later merged with another magazine, *Today*, dropped the hyphen to become *Newsweek*, and evolved from a mere news digest into a magazine providing background explanation, interpretation, and signed opinion. Circulation rose steadily during the 1950s, and continued to rise after the magazine's purchase in 1961 by Philip Graham, owner of the *Washington Post*. Its mid-1980 circulation of 3.0 million placed *Newsweek* second in readership to *Time* among weekly news magazines.[95]

Founded by Washington political reporter David Lawrence, *U.S. News*, as its name suggests, focused on national news, and from its inception in 1933 provided analysis and forecasts. In 1946 Lawrence introduced a sister publication, *World Report*, and two years later merged the two into a single magazine. *U.S News and World Report* focused on economic and political affairs, and its reporting often reflected Lawrence's political conservatism. After lagging well behind *Time* and *Newsweek*, its circulation increased 244 percent between 1950 and 1962, to over 1,250,000, and advanced further in the mid-1980s, to over 2.1 million.[96]

Business Week, one of many magazines serving a specific profession or interest group, is treated here as a national weekly news magazine because of its wider coverage of national and foreign news.[97] Founded in 1929 by McGraw-Hill, *Business Week* survived the Depression and in 1980 became one of the ten top magazines in total advertising revenue.[98] Its mid-1980s circulation exceeded 850,000.

Two additional magazines included in this case study reflect map use among weeklies in Canada and the United Kingdom. Founded in 1843 by banker and politician James Wilson, Britain's *Economist* originally addressed the concerns of businessmen dealing in commodities, railroads, and other investments.[99] Its use of illustrations other than business charts was minimal until the 1930s, but its departmental organization, broad scope, and use of graphics is similar in the late 1980s to American news weeklies. Indeed, the *Economist* might even be considered an American magazine, for not only is it less British in outlook than two decades ago, but since 1985 a North American edition has been printed in Connecticut, using plates made from film flown across the Atlantic on the Concorde.[100] In the mid-1980s, less than a third of the *Economist*'s worldwide circulation of over 250,000 was in Britain. Similar in content to its European counterpart but with the "American Scene" department moved to the front of the magazine, the North American edition had 100,000 regular readers. Still providing a business intelligence service, The Economist Publications Ltd. supplements its weekly "newspaper" (in staff jargon) with several highly regarded, more expensive newsletters, including the monthly *Development Report*, the biweekly *Financial Report*, the biweekly *Connections* (about communications), and the confidential weekly Foreign Report, as well as a variety of topical "Special Reports" from the Economist Intelligence Unit. All Economist newsletters and reports use maps and information graphics.

Self-proclaimed "Canada's national newsmagazine," *Maclean's* has been a member of the *Time* genre only since 1978, when it announced the "intention to humanize the endless flow of decisions that govern our lives from a point of view that no newsmagazine has ever attempted before: an authentically Canadian perspective. . . [as] Canada's first newsweekly."[101] *Newsweek* and *Time* compete in the Canadian market, but Canada's tax laws have discouraged advertising in foreign publications. Older than its competitors, *Maclean's* was founded in 1905 as *Busy Man's Magazine*, and received its present name in 1911. It published fortnightly through 1966, and then monthly until 1975, when it became a semi-monthly for three years. Its format was similar to that of *Life* and *Look* from the 1930s through the 1960s, and akin to that of *Harper's* and the *Saturday Review* while it was a monthly. As a weekly covering world and national news with a distinctly Canadian outlook, *Maclean's* had a mid-1980s circulation of more than 640,000, including sales outside Canada of fewer than 16,000 copies.[102] Within Canada, it outsold *Time* and *Newsweek*, which had Canadian circulations of approximately 350,000 and 90,000, respectively.

A sample of all January and July issues, similar to the sample for the five elite newspapers but beginning in 1930, provides an empirical base for examining map use by these six weekly news magazines. Two indexes address the frequency and consistency of their use of maps. The average number of maps per issue, obtained by dividing the number of mapped articles and stand-alone maps for the year by the total number of issues with a January or July cover date, is a measure of average cartographic effort comparable to the average daily mean rates computed for daily newspapers. Because a number of magazines published issues without maps, even for the more recent sample years, a second index—the percentage of issues with one or more maps—is used to measure the regularity of map use. That these two indexes complement each other is demonstrated by *U.S. News and World Report*, which in 1985 had the highest per-issue rate (more than 3.1) but only the third highest percentage of issues with maps (75 percent). In 1980, it registered the highest rate of the six magazines for any year (3.6)—a rate more than three times higher than *Time*'s (1.1) for that year. Yet both *Time* and *U.S. News and World Report* employed at least one map in every issue sampled for 1980.

As shown in Figure 2.22, a time-series graph of the per-issue rate, weekly news magazines have substantially increased their use of maps since 1930. Of the four magazines publishing in 1930, all had rates below one mapped article every two weeks, in contrast to average rates in 1985 for four of the six magazines of at least one mapped article each week. But although generally upward, the trends have been erratic, perhaps

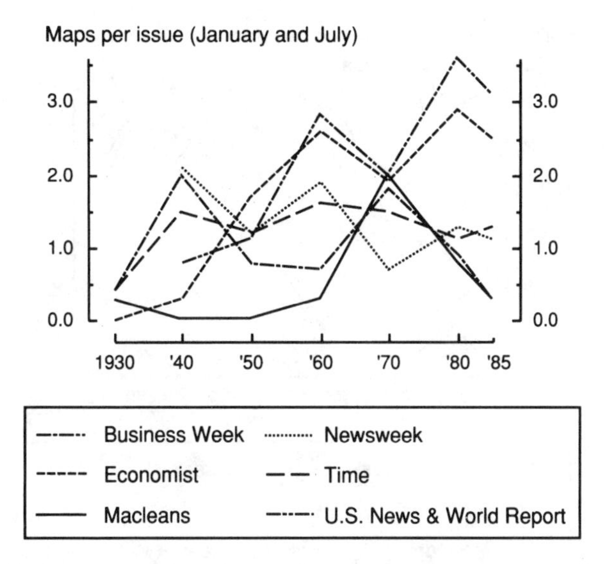

Figure 2.22. Average number of mapped articles per issue, 1930-1985, for *Business Week*, the *Economist*, *Maclean's*, *Newsweek*, *Time*, and *U.S. News and World Report*. Means are based on all issues with a January or July cover date for the years sampled.

because of temporal variations in the significance in the news of geographic relationships or locations. Peaks for 1940, 1960, and 1980 reflect World War II and presidential elections that coincided with the 10-year sampling interval.

More striking is the variation among the six magazines. The *Economist* and *U.S. News and World Report* show noteworthy and consistent increases, except for comparatively slight declines from 1960 to 1970 and from 1980 to 1985. After a significant rise between 1930 and 1940, *Time*'s rate has been generally stable, in a range between 1.1 and 1.6 mapped articles per week. Its principal competitor, *Newsweek*, which led all news magazines in 1940 with a per-issue rate of 2.1, has fluctuated more widely, with a drop to a low of 0.7 for the 1970 sample. For 1980 and 1985, though, *Newsweek* and *Time* had similar rates, between 1.1 and 1.3. *Business Week*, which registered the relatively high rates of 2.0 and 1.8 for its 1940 and 1970 samples, respectively, declined to 0.9 mapped articles per issue for 1980, and to a mere 0.3 for 1985. *Maclean's*, which was not a typical weekly news magazine until 1978, nevertheless reached its peak per-issue rate of 2.0 in the 1970 sample as a monthly, and recorded the much lower rates of 0.8 and 0.3 for 1980 and 1985. Three clusters emerge in the 1985 sample: *Business Week* and *Maclean's* with low per-issue rates, below

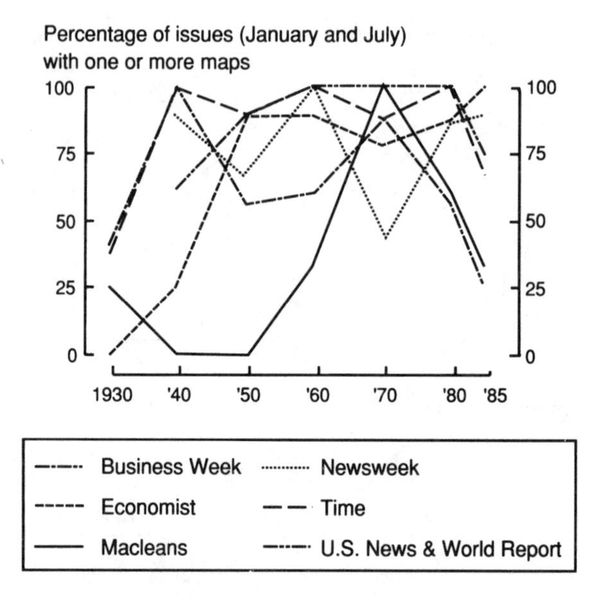

Figure 2.23. Percentage of January and July issues with at least one map, 1930 to 1985, for *Business Week*, the *Economist*, *Maclean's*, *Newsweek*, *Time*, and *U.S. News and World Report*. Percentages are based on all issues with a January or July cover date for the years sampled.

0.4; *Newsweek* and *Time* with rates slightly above 1.0; and the *Economist* and *U.S. News and World Report*, with comparatively high rates, above two mapped articles every week.

As shown in Figure 2.23, news magazines with higher per-issue rates generally have been more consistent in their use of maps. Among the most regular map users are the *Economist* and *U.S. News and World Report*, both of which employed maps in at least three-quarters of their issues from the 1950 sample onward. For 1940 through 1980 *Time* used maps in over four-fifths of the issues examined, despite a drop in its consistency index from 100 percent for 1980 to 67 percent for the 1985 sample. *Newsweek*, which used maps in only 37 percent of its January and July issues in 1970, presents a more uneven pattern, marked by a recovery in its consistency index to 87 percent for the 1980 sample. It seemed to have achieved a stable regularity, and employed maps in 89 percent of the issues sampled for 1985—a consistency exceeded that year only by the *Economist*. Not surprisingly, the two magazines with the lowest per-issue rates for 1985 also had the smallest percentages of issues with maps. *Maclean's*, which used no maps during January and July in 1940 and

1950, when it published only twice a month, carried mapped articles in both monthly issues in the 1970 sample, yet as a weekly fell in cartographic consistency to 62 percent for 1980 and 33 percent for 1985. *Business Week*, which experienced a similar decline for these two most recent samples, had been among the most regular users of maps in 1940 and 1970. Perhaps the only trend suggested by Figure 2.23 is that news magazines have used maps with modest but variable frequency after shifting in the 1930s from a more limited pattern of journalistic cartography.

Unlike newspapers, news magazines appear not to have increased greatly their use of maps in the transition from hot type and letterpress to cold type and offset lithography. Because of large circulations and less frequent deadlines, magazines of the *Time* genre have long had richer graphics than daily newspapers. But their concise verbal accounts and inherently limited length have constrained the magazines' capacity for cartographic expansion—and for an increased use of pictures in general. Having found their niche in the print media, the news magazines have been reluctant to abandon a successful formula. Unlike the newspapers, they could not add sections or produce a massive Sunday edition. Subscribers valuing the magazines' condensed and selective presentation of the news would not tolerate much longer issues, with more and longer stories and more graphics. Besides, *Time* already had a pictorial sister publication, *Life*, and for readers narrowly interested in a specific kind of news, there were *Sports Illustrated*, *People*, *Money*, and numerous publications with emphasis on business, politics, the arts, and other topics the weekly news magazines treated with a single department.

Ironically, one effect of the recent design revolution on news magazines seems to have been to inhibit their use of maps. Without question, weekly news magazines did indeed become "more graphic" in the late 1970s and 1980s—but not necessarily by adding more graphics. Although the number of illustrations per page might not have increased substantially, art directors and executive editors became much more conscious of the need for graphic appeal. The magazines are now using color photographs more frequently, at the expense of black-and-white photos, and most maps are now in full color. Most notable, though, has been an increased use of the information graphic—the statistical diagram—which seems more widely used than the map as a stand-alone graphic or sidebar. Histograms, pie charts, and time-series "fever charts" once were far more mundane than most maps. But the addition of drop shadows, colored backgrounds, pictorial icons, and other "creative" flourishes can easily convert a straightforward data graph into a catchy (if not actually "hyped") page decoration.[103] Apparently most news stories

Table 2.2 Average Number of Map and Nonmap Data Graphics, for January and July Issues, in Six Elite Weekly News Magazines, 1975, 1980, and 1985.

	1975		1980		1985	
Magazine	Map	Nonmap	Map	Nonmap	Map	Nonmap
Business Week	0.4	6.4	0.9	5.9	0.3	8.8
The Economist	3.8	8.5	2.9	9.6	2.5	16.6
Macleans	0.5	0.0	0.8	0.4	0.3	0.1
Newsweek	1.6	1.3	1.3	1.8	1.1	1.3
Time	1.0	1.6	1.1	1.4	1.3	1.1
U.S. News and World Report	2.2	4.5	3.6	3.7	3.1	5.6

can be dressed up more easily with a statistical diagram than with a map, and as Table 2.2 demonstrates, charts—rather than maps—account for most of the recent growth in information graphics.

Numerous technological advances, most notably photoengraving and photomechanical page makeup, underlie the increased prominence of news maps. But the rise of journalistic cartography reflects two other currents in the history of publishing: broader reporting and improved visual appeal. Early increases in the number of news maps reveal the effects of photoengraving technology and wartime journalism, whereas more recent increases—and the use of maps to illustrate a more diverse array of news and features—reflect a wider journalistic scope, including stories about travel and local economic development, as well as a heightened concern with the organization and appearance of news publications. Chapter 3 explores the role of wire services and syndicates, journalistic institutions particularly important in providing nonlocal news maps to small and medium-size daily newspapers.

3 Wire Services, News Syndicates, and Graphics Gatekeepers

Smaller daily newspapers, which can afford neither foreign correspondents nor staff artists, have long relied upon press syndicates for nonlocal news and whatever maps might be needed to describe foreign wars, national elections, and the sites of devastating accidents and disasters. Economies from sharing the reporting costs for news stories also accrue to the drawing costs for information graphics: a map explaining a terrorist attack in the Middle East is produced more efficiently in one location by a single artist for distribution to several hundred newspapers than by each paper's own illustrator. The wire-service map should be more accurate as well, for a press association's map maker generally would have not only more complete source materials but also ready access to correspondents in the field. Without syndicates and wire services, few daily newspapers would have timely news maps illustrating stories of more than local importance.

Syndicate design strategies, operating practices, and distribution systems have promoted a homogeneity in news graphics much more marked than in news stories. An editor can readily shorten a wire-service news story, incorporate a local angle, change the emphasis, and even prepare a composite story from accounts supplied by two or more news syndicates. Yet rarely has a local newspaper altered or redrafted a wire-service news graphic, except to change its size. For small and medium-size newspapers, the news-graphics service has been both an indispensable facilitator and a rigid gatekeeper.

This chapter explores the rise of news syndicates and the development of the electronic technology that permits the timely distribution of centrally produced maps and information graphics. Perhaps the single most significant advance has been photofacsimile transmission, which is almost as old as the electronic telegraph. Despite the sporadic success in the late nineteenth and early twentieth centuries of several independent inventions, photofacsimile did not begin to have a wide effect on journalistic cartography until 1935, when the Associated Press opened its Wirephoto network. Substantial progress in facsimile transmission in the early 1930s reflected not only the vision of individual engineers but the organizational efficiency of the American Telephone and Telegraph

Company's Bell Laboratories and the AP, which perceived a strong demand for timely news photographs. Maps merely went along for the ride, as they had decades earlier, when stereotype engraving enabled news and feature syndicates to market line engravings and halftones to news publishers eager for low-cost illustrations of decent quality. This chapter examines briefly the histories of the Associated Press, United Press International, and several syndicates that are prominent providers of news maps. The recent revolution in newspaper design has led to a greater appreciation of the maps and other line art of graphics syndicates and graphics networks. Only with the microcomputer revolution of the mid-1980s, though, are newspapers beginning to overcome the graphic limitations of transmission equipment designed primarily for continuous-tone photographs.

News Services, Postal Systems, and Telegraph Lines

Journalism has had a long and close relationship with the telecommunications business. Before the commercial telegraph in 1844 and the telephone much later in the nineteenth century, newspapers relied on postal services not only for dispatches from distant correspondents but also for delivery of their printed product itself. Indeed, the first postal services were royal courier systems carrying news to an emperor and orders to his provincial administrators. With government sanction, private postal systems arose to meet needs for business intelligence and commercial transactions. In the sixteenth century the Taxis family operated a private mail service that served much of Europe, from the German states westward to Spain.[1] In the latter part of that century the Fugger family of Austrian bankers and traders established a network of couriers and correspondents in Europe, South America, North Africa, and East Asia. Surviving manuscripts of the Fugger news-letters provide a chronicle for modern historians of the often confused and untimely news reaching European businessmen and officials.[2] In 1840 British tax reformer Rowland Hill initiated the first modern postal system by introducing prepaid postage stamps and a uniform, countrywide letter rate, which simplified mail handling by eliminating a complex, time-consuming schedule of individual point-to-point rates for domestic mail.[3] In the American colonies a number of newspapers, including John Campbell's *Boston News-Letter* and Benjamin Franklin's *Pennsylvania Gazette*, were operated by postmasters, and persons receiving letters from distant places often shared news with these postmaster-editors and their readers.[4]

News generally traveled at the same speed as mail, which was carried with passengers on stagecoaches or ships. As transportation improved early in the nineteenth century with the development of the railway and the steamboat, news arrived sooner. In 1790, for example, European news generally was two months old when printed in Philadelphia newspapers, but by 1840 the time lag had been reduced to a couple of weeks.[5] Over this 50-year period, the average time lag for news from Baltimore dropped from 6.0 to 1.1 days, and for news from Boston, from 12.0 to 2.7 days.

With the development of inexpensive, mass-circulation daily newspapers in the early 1830s, publishers became more concerned with timeliness. A variety of signaling systems and special courier systems was used, including carrier pigeons and relay messengers on horseback. Perhaps the most ambitious preelectronic communication system was the semaphore network developed in France, starting in 1794 and operating until 1852. With the aid of telescopes, operators could read coded messages sent between stations about 15 km (10 mi) apart, which used movable signal arms or the "clockface" dials shown in Figure 3.1, mounted on towers to improve visibility. In clear weather a semaphore telegram could travel the 800 km (500 mi) from Toulon to Paris in an hour.[6] In 1825 Charles Havas organized a network of correspondents in European capitals, and in the late 1830s he added newspapers to his diplomatic and business clients. Havas used both the semaphore telegraph and carrier pigeons to gather and distribute news.[7] His company, Agence Havas, was the first modern news agency; it also dealt in advertising. In the early nineteenth century, though, the *Times* of London relied heavily upon mailed bundles of newspapers and posted reports from correspondents for news from Europe. After 1809, when Napoleon halted mail service from all continental ports, the *Times* started its own packet service to smuggle news across the Channel.[8]

No other messenger system could match the electric telegraph for speed or efficiency, and after 1844 important news moved wherever possible by wire. As with many technical advances, a number of inventors experimented with the telegraph in various forms years ahead of its commercial development. Among the many electronic signaling devices proposed or tested in the late eighteenth and early nineteenth centuries were several systems requiring one wire for each letter of the alphabet.[9] Particularly important was Hans Christian Oersted's discovery, in 1820, that an electric current flowing through a wire could deflect an adjacent magnetic needle. Oersted's observation led William Fothergill Cooke and Charles Wheatstone to develop a five-needle, five-wire telegraph in 1837, and Samuel F. B. Morse and Alfred Vail to develop a long-distance relay system using a dot-dash code in 1838. So practical was the electric telegraph and so great the demand for rapid communications that the successful demonstration of the two-wire Morse telegraph in 1844, over the 65 km (40 mi) between Baltimore and Washington, led in two decades to the installation in the United States alone of over 160,000 km (100,000 mi) of telegraph line.[10] In 1851 the first successful submarine cable, across the English Channel, reflected advances in the manufacture of insulated

Figure 3.1. An experimental telegraph demonstrated in 1791 by Claude Chappe employed 16 visually distinct signs arranged on a clock face, on which a single hand pointed to specific characters. Observers could take readings whenever the hand paused. A continuously revolving hand was also tried, with readings taken whenever a boom or a gong sounded, to indicate that the hand was pointing at the correct symbol. Alphabetic characters, numbers, punctuation, and other commands could be transmitted as sequences of two of these hexadecimal characters. [Source: Alexis Belloc, *La Telegraphie Historique* (Paris: Librarie de Firmin-Didot et Cie, 1889), 69.]

wire. In 1866 the first successful trans-Atlantic cable, between Trinity Bay, Newfoundland, and Valentia, Ireland, became a monument to electronics manufacturing, shipbuilding, and the enthusiasm and determination of Cyrus Field, the youthful "retired" manufacturer who had directed and secured financing for the project.[11] Despite service interruptions because of damaged insulation, the 1866 cable reduced the trans-Atlantic news lag from a week or more to a matter of hours. As shown in Figure 3.2, for instance, in 1870 an afternoon newspaper in Ithaca, New York, could report the day's developments in the Franco-Prussian War. Reliability increased and prices declined in the 1870s, when several other cables crossed the Atlantic, including one linking Portugal and Brazil in 1873. In a few decades improvements in cable manufacture and telegraph instruments and an increasingly more dense transoceanic cable network diminished a geographic isolation produced by millions of years of continental drift.

Early trans-Atlantic messages cost $5 a word, and cable news dispatches were necessarily brief. Even land-line telegraphic charges could strain the news-gathering budget of a mass-circulation daily newspaper.[12] Publishers economized by dropping "a," "the," and other easily reinserted words, and by collapsing words and phrases to coin abbreviations similar to the acronyms of present-day computer programs and federal agencies. Thus, "pwf" meant "powerful," and "potus" stood for "President of the United States."[13] But these savings were puny when compared to the scale economies of a press association, in which newspapers not only shared telecommunications charges but also the cost of a large network of reporters and editors. No single paper could afford the news network of even a modest-size press association.

Although accelerated by electronic communications, cooperative news-gathering arose when news publishers recognized that the steady advantages of comprehensive national and foreign reporting outweighed the ephemeral and uncertain benefits of an occasional scoop. The Harbor News Association, precursor of the AP, began in 1848, when six New York City newspapers pooled their efforts at sending newsboats out to steamships arriving from Europe with dispatches and newspapers.[14] After developing a communications system for their own needs, the members sought to sell their foreign reports to newspapers outside New York. They also organized the New York Associated Press as a cooperative to collect domestic news. Several subsidiary regional press associations formed, such as the New York State Association and the Western Association, and these groups collectively became the Associated Press.[15] By 1872 the New York Associated Press provided telegraphic news to more than 200 daily newspapers, at an annual cost of over $200,000 for cable service alone. In

1880 the AP supplied foreign and domestic news reports to 355 newspapers—30 percent of the nation's dailies—and sent 611 million words by Western Union, at a cost of almost $393,000.[16] Because of its size, the AP was able to negotiate a somewhat favorable "press rate."

In the second part of the nineteenth century, news agencies formed in most of the world's countries, following the examples of the Agence Havas, the AP, and Reuters, an independent agency founded in 1851 by

ITHACA, N. Y., FRIDAY, JULY 22, 1870.

By Telegraph

TO THE

ITHACA DAILY JOURNAL.

TO FOUR O'CLOCK, P. M.

Battle Fought!

The French Defeated.

200 PRISONERS CAPTURED!!

FOREIGN INTELLIGENCE.

A BATTLE AT SAARBACK.

A battle is reported to have been fought near Saarback, between the French and Prussians, the former being defeated and pursued. The fight was occasioned by two hundred French reconoitering across the frontier. After a little fighting all were taken prisoners. Some were killed.

Figure 3.2. The *Ithaca Daily Journal* displayed a daily telegraphic dispatch at the top-center of its front page. Dispatch for July 22, 1870, reported a Prussian victory in the Franco-Prussian War, which led to the unification of Germany in 1871. [Source: *Ithaca Daily Journal*, 22 July 1870, 1.]

Julius Reuter, a former bank clerk who had once operated a carrier pigeon service for Havas. Reuters developed the most extensive worldwide coverage of any news service, and for decades was the principal source of news from Africa, East Asia, and the Middle East.[17] Many governments formed an official or semi-official news agency, and others endorsed an affiliate or branch of a worldwide service, such as Spain's arrangement with Havas.

Telegraphic news services have had a profound effect on the style as well as the timeliness of news stories. Before telegraphic dispatches, nonlocal news stories often were rambling narratives that established a background for the correspondent's observations and opinions but concealed until the end many important facts about the central event and its effects. Because telegraphed verbosity was costly and line failure could interrupt the dispatch, editors encouraged a writing style termed the "inverted pyramid," in which the climax follows the lead, the first few paragraphs present the most significant facts, and subsequent paragraphs become less and less important. American journalists first used the inverted-pyramid story in reporting the Civil War, and the AP perpetuated the practice by telegraphing full versions of a story to some papers and truncated versions to others, according to each member's needs and assessment.[18]

Wire reporting has also made news stories more bland. In the early 1800s American newspapers were highly partisan, and political invective entered news accounts as well as editorials. Because wire stories had to serve papers with a variety of viewpoints without alienating a substantial number of members, press-association reports followed a middle course, with a focus on fact and objectivity. The news services thus led a trend toward uniformity, broken only on the editorial page or by the individual paper's "rewrite man."[19]

In the 1950s, the teletypesetter saved publishers money by linking the news wire to the linotype through paper tape—and added further blandness at newspapers that considered editing little more than an uncomplicated matter of selecting a story and cutting it off at the end to make it fit the page.[20] In the 1970s the computerized "front-end" editorial composition system made easier the local editor's task of modifying wire-service stories. Yet not until the late 1980s did computer graphics systems begin to offer local editors a similar ready control over the design and content of wire-service graphics.

The greatest effect of telegraphic news on journalism was a substantial increase in the variety and timeliness of news stories. Before telegraphy the news editor often had little copy from which to choose, and was at the mercy of a slow, erratic transportation system, which delivered news in "geographic bundles," as when a shipment of Philadelphia newspapers arrived in New Orleans. The importance of these dispatches lay in the significance of their place of occurrence, not in how recently the events described had occurred, or even in their inherent newsworthiness.[21] With telegraphic news delivery, though, editors and readers learned to appreciate news for its timeliness, and an event that had occurred three weeks earlier in Boston or New York was no longer newsworthy. Newspapers improved markedly because editors could select stories thematically, for their inherent significance.

SCENE OF SATURDAY'S BATTLE NEAR LADYSMITH

Figure 3.3. Caption for map of Ladysmith and vicinity, printed in the *New York Times* on Tuesday, 9 January 1900, reads: "The above map shows the section in the immediate vicinity of Ladysmith and is a part of a map printed Dec. 13 in *The London Daily Chronicle*. Ladysmith lies in a hollow surrounded by kopjes. The principal Boer positions are near Pepworth Hill on the north, Lombard's Kop and Islmbulwhana Mountain (generally called Bulwaan or Bulwana in dispatches) on the east, while on the south they run along the kopjes along the railway, and in the west along the hills between Bester's Sprult and the town. The 'camp' marked northwest of Ladysmith is the old camp used as a kind of Natal Aldershot before the war." [Source: *New York Times*, 9 January 1900, 3.]

Toward the end of the nineteenth century, a newspaper with a photoengraving plant, an illustrator, and an atlas could with modest effort produce maps to accompany stories about rail disasters, ship wrecks, riots, and other sensational events. For breaking news in distant, enigmatic places, wire-service reports usually provided geographic details sufficient for at least a simple locator map. Yet in the early twentieth century, wire-service stories seem to have inspired few news maps. In January 1900, for instance, the *New York Times* used three maps to report events in the Boer War, but one (Figure 3.3) was a map printed four weeks earlier in the *London Daily Chronicle*, and the other two were identical and probably had been published elsewhere. The four maps used that July by the *New York Times* addressed colonial military expeditions in China; although the stories accompanying these locator maps clearly were breaking news, none of the maps seem to have been altered a few hours before presstime to reflect the latest news. In contrast, the *Times* of London quite likely produced the ten maps used in January 1900 especially to chronicle the most recent developments in Britain's campaign against the Orange Free State. Telegraphic reports seem to have influenced the selection, if not the content, of these maps; after all, the story was of particular interest to the British, and the newspaper had correspondents on the scene. But that July the *Times* had only two mapped articles, and in neither case do the illustrations seem to have been produced hours before. A cartographically illustrated story on Egypt's water supply was hardly breaking news, and the maps of Peking accompanying a story on the fighting in China were from the *Times Atlas*. As noted earlier in this chapter, before the wire services provided news pictures, they had little direct effect on journalistic cartography.

Stereotyping and Syndicate Art

Newspapers have never given their readers a pure diet of timely, breaking news. Colonial papers, for instance, offered commentary and correspondence that often were more human-interest features than hard news. Moreover, feature articles, short stories, cartoons, and crossword puzzles are at least as old as the mass-circulation daily newspaper. Much of this material has come from news syndicates, which have supplied feature material to complement timely nonlocal news from wire services. Because syndicates originally used the mails or railway express for their less perishable information, they could supply illustrations as well as written matter, and sometimes provided maps.

Among the earliest news syndicates was Tillotson and Son, publishers of daily and weekly newspapers in Bolton, England, which in the early

1870s began to sell serialized fiction and short stories to other newspapers.[22] Demand was sufficient to warrant opening an office in the United States in 1889, but by that time many other news syndicates were supplying fiction, features, pictures, and advertising "cuts" to American weekly newspapers and to small and medium-size daily papers in particular. Like Tillotson, the American syndicate organizer would contract with a well-known writer to prepare a series of stories. The syndicate would then attempt to sign up a number of newspapers, each in a different city. Serials were much in demand because a reader enticed by the first episode could more easily be converted to a subscriber or regular purchaser than a reader attracted by a single short story. In 1884 Charles Dana, editor of the New York *Sun*, offered original stories by Bret Harte and Henry James, and the success of Dana and Tillotson inspired Samuel Sidney McClure to start a similar service the same year. From an initial weekly mailing of a single 5,000-word short story, McClure increased his service to 20,000 words per week in 1885, and with the addition of general articles, to 30,000 words per week in 1886, and 50,000 words the following year.[23] To accommodate additional material from his writers, he established *McClure's Magazine* (1893-1933), which later published the stories of Lincoln Steffens, Ida Tarbell, and other famous "muckrakers" and became one of the nation's leading mass-circulation magazines.

Historians of North America journalism recognize Ansel Nash Kellogg as the "Father of the Newspaper Syndicate."[24] Kellogg was editor of the Baraboo, Wisconsin, *Republic* when, in 1861, his journeyman printer and his principal assistant both enlisted in the Union Army. Their loss forced Kellogg to order paper with war news printed on the "inside" by the Madison *Wisconsin State Journal*. The "outside" of his four-page paper contained local news and advertising he managed to print himself. Several other weeklies in the region soon adopted the *Journal*'s "ready-print" service, which began to carry legal and other advertising the same year. As more printers left for military service, the Milwaukee *Evening Wisconsin* began a similar service, and improved rail transport boosted its roster of clients to over 30. Kellogg recognized the potential profit in a centrally located newspaper syndicate, and after the Civil War he sold his paper, moved to Chicago, bought the weekly *Railroad Gazette*, and started his own ready-print business. By the beginning of 1870, 193 papers in 13 states bought paper printed on the "inside" with news, advertising, and fiction that Kellogg chose and edited for rural newspapers. His principal competitor, the Milwaukee *Evening Wisconsin*, carried more advertising, offered a lower price, and could boast a more impressive list of 294 customers. A convincing salesman, Kellogg increased his own advertising by impressing several Eastern advertising agencies with his broad circulation. Competition increased further when two other "newspaper unions" formed to share the expanding ready-print market. After losing much of its equipment and supplies in the Chicago fire of 1871, Kellogg's firm recovered rapidly, and he introduced serialized fiction in 1871 and illustrated articles in 1872. He opened a St. Louis branch in 1872, St. Paul and Cincinnati branches in 1874, and a Cleveland branch in 1876. By 1880 Kellogg was supplying over 800 customers, and when he died in 1886, the firm's clients numbered almost 1,400.

Shipment by rail of half-printed sheets would have been too costly for medium-size and large newspapers, and the competitive urban newspaper market called for greater freedom in the integration of syndicate material with breaking news, local news, and local advertising.[25] The rapid growth of the news syndicate in the 1880s was possible largely because of the stereotype plate, which earlier had made practicable the high-speed rotary press. Adoption of stereotyping by Kellogg and other ready-press news syndicates was an occasion not only for expanding their business to larger clients but also for developing a lighter stereoplate that was less expensive to ship. As early as 1858, in Sheffield, England, type-high solid metal stereotype plates had been produced at a central stereotyping plant from type frames supplied by individual customers, but high transport costs restricted this service.[26] The weight of conventional stereoplates would have severely limited the use of stereotyping by news syndicates in the 1880s had not a thin metal stereoplate been devised to lower the costs of both transport and metal casting. Kellogg provided client newspapers with standard blocks into which column-width strips of stereotyped "boiler plate" could easily be locked.[27] In 1875 Kellogg's firm patented a system in which plates drilled for screws or tacks could easily be anchored to a wooden base, and in 1878 it developed the "butterfly plate," with an X-spring on the back that could be pinched for easy insertion of the plate into a slot on a metal base.[28] Returning the plate lowered further the cost to the customer. A later improvement was a still lighter, celluloid plate.

Syndicate material had a strong influence on the content and appearance of daily and weekly newspapers, especially those in smaller cities, with limited competition. The syndicate emerged as a primary gatekeeper in the flow of information, but competition seems to have promoted variety and volume, if not quality. Kellogg's paternalistic pitch promised both economy and special attention: "By giving me *carte blanche* in the selection of matter of the various classes, you will get the reading fresher than if I send out a proof sheet. . . . [and] my long experience . . . is sufficient guarantee that nothing of objectionable character will be used . . . Special care will be taken to avoid sending any matter which would interfere with, or duplicate, that furnished to other parties in the vicinity."[29] But there was much to select from, and a publisher could manufacture at low cost a newspaper with minimal local coverage and a selective potpourri of prominently displayed yet shallow "telegraphic dispatches." Some editors were correctly criticized for "editing their paper with a saw," whereas others shunned all syndicate material with a snobbish obsession.[30] So while readers in general received more fiction and less political commentary, they also were offered a variety of practical, educational, and entertaining articles—many embellished with line drawings (Figure 3.4), and sometimes with maps (Figure 3.5). Whereas the newspapers of 1870 had comparatively few illustrations, those of 1880 and thereafter offered abundant centrally produced boiler-plate art—little of it cartographic, however.

Another technical advance promoting syndicated artwork was the dry *mat*, or *matrix*, a paper-mache or cardboard mold, which could be shipped or mailed less expensively than a thin metal stereoplate. A newspaper with a *flat casting box* and melting pot could use the mat to cast its own stereoplates (Figure 3.6). The engraver carefully placed the mat on the horizontal steel casting table, covered it, and rotated the casting table to a vertical position. He then poured molten metal into the casting box and allowed it to cool. When rotated back to the horizontal and opened, the casting table held a flat, 3-mm-thick (1/8-in.-thick) cast, ready for mounting type-high on a standard wood base.[31] After 1895 newspapers used flat stereotyping extensively for syndicate stories, artwork, news pictures and other halftones, and advertising displays.[32] Syndicates continued for several decades to ship metal plates to smaller papers unable to afford flat stereotyping, but many of these newspapers failed when better-financed competitors set up photoengraving plants and bought linotypes in the years following World War I.[33]

In the early twentieth century capitalists and entrepreneurs in the news business had two opportunities for expansion: the news syndicate, which sold news and features, and the "chain," which owned newspapers in several cities. Because the scale economies of common ownership extended beyond the purchase of newsprint and other supplies to the preparation of news stories and feature articles, the larger chains also functioned as syndicates, and even sold their news and feature services to noncompeting papers. Among the earliest American newspaper chains was that started in the 1880s by Edward Wyllis Scripps and his partners. Scripps controlled nine daily newspapers by 1900, and expanded to 22 by 1910.[34] His enterprises included the Scripps-McRae Press Association, founded in 1897 and operating largely in the Midwest, and the Scripps News Association, organized that year from a loose cooperative arrange-

Figure 3.4. Pencil sketch of an improved mail collection system, accompanying a story entitled "Up to Date Mail Collecting," is typical of the line illustrations used by feature syndicates in the late nineteenth and early twentieth centuries. [Source: *Cortland Evening Standard*, 20 January 1900, 8.]

Figure 3.5. Hand-lettered map of the southeastern states, accompanying a story on a solar eclipse expected on 28 May 1900, illustrates the crude pencil-drawn maps supplied as line photo-engravings or mats by feature syndicates. Note the spelling "Illinonis." [Source: *Cortland Evening Standard*, 23 May 1900, 3.]

ment among Scripps's papers on the West Coast. These and other press associations provided telegraphic news to newspapers excluded by the AP, whose members at that time could veto service to nonmember competitors. Started in 1902 to supply features to Scripps-McRae papers, the Newspaper Enterprise Association expanded in 1909 to offer features, comics, and news pictures to papers outside the chain. Printed copy and mats delivered by mail and express to clients included timely illustrated news features chosen to anticipate likely "spot news" about international conflicts, elections, sporting events, and explorers' expeditions.[35] NEA

Figure 3.6. Flat casting box used for casting metal plates from mats, or matrices, supplied by news and features syndicates and advertising agencies. The casting frame was rotated into a vertical position to receive the molten metal, and back to the horizontal to remove the plate. Job printers and small newspapers used these metal plates in flat form on their flatbed printing presses. At larger newspapers flat metal plates cast from mats were laid out with the type in a form for an entire page. Printers would then press a paper mat from this flat format and use it to make a round stereoplate for a high-speed rotary press. This Goss flat vacuum casting box was an improved model manufactured in the 1930s. Few were made after that time because rotary presses replaced most flatbed presses. [Source: Courtesy of Graphic Systems Division, Rockwell International Corporation.]

grew to serve nearly 400 clients by 1920, and 700 papers by 1930. As early as 1924, NEA used air express and air mail to serve clients from its Cleveland headquarters. In the 1930s a New York subsidiary, Acme Newspictures, provided timely news photographs to over 100 papers. During the 1930s, 1940s, and 1950s, NEA and Acme were the principal source of cartography for many small and medium-size newspapers.

Cleveland was the home of another "budget" feature service, the Central Press Association, founded in 1910 by V. V. McNitt with backing from publishers of the Cleveland *Leader* and the Cleveland *News*.[36] The Central Press at first offered only whatever features, cartoons, and news pictures these Cleveland papers could provide, but later the syndicate developed most of its own material. In 1912 the Central Press bought the North American Press Syndicate, in Chicago, and acquired 180 additional clients. McNitt established a separate Central Press Association in New York and absorbed several similar firms until his own company was purchased in 1930 by King Features, owned by William Randolph Hearst. But until the 1970s the Central Press retained its identity, and its Cleveland office supplied features and artwork. During the 1930s, 1940s, and 1950s the mat service of the Central Press was an important source of news maps for small daily papers, and for a while in the 1950s the syndicate even operated its own photographic wire service. Figure 3.7 is one of many Central Press maps illustrating developments in the cold war.

Clichés such as "dynasty" and "empire" seem fully appropriate when describing William Randolph Hearst's news publishing business. In the late nineteenth century Hearst expanded from San Francisco to New York, Chicago, and other cities, and built huge circulations through sensational stories and blatant headlines.[37] Hearst's holdings included magazines, several paper companies, and King Features Syndicate, which hired salaried artists and controlled many popular comic strips. Hearst linked his newspapers with private telegraph wires leased from Western Union, and began to sell his news service to other papers through several companies, which he combined in 1911 into the International News Service. The INS provided not only telegraphic news but also comics, features, and news pictures. In the 1910s it had both domestic and foreign clients, and had agents abroad who dispatched breaking news by cable and photographs by ship. The INS provided photographs, illustrations, and even complete pages as mats sent by rail express, and its Dry Plate Department in New York operated branches with staff photographers in Atlanta, Boston, Chicago, Los Angeles, and San Francisco.[38] Its illustrators, who decorated the borders of fashion, sports, and other photographic layouts, occasionally drew maps, which stood alone or formed part of a display picture. Hearst later reorganized International News Photos as a

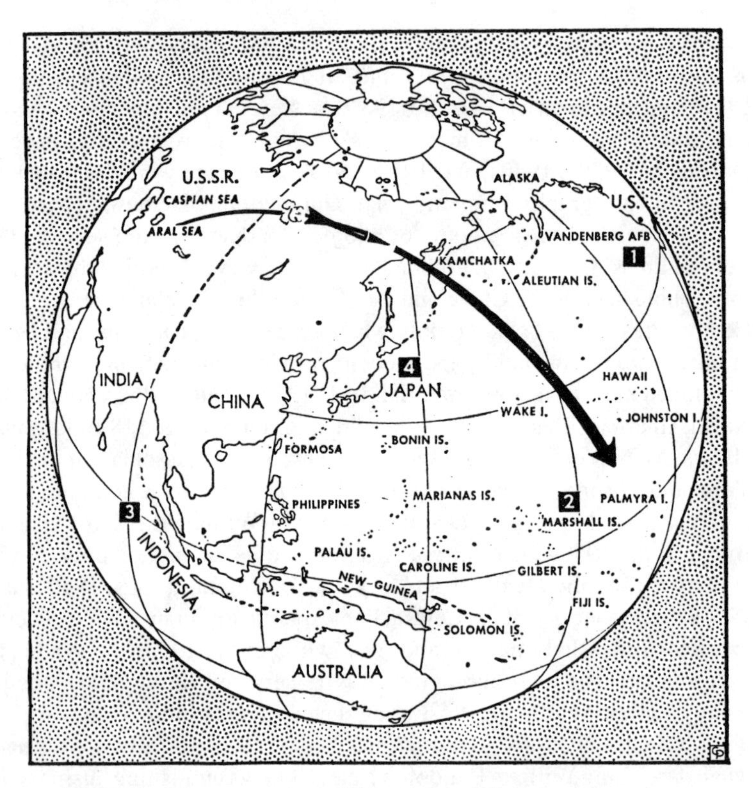

Figure 3.7. Central Press map uses oblique azimuthal projection to show a planned 8,000-mile test flight of a Soviet missile from the Aral Sea to an impact area east of the Marshall Islands (2). U.S. tests, launched from Vandenberg Air Force Base (1) used the same area. Other numbers locate Indonesia (3), which Khrushchev planned to visit in early 1960, and Japan (4), which protested the Soviet test. [Source: *Norwich Sun*, 21 January 1960, 4.]

separate company, and entered the newsreel business with Hearst Metrotone. In 1930 *Fortune* acclaimed him "the undisputed King of Newspictures."[39]

Three of Hearst's four principal competitors in the news photo business of the early 1930s were also newspaper publishers: the Scripps-Howard chain operated Acme News Pictures; the *Chicago Tribune* and the New York *Daily News* owned Atlantic and Pacific Photos; and the *New York Times* owned Wide World Photo. Hearst's principal challenger, the Associated Press Photo Service, was the youngest, initiated in 1928 with the blessing of several AP members also in the news photo business. Hearst, Adolph Ochs of the *New York Times*, and Joseph Medill Patterson of the New York *Daily News* approved the AP's venture because the benefits of lower, shared costs and a richer variety of news pictures apparently offset whatever revenue might be lost by their own picture services.[40] AP general manager Kent Cooper's announced intention of developing a photographic wire service was particularly attractive to Ochs. Apparently the costs of a timely news photo service were straining the budgets of the others. In 1930 *Fortune* estimated that the AP was breaking even, whereas Ochs's Wide World Photos was earning a slight profit, and the other three were each losing about $250,000 a year.[41] That year Atlantic and Pacific News Photos sold its European service to AP and its American operations to Acme. The *New York Times* later downgraded Wide World Photos to a photo feature service, and then sold it to the AP in 1940.

AP's Photo Service, initiated in 1928, had two precedents: the Feature Service, started a year earlier, and the General Mail Service, sent to members as a less timely supplement to the wire service. The mimeographed copy provided by the General Mail Service lacked illustrations and appeared dull when compared to the neatly printed and illustrated offerings of commercial feature syndicates. The first illustrated AP features released in 1927 had printed text and matted pictures, and the mailing included short stories, news briefs, and daily and weekly illustrated columns addressing sports, women's fashions, movies, the arts, New York and Washington topics, and foreign affairs. Biographical sketches of prominent persons accompanied column-wide portraits, which member newspapers could file away until they needed them for obituaries or news reports.[42] All 1,228 members received the first mailing free, and only 78 declined to take the service. The weekly rate varied from $1 to $6, depending on circulation, and for an additional $3.50 the AP supplied illustration mats; commercial syndicates charged much more for similar service.[43] This picture service gave the AP a year of useful experience in providing illustrated material, and it whetted members' appetites for the AP Photo Service inaugurated in 1928. AP artists supplied both the Feature Services and the Photo Service with maps and other line drawings. Later called AP Newsfeatures, or merely APN, the Feature Service during the 1940s mailed member papers two maps each week, together with a 600-word explanation.[44] Member newspapers unable to afford the AP Wirephoto service could receive APN maps to portray the threats and drama of World War II. In the 1980s several news syndicates provided maps to newspapers by mail. Most notable, in addition to AP News Graphics, were: NEA; Tribune Media Services, a subsidiary of the *Chicago Tribune*; and the News America Syndicate, owned by Rupert Murdoch.

Since the development of halftone engraving in the late nineteenth century, newspapers have used more photographs than line drawings—

with the possible exception of some smaller newspapers, when cartoons are included. The photograph, after all, is an efficient and visually effective means for communicating images of people—individually or in groups, in action or at rest. Photojournalism emerged in the early 1930s as a new profession, concerned with art and ethics, graphic technique, and recognition of achievement.[45] Press associations and syndicates strove to satisfy news publishers' demands for low-cost photographs, and provided portraits and action shots of nationally and internationally prominent people and events even to small newspapers without photoengraving. Although maps and other data graphics clearly were secondary to the photograph, distribution systems developed for the demanding graytones of the more-in-demand photo made possible an unprecedented application of journalistic cartography.

Photos (and Maps) by Wire

As soon as they could economically illustrate their newspapers in the late 1800s, news publishers sought ever more timely new photographs, and a few even experimented with electronic transmission. Although the rapid and systemic movement of words by wire preceded by nearly a century the development of a practical electronic system for distributing news pictures, successful experiments in the mid-nineteenth century demonstrated the possibility and promise of facsimile transmission. Moreover, proposals for an electric telegraph that would transmit and receive printed or hand-written messages had preceded these demonstrations by nearly a half century. Although interest in the "telautograph," or "writing-telegraph," waned early in the twentieth century when the telephone and teletypewriter proved more useful, numerous inventors sought improved methods for the electronic transmission of graphic symbols, drawings, and photographs.[46]

Mid-nineteenth century experiments in graphic telegraphy anticipated the two fundamental data structures of present-day computer graphics—*vector* and *raster*. The telautograph, which used a stylus or pencil to transmit and a pencil or ink-pen to record, was a vector system that encoded electronic signals to direct a writing instrument from point to point, in sequence.[47] As Figure 3.8 illustrates, an operator writing on the "transmitting paper" with pencil A would vary the electrical resistance of two rheostats R and R′, and thereby control the current transmitted to a pair of galvanometers G and G′, linked in turn to the recording pen B. Pen B thus reproduced many miles away the message traced with pencil A. Early telautograph systems included an apparatus demonstrated in 1874 in

England by E. A. Cowper, but never used commercially, and several systems based on Cowper's work, developed in the United States around 1884 by J. Hart Robinson and H. Etheridge, and used for business communications and reporting baseball news.[48] Earlier still were a number of raster systems that represented a message with a series of evenly

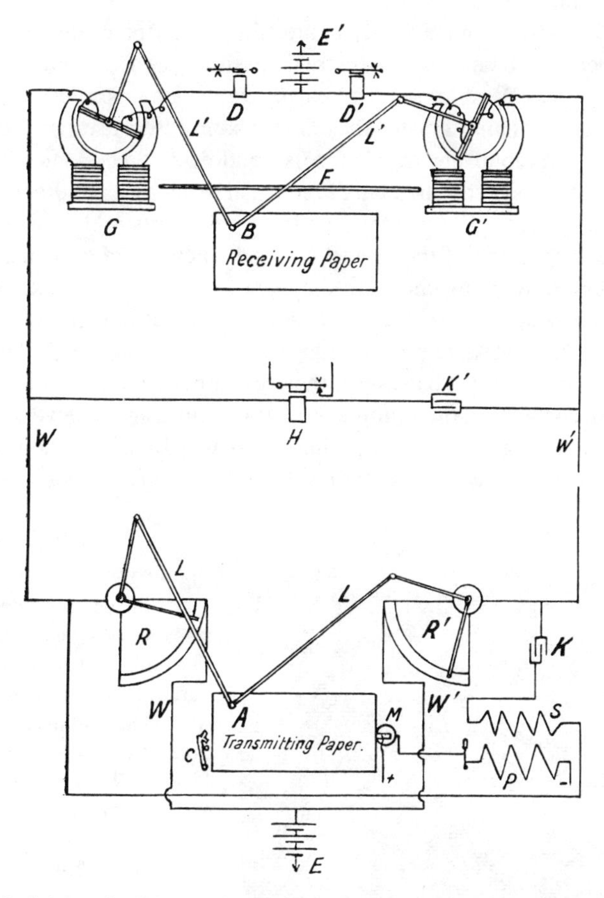

Figure 3.8. Schematic diagram describing the two-wire telautograph developed by Foster Richie at the turn of the century. A message written with the transmitting pencil (A), linked to a pair of rheostats (R and R′), is reproduced by the recording pen (B), driven by a pair of galvanometers (G and G′). At the receiving station a lifting magnet (not shown) raises the recording pen when the pencil (A) is raised between characters and thus not pressed against the transmitting paper. [Source: "The Telautograph," *Nature* 64, no. 1648 (30 May 1901), 107-9; Figure 2 on p. 109.]

spaced parallel lines, parts of which were darkened, as in Figure 3.9; this image thus is similar in structure to the raster image on a television screen. A distant receiver, synchronized with the transmitter to canvass a similar series of lines at the same speed, reproduced the message or drawing line-by-line. The receiver darkened only those parts of the line addressed while current was flowing, and the image appeared as a set of closely spaced parallel-line segments.

Alexander Bain, a Scot, patented the first graphic telegraph in 1843, a year before Morse's successful transmission of coded messages between Baltimore and Washington. Bain's apparatus demonstrates several principles and problems of raster-facsimile systems. Because his system could not store and replay electrical signals, sending and receiving instruments had to be fully synchronized in order to start each row simultaneously and to sense and record corresponding image elements at the same distance from the beginning of the row. Bain used a pendulum to scan across the rows, a clockwork mechanism to advance the scan from one line to the next, and a set of electrical contacts and electromagnets to enforce synchronization—if one pendulum were to move ahead of the other, a magnet would hold it back until the second pendulum reached the same position, so that the pendulums at both transmitter and receiver could start a new stroke together.[49] Equally ingenious was Bain's method of detecting image cells: the letters of a message were set in a block of metal type,

Figure 3.9. Example of a raster image, as scanned and transmitted by a telegraphic facsimile system. Note the parallel scan lines that vary in thickness in inverse proportion to image brightness. [Source: T. Thorne Baker, *Wireless Pictures and Television* (New York: D. Van Nostrand, 1926), 16.]

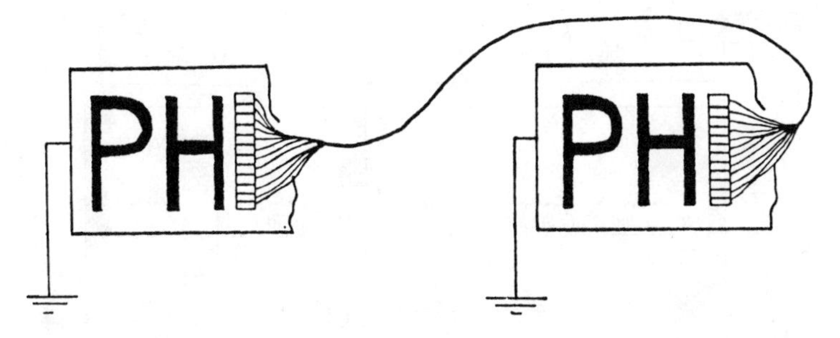

Figure 3.10. Principles of Bain's 1843 telegraphic facsimile system. The transmitter was based upon a metal brush of several narrow springs that passed over a metal plate bearing large metal letters. The springs could touch the letters but not the plate on which they rested. Each spring on the transmitter was connected to a wire linked to a corresponding spring on a similar brush at the receiver. The receiver's brush passed over a piece of chemically treated paper (also resting on a metal plate) at the same speed the transmitter's brush passed over the raised metal letters. When a transmitter spring sensed a letter—and thus completed the electrical circuit between transmitter spring and receiver spring—the receiver made a corresponding mark by passing an electric current through the paper. [Source: T. Thorne Baker, *Wireless Pictures and Television* (New York: D. Van Nostrand, 1926), 11.]

and raised to make contact with a small metal stylus attached to the pendulum. As Figure 3.10 shows, this contact should close an electrical circuit and send current to the recording pendulum, which carried a light metal point across a sheet of conductive, electrochemical recording paper. When the circuit was closed at the transmitting station, the current discolored the recording paper to mark the overlap between the scan line and the letters on the "message block" of raised metal type.

A few years later Frederick Bakewell, another Englishmen, significantly improved Bain's design. Bakewell's "copying telegraph," patented in 1848, employed a simpler and more efficient scanning method, in which the message was mounted on a cylinder and a thin electrical contact was moved slowly along a long lead screw, parallel to the cylinder's axis, as the cylinder was turned by a clockwork mechanism powered by a weight. Bakewell's cylinder-and-screw approach is the basis of present-day rotating-drum scanners, which examine an image in a single, continuous spiral, as Figure 3.11 shows. The recording cylinder held a chemically treated, electrically sensitive paper, similar to that used by Bain, whereas the transmitting cylinder carried an image scribed with a stylus on a sheet of foil coated with a nonconductive varnish. Where the stylus had removed the varnish, the foil was exposed to close the circuit through the metal contact in order to transmit an electric current to discolor the

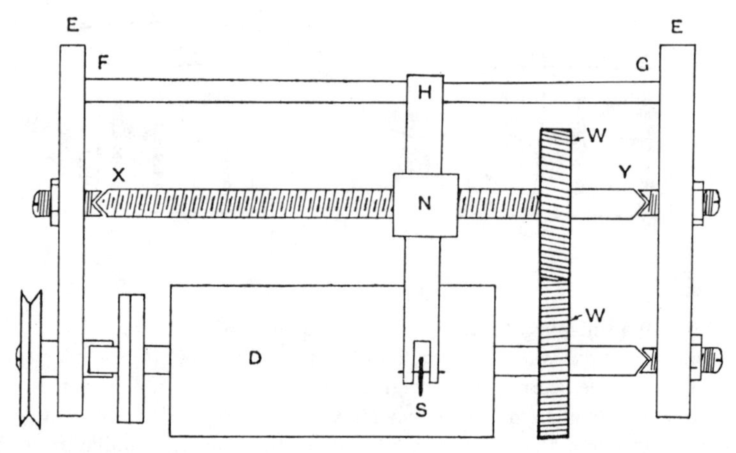

Figure 3.11. The cylinder-and-screw scanning mechanism, first employed in 1848 for Bakewell's telegraphic facsimile system. Drum D on the transmitter carried the image to be scanned by stylus S. Carrier arm H (supported by cylindrical steel bar FG, between vertical end plates E and E) was linked to threaded shaft XY by nut N. The shaft of the drum was connected to the motor through a pulley at the left and to threaded shaft XY by the gear wheels W. As drum D turned, stylus S advanced slowly in a lateral direction, parallel to the axis of the drum, to trace a spiral over the revolving drum and thus approximate a set of closely spaced scan lines across the image. [Source: T. Thorne Baker, *Wireless Pictures and Television* (New York: D. Van Nostrand, 1926), 87.]

corresponding image element on the recording paper. Although this apparatus proved impracticable for commercial service, Bakewell demonstrated the ability of the telegraph to send drawings as well as words.

Development of reliable electric motors in the second half of the nineteenth century led to further improvements in facsimile transmission, especially in synchronizing the sending and receiving cylinders. The most notable advance followed telegraph engineer Willoughby Smith's discovery, published in 1873, that selenium changed its electrical resistance when exposed to light. Smith observed that selenium resistors exposed to sunlight retarded an electrical current more effectively at night than during the day. Around 1880 Shelford Bidwell, an eccentric British scientist, suggested that selenium cells might be useful in facsimile transmission to distinguish between the light and dark parts of an image. And in 1891 Noah Amstutz, who had experimented with facsimile transmission in Cleveland, demonstrated the use of a selenium detector for transmitting halftone photographs. Yet photofacsimile remained little more than a promising curiosity until 1902, when Professor Arthur Korn demonstrated in Germany the first practical photoelectric sensing and transmission system.[50] Korn mounted a transparent, film-positive photographic image on a glass cylinder, inside which a selenium cell was positioned to receive a thin beam of light aimed through the glass. A cylinder-and-screw apparatus provided a systematic scan of the photograph. Where the image was relatively light, the narrow beam penetrated the film and the glass cylinder to lower the electrical resistance of the selenium cell and thus increase the strength of the current sent to the receiving station. Where the photograph was comparatively dark, little or no light penetrated and the transmitted current was much weaker. The receiver amplified the current, which controlled the intensity of a light focused through a set of lenses onto a small portion of a sheet of photosensitive film mounted on the synchronized recording cylinder. When developed, the exposed image reproduced the original picture with modest success. Other inventors designed similar receivers using photographic film and a focused light beam either powered by the amplified incoming signal or deflected onto or away from the film by a galvanometer that served as a *light valve*. Korn made significant improvements to his system over the next two decades and advanced photographic facsimile through both his research and his successful demonstrations of graphic telecommunications.

News publishers were early and often enthusiastic users of facsimile technology. On June 21, 1895, in what might have been the first journalistic use of a telegraphically transmitted illustration, the Chicago *Times-Herald* printed a profile sketch of inventor Elisha Gray transmitted from Cleveland with facsimile equipment from Gray's Telautograph Company.[51] A few years later Ernest Hummel, a German immigrant to St. Paul, Minnesota, developed his "Telediograph." In 1899 Hummel demonstrated his system in a successful exchange of pictures between the Boston *Herald* and the New York *Herald*.[52] Hummel's transmitter, which used varnish-coated foil scribed with a stylus, was suitable only for line drawings. Photoelectric systems such as the Telediograph could transmit halftone photographs, but because of transmission noise and other distortions, image quality was poor unless the halftones were enlarged photographically before transmission and then reduced before use in the paper.[53] A few British and French daily newspapers and illustrated weeklies were interested in Korn's selenium-cell facsimile system, and sought exclusive rights for its use within their countries.[54] In 1908 the London *Daily Mail* and *L'Illustration*, in Paris, used Korn's system to exchange photographs via land telephone lines and a telephone cable under the English Channel.[55] In 1909 the London *Daily Mirror* employed a "Telectograph" system between London and Paris and between London and Manchester, and was considering trans-Atlantic facsimile.[56] In 1912

L'Illustration operated a photofacsimile service that provided Paris morning newspapers with photographs taken the previous afternoon on the Riviera.[57] In 1920 Edouard Belin, a Paris experimenter who had transmitted photographs by wire between major cities in Western Europe, demonstrated his "Telestereograph" system in the United States, in a test exchange between the New York *World* and the St. Louis *Post-Dispatch*.[58] In the early 1920s, however, picture quality was variable and often poor, the equipment was not reliable, and no news publisher, syndicate, or press association was yet ready to invest substantially in an unproven, still largely experimental technology.

In the early 1920s the American Telephone and Telegraph Company made the significant institutional commitment required for the commercial development of facsimile telecommunications. AT&T, which had not only the incentive of more business for its nationwide long-distance telephone network but also the resources of its Western Electric manufacturing facility and its Bell Laboratories research arm, wanted to explore the commercial possibilities of facsimile transmission. In 1924 AT&T demonstrated its system by transmitting news photos to New York and Chicago from the Republican presidential convention in Cleveland.[59] Its photoelectric drum-and-screw transmitter, which scanned a transparent positive image made from a negative provided by the customer, could accommodate a picture as large as 10.8 by 16.5 cm (4.25 by 6.5 in.).[60] Scanning at 39 lines per cm (100 lines per in.), the system in five minutes could simultaneously recreate the image at the receiving station as a pattern of equally dense variable-width parallel lines. AT&T initiated a telephotograph message service in April 1925, with offices in New York, Chicago, and San Francisco, and in the next two years extended its facsimile service to Atlanta, Boston, Cleveland, Los Angeles, and St. Louis. Typical rates for sending a picture from New York were $15 to Boston, $20 to Chicago, and $45 to San Francisco.

By 1930 AT&T recognized the limited commercial potential of its expensive first-come, first-served common-carrier facsimile service. Drawbacks were both technical and competitive, and included: the delay at the transmitting station of preparing the positive film transparency from the customer's negative; the need to transmit a large or medium-size picture or drawing as a series of overlapping smaller pictures and to assemble these as a mosaic at the receiving station; fair-to-poor image quality; a limited number of access points to the network; and competition with the U.S. Post Office's improved air mail service. A growing economic depression promised additional losses, and AT&T was unwilling to invest further in its faltering facsimile service, which it discontinued in 1933. But the AP, which in the late 1920s had foreseen the advantages of a dedicated photofacsimile network, in 1930 asked AT&T engineers to study the feasibility of a leased-wire facsimile system with simple, reliable transmitters and receivers meeting newspaper standards for picture quality. In 1933, AT&T, eager to recover the losses of its defunct commercial facsimile service, expressed a willingness to develop a system to meet the AP's needs. AP general manager Kent Cooper was concerned that the AP have exclusive use of the service for at least five years, and advised AT&T on the wording of a letter sent to competing picture services, to solicit commitments to an AT&T system serving a minimum of 25 offices at an annual cost of almost a million dollars.[61] The AP's competitors—Acme, International, and Wide World—declined, and Cooper made an oral agreement to the terms he had negotiated with AT&T.

Although he had the tentative approval of the AP Board of Directors, Cooper faced two challenges: signing up 25 members at prices three to fifty times what they were then paying for pictures by mail, and overriding the objections of William Randolph Hearst and Roy Howard, influential AP members with competing photo services. In addition, some members were concerned that nonsubscribers would subsidize the electronic facsimile service, discussion of which dominated the AP's 1934 annual meeting.[62] After lengthy debate, Cooper gained not only the approval of a majority of the members but also the indispensable contracts with the publishers of 39 newspapers in 25 cities. Particularly important were three California papers, which made the network truly national in scope. Its trade name "Wirephoto," coined between the signing of the contracts and the system's official opening on January 1, 1935, aptly describes this AP triumph in both telecommunications and cooperative news-gathering.

Designed especially for news pictures, the AP Wirephoto service was a marked improvement on earlier facsimile systems. Subscribing newspapers had both a receiver and a transmitter, and timely pictures could readily enter the network at major American cities. The new system required only a paper positive print, which was easier to produce and edit than the film transparency used with AT&T's commercial system. Operating in daylight instead of a darkroom, the transmitter could accommodate a print as large as 28 by 43 cm (11 by 17 in.), and measured the intensity of a beam of light reflected from the print mounted on a drum with a 28-cm (11-in.) circumference.[63] Most pictures were 8 by 10 inch prints described in a one-inch caption. Scanning at 39 lines per cm (100 lines per in.), the AP system transmitted the 880,000 picture elements of a typical image in 8 minutes. Newspapers received the Wirephoto as a negative, ready to be developed and photoengraved. Because of improvements in noise filters and the light-valve and synchronization mechanisms, halftones made from the original photo and its transmitted Wirephoto were comparable in quality.

Through a voice circuit the network controller in New York scheduled individual transmissions and mediated requests for photographs of particular interest to specific subscribers.[64] When the negative proved unsatisfactory because of transmission error, an important picture could be retransmitted. The network could also be divided for the limited distribution of political, sports, and other photos of only local or regional interest. Available 24 hours a day, the network normally operated only 16 hours a day, 7 days a week. Portable transmitters were available for covering special events, as well as for connecting to the leased-wire network additional papers receiving a more limited service at a lower cost.[65] When the Wirephoto network opened in 1935, ten additional newspapers received "expedited delivery" of photos from the nearest full-service node. By 1940, the AP Wirephoto system was serving newspapers in 72 cities, as shown in Figure 3.12.

Like feature syndicates and the major news photo services, the AP's Wirephoto service included a few maps among a much larger array of pictures dominated by news and sports photos. The 53 pictures moved on January 5, 1935, for instance, included two weather maps, sent from Washington at 10:14 A.M. and 10:20 P.M.[66] The Wirephoto log for that day also listed two unidentified graphs moved from New York. But sports, disaster, and cheesecake photos from a range of origins clearly dominated that Saturday's pictorial traffic. Among the network's early cartographic transmissions was Figure 3.13, a map showing the Saar Basin, scene of the 1935 plebiscite that started the expansion of Nazi Germany. Cartography had a comparatively minor role in the Wirephoto service, as apparent in an *Editor and Publisher* story noting that "the service is, of course, not limited to news photos, but will include charts with criminal stories, and the daily national weather map prepared by government experts in Washington, D.C."[67] Prepared by the Weather Bureau and ready for immediate transmission, the timely weather chart was a distinctive service, particularly attractive to some AP members.

Except for the daily weather map and an occasional locator map or government handout, Wirephoto was insignificant as a source of cartographic material until World War II. The AP had designed Wirephoto, institutionally as well as electronically, for supplying photographs, not maps. Because of difficulty in transmitting type labels, maps for Wirephoto had to be specially drafted, with few place labels and blatant, bold type. Preoccupied with retouching photos and serving the illustration needs of the mail features and picture services, the small AP art staff had

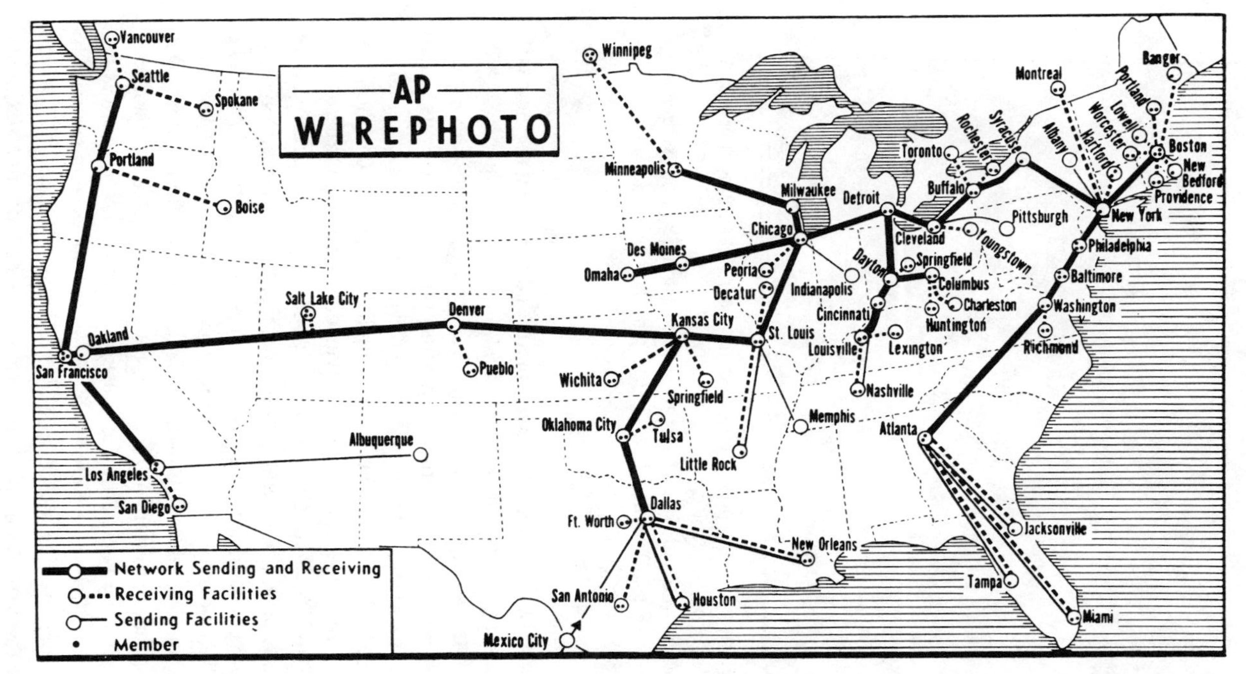

Figure 3.12. Map of the Wirephoto network, prepared on its sixth anniversary by the AP for *Editor and Publisher*. Solid black lines identify the leased-wire network, which provided direct service to 55 member newspapers in 29 cities. “Abbreviated service” provided Wirephotos on demand through long-distance telephone circuits to 51 additional papers. In addition, 31 portable transmitters provided fuller nationwide coverage, particularly important for the southern states and other parts of the country with few Wirephoto subscribers. [Source: *Editor and Publisher* 74, no. 1 (4 January 1941), 9.]

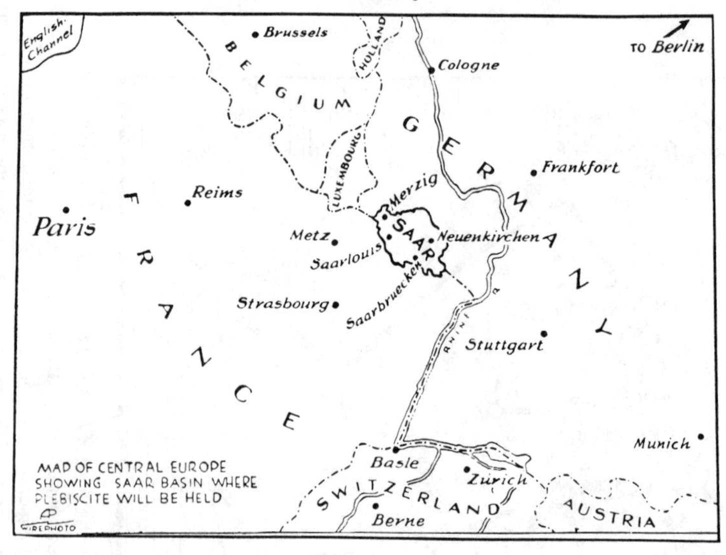

Figure 3.13. Early AP Wirephoto map shows the Saar Basin region, between France and Germany. A plebiscite held in early 1935 reincorporated the region into Germany and initiated Hitler's territorial expansion. [Source: *Syracuse Herald*, 13 January 1935, section 1, 8. Courtesy of the Onondaga Historical Association, Syracuse, N.Y.]

little time for drawing maps. Moreover, picture editors hired for their skill with photos, not artwork, had little inclination to provide maps.

Wirephoto's demonstrated success attracted competing picture services into electronic facsimile transmission. Particularly useful to these competitors were transmitters that converted electronic signals to acoustic pulses which could be sent through an ordinary telephone, at standard long-distance rates, to a receiver equipped to restructure sequences of high-pitched chirps into a photographic image.[68] These "soundphoto" machines provided a flexible, low-cost photowire system, without the expense of a leased-wire network. Arrangements could be made for subscribers to receive only those photos they requested or were likely to use, and even though transmission time for individual photos was about double that of Wirephoto service, the overall cost was much less. Electronic facsimile was particularly useful in transmitting pictures between regional offices of a picture syndicate, for subsequent delivery to many subscribers by rail, air, or special messenger. By late 1937, Acme Newspictures was operating its "Telephoto" service with nine stationary and four portable facsimile units; Hearst's International News Photo had a "Soundfoto" network served by 15 stationary and 12 portable units; and the *New York Times*'s Wide World Wire Photo had 15 fixed units and 40 portable transmitters. In addition, Western Union operated a common-carrier telegraphic facsimile service between New York, Buffalo, and Chicago, and Finch Telecommunications operated a similar service between New York, Chicago, Los Angeles, and San Francisco. Although the AP bought Wide World in 1940, competition increased when the NEA picture syndicate began to distribute mats for Acme Telephoto, and the Central Press began to distribute matted International Soundfotos.[69] In the 1950s, though, a single strong competitor challenged the AP and its Wirephoto, after the United Press bought Acme in 1952 and merged with International New Service in 1958 to form United Press International.

From Copper Wires to Dish Antennas

The half century following Wirephoto's inauguration in 1935 has seen substantial improvements in facsimile technology and significant growth in the number of newspapers with photowire service. From only 39 American daily newspapers in 1935, the list of newspaper and other clients with direct AP photowire service had risen by the early 1980s to about 1,500, roughly half in the United States.[70] UPI served roughly similar numbers of domestic and foreign news picture clients in 1980, and most of the country's larger newspapers received photowire service from both the AP and UPI.[71] Although a few smaller American dailies, typically with circulations under 10,000, have no photowire, most newspapers need and can afford timely, electronically delivered news photos. Moreover, satellite facsimile transmission has eliminated the need for an expensive leased-wire network, and improved photoreceivers and more efficient photoengraving techniques have reduced the cost and delay of putting photowire illustrations in the paper. In 1935 the AP paid $16,000 each for receivers and transmitters, and would not have considered providing Wirephoto service to a small newspaper. But in 1985 the added charge for providing Laserphoto I service to an AP member with a daily circulation of only 10,000 was less than $10,000 a year, including materials.[72]

In 1935, even if the AP had provided the service free, the typical small newspaper would have had difficulty using an on-site Wirephoto receiver. Although a small paper could use its casting box to make stereoplates from matted photos received from a picture syndicate, to prepare a stereoplate from a Wirephoto film negative the newspaper would have had to buy expensive photoengraving equipment and hire additional staff. As a result, few small papers had timely news photos—local or national—during the 1930s and 1940s, and few had a staff photographer. Local photos, of course, were a more pressing need than Wirephoto service; after all, picture syndicates provided useful if not highly recent photos of movie stars, tanks, and politicians, but local people were interested principally in seeing pictures of themselves, their neighbors, and their own athletically inclined sons and daughters. Small news publishers would make substantial investments in photoengraving principally to improve local coverage, not to accommodate Wirephoto.

Technical advances serve needs at many levels, and, unsurprisingly, the electronic scanner-engraver—an important catalyst for local photojournalism—was an improved version of a mechanical engraver that the AP's research laboratories built in 1941 to promote Wirephoto.[73] AP engineers had hoped to develop an engraver they could connect directly to the Wirephoto network so that smaller papers could obtain relief plates without investing in a costly photoengraving plant. But after the prototype proved too expensive to manufacture, AP researchers experimented with an inexpensive electronic engraver deemed useful to member newspapers. The AP later turned its development over to the Fairchild Camera and Instrument Corporation, which in 1949 introduced the Fairchild Photo-Electric Engraver (described in Chapter 2), a scanner that made halftones from photographic prints. The Scan-A-Graver, as it was later called, was like a Wirephoto system without the Wirephoto network: light reflected from a positive print mounted on a rotating drum controlled the generation of a raster image with 26, 33, or 47 dots per cm (65, 85, or 120 dots per in.), depending on the model. Instead of producing a negative image hundreds of miles away, the Scan-A-Graver formed a relief image on a plastic plate mounted on a second drum, a mere inches away and sharing the same axle.[74] A pressroom worker using double-stick tape could easily mount the resulting "plastic cut" onto a metal stereoplate with the type for the page. The Scan-A-Graver enabled many small-circulation daily newspapers, and even some weeklies, to introduce timely local photos.[75]

With a Scan-A-Graver to make press-ready relief cuts, the small newspaper needed only a low-cost Wirephoto connection and a facsimile receiver producing paper positives, instead of film negatives. Because each additional connection to the network lowered the unit cost for all users, and because the population served largely governed AP rates, an improved photoreceiver became a critical need. The AP met this need with the Photofax receiver, which it first demonstrated in 1952 and modified in the mid-1950s.[76] The Photofax receiver (Figure 3.14), which used a roll of paper sensitive to electrical current but not to light, could operate unattended in the newsroom. It required little more attention than a teletypewriter—a daily change of paper and a periodic inspection and

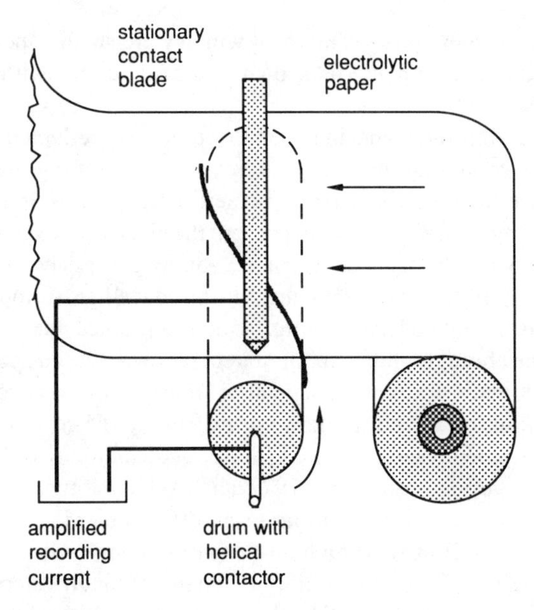

Figure 3.14. Principles of the Photofax receiver. Current from the recording amplifier completes a circuit from the helical contactor on the rotating drum through the electrolytic paper to the stationary contact blade, and with continued rotation of the drum describes a pattern of parallel lines on the paper. When the current is flowing the paper is darkened at the point of intersection between the helical contactor and the blade. The drum rotates rapidly relative to the speed of the paper to provide a high-resolution image of closely spaced parallel lines.

cleaning. As the paper advanced slowly past a stationary printing blade, a helical contactor on a rotating drum swept across the blade on the other side of the paper, and in one "stroke" addressed all the picture elements along a single scan-line. The moist electrolytic paper, which carried a current from the contacts to the blade, darkened the paper in proportion to the current, which by electrolysis transferred some of the metal from the printing blade to the electrolyte-impregnated paper.[77] Although picture quality was not as good as that of older, photographic Wirephoto receivers, which some papers retained for reproduction copies, Photofax prints emerged dry from the recorder and could be ripped off and examined as soon as they were received. By 1956, 200 of the AP's 500 Wirephoto subscribers had Photofax units, and the network had grown from 10,000 to 30,000 miles in two decades.[78] UPI, which two years earlier had introduced its own electrolytic facsimile receiver, called Unifax, by 1956 was serving 142 newspapers, many of which were small papers that only recently had begun to make their own photoengravings with a Scan-A-Graver.[79]

Further improvements in facsimile equipment reflected a sustained interest in news photographs and an increased concern for graphic quality, particularly among smaller papers with new offset presses. In 1956 the AP introduced a moisture stabilizer to prevent the electrolytic Photofax paper from drying out in the dry climates of the Southwest or in heated buildings in the North.[80] In the mid-1960s the AP introduced an Automatic Wirephoto receiver, dubbed AutoPhoto, which combined the reproduction quality of the photographic Wirephoto receiver with the automatic operation of Photofax.[81] Twenty seconds after transmission was complete, an "automatic guillotine" cut a glossy print from a 165-m (500-ft) roll of paper and dropped it into a bin. UPI introduced an improved, automatic portable Telephoto transmitter in 1971, and a fully automatic electrostatic Unifax II receiver in 1975.[82] Similar to an office copier in its use of a dry-ink toner, Unifax II provided a high-definition image with 53 scan lines per cm (135 lines per in.) and a similar density of picture elements along each scan line.[83] In the late 1970s the AP, working with engineers at the Massachusetts Institute of Technology, developed the Laserphoto receiver, which used a precise beam of laser light to record a detailed latent image on dry-silver paper.[84] Heating the paper developed automatically the latent image, recorded at 44 scan lines per cm (111 lines per in.). The AP's "Electronic Darkroom," a computer for storing digital images of news pictures and for cropping, enlarging, reducing, and sharpening contrast, further improved the quality of Laserphotos.[85] The Laserphoto II receiver, with a still finer resolution of 65 scan lines per cm (166 lines per in.), provided sharper, more highly detailed pictures for newspapers using halftone screens with densities of 33 lines per cm (85 lines per in.) or finer.[86] With Laserphoto II, the AP introduced a second, high-speed photowire providing not only high-resolution images but also color separations—four for photos to be printed in "full-color" with black, yellow, magenta, and cyan inks.[87]

Satellite transmission was a further boon to AP and UPI photowire service. Despite occasional, generally predictable interruptions caused by electromagnetic storms or "solar noise," sending signals to a telecommunications satellite for rebroadcast to an unlimited number of earth stations generally was more reliable than land-line transmission, which was subject to unpredictable signal distortion as well as service interruptions because of severe weather.[88] But the greatest incentive for the shift to satellite communications was the threat of greatly increased cost for local and regional service provided by the Bell System's regional operating companies, severed from AT&T in the court-ordered divestiture of the early 1980s. Between 1983 and 1984 the AP's annual telecommunications charges might have risen from $12.7 million to $21.5 million had it not adopted satellite transmission.[89] UPI, with telecommunications accounting for 30 percent of its total operating expenses, hoped to save $6 to $7 million annually by shifting from land lines to a communications satellite.[90]

Despite the cost and the complicated logistics of installing hundreds of receiving dishes, the changeover to satellite communications was rapid. In 1981 the AP leased a transponder on Western Union's Westar III satellite, and by the end of 1982 had installed dish antennas at 450 newspapers and broadcasting stations.[91] Also using Westar III, UPI had by the end of 1982 installed receiving antennas at 315 of its 900 planned earth stations. By 1985, the AP and UPI had each installed over 2,000 receiving dishes, the AP had bought one of Westar III's 24 transponders, and UPI had established its own ground transmitter, or "satellite uplink." Both wire services had not only reduced substantially their dependence on the telephone system but were selling data transmission and facsimile services to business clients and smaller news syndicates. In the mid-1980s, for example, UPI carried a package of Accu-Weather forecasts and maps, which UPI clients could use for an additional fee.[92] In addition to providing commercial telecommunications services, the AP manufactured Dataphone receivers and transmitters for sale to nonsubscribers desiring Laserphoto II picture quality.[93]

The Associated Press Managing Editors Association, composed of news executives from AP-member newspapers, has been a constructive critic of Wirephoto operation. The Photo and Graphics Committee has been one of the more active of the APME committees monitoring numerous aspects of the cooperative's performance and presenting study reports and recommendations at the annual meeting. In addition to surveying the photographic needs of AP members and agitating for improved image quality, the Photo and Graphics Committee occasionally has called for more and better maps. In 1949, for instance, the committee noted a need for a forecast weather map, instead of a map showing past conditions.[94] In 1974 the committee requested that the weather map be redesigned and called for regional weather maps as well.[95] Management responded with a new design but noted that "the more striking the design is, the less information the map carries."[96] Responding to requests for maps of the Korean conflict, the AP in 1950 offered four "spot maps" (that is, timely news maps) a day, two for morning papers and two for evening papers. As demonstrated by Figure 3.15, these AP war maps could be as dramatic and informative as most action photographs. According to the committee's

Figure 3.15. In July 1950 many newspapers ran a daily map on the front page to show how well U.S. troops and their allies were resisting an onslaught of North Korean and Chinese forces, shown on this Associated Press map by ominous dark arrows. [Source: Syracuse *Herald-Journal*, 29 July 1950, 1.]

report for that year, "Never has AP experienced [a] heavier call for play-by-play illustration of any situation in map form."[97] In 1954 a request for color separations by wire mentioned maps as well as photos.[98] In 1964, the committee noted that more maps would have improved coverage of the Kennedy assassination, and called for "up-to-date explanatory maps which would quickly illustrate stories."[99] "Constructive suggestions" offered in 1965 included "more imaginative techniques in maps, apart from basic locator maps."[100] The following year the committee reported a need for more detailed locator maps, less lag in transmission between the story and accompanying maps, and the use of "imagination and skill" in preparing maps that tell the story.[101] In 1968, in response to persistent complaints, Hal Buell, manager of the AP photo department, promised improvements. "We're presently seeking ways of sophisticating our maps and graphs," Buell asserted, "and I think you'll be seeing more of the complete map that within itself tells the story, with boxes and lettering included."[102] But after reviewing 64 maps and charts offered over a 53-day period, the committee in 1971 concluded that "the AP needs significant help in the cartography department."[103] Although praising the redesign of the Vietnam map to show terrain, the report found that "most of the base maps used by the AP remain cluttered and antiquated." The seven years following this stinging analysis were comparatively quiet on the cartographic front, though, with the annual committee report criticizing only the AP weather map.

By the late 1970s the revolution in newspaper graphics was underway, and the APME again looked at AP cartography. At a workshop on maps at the 1978 annual meeting, art supervisors from three member newspapers discussed the organization and operation of their own map departments.[104] The committee report noted that Frank Peters, of the *St. Petersburg Times*, normally used AP maps only as data for redrafting "into *Times* style." The following year the committee observed that AP graphics were "frequently below the standards of its members."[105] The report cited transmission problems as a limitation but noted that the AP "often lacks versatility in its work." In 1980 a similar but hopeful verdict concluded that "maps and graphics from the Associated Press are better than they used to be, but they still aren't very good."[106] The 1980 report also suggested a reason: "Nobody at AP seems to be really interested. Graphics are an afterthought. Coordination between graphics and stories is often worse than coordination between photos and stories." A workshop at the 1981 annual meeting discussed several examples of "how wire graphics that are less than attractive can be dressed up dramatically."[107] But at least one committee member was sufficiently impressed with the adoption of Helvetica type and an "open, cleaner style" to conclude that "the overall appearance of AP graphics no longer is 'dowdy,' as was charged in the 1980 study, and the new-look maps and charts generally are 'finished' enough for immediate use."[108]

In 1983 the committee announced the most important improvements since Wirephoto's inauguration: not only had the new Laserphoto II facsimile system improved image quality, but the AP was "placing increasing emphasis on graphics, especially informational graphics."[109] In addition to engaging graphics expert Robert Lockwood as a consultant, the AP announced plans to establish a graphics department. At his News Graphics studio in New Tripoli, Pennsylvania, Lockwood and his several

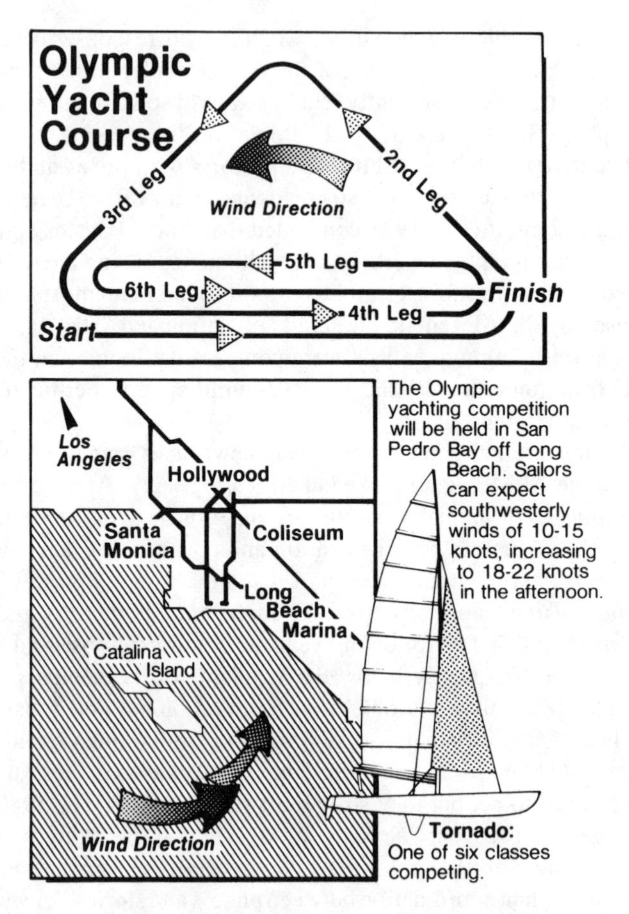

Figure 3.16. Anticipatory graphic prepared by Robert Lockwood's News Graphics firm for graphics package mailed to AP members several weeks before the 1984 Summer Olympics. [Source: Courtesy of the Associated Press.]

employees prepared special weekly packages of graphics to be mailed to AP members.[110] Figure 3.16, which illustrates Lockwood's style, is a pair of maps included in a package of anticipatory graphics mailed to AP members several weeks before the Los Angeles 1984 Summer Olympics. Although not established as a separate unit until late 1985, the new graphics department at AP headquarters in New York was one of three significant improvements to AP service announced in 1985 by AP president and general manager Louis Boccardi. With a "director of graphics" and three additional artists, the new department's mission included the preparation of graphics for state news reports, as well as for national and foreign stories of interest to all AP members.[111]

Although an increased awareness among newspaper editors of the value of graphics explains much of the recent improvement in the AP's cartographic art, a second and perhaps more important influence is competition with the cooperative's principal rival, United Press International. Indeed, the AP fulfilled its commitment to a graphics department only after UPI announced the expansion of its UPI graphics staff in mid-1984.[112] Yet the recently widened acceptance of and increased demand for newspaper graphics reinforced whatever role might have been played by competition among news agencies. With a director and ten graphics specialists, UPI's graphics department represented a significant investment by this financially vulnerable wire service, which never has attained the strength, prestige, and stability of the AP. Both the AP and UPI have lost business since the 1950s as a result of closings and mergers in the newspaper industry. Many publishers no longer facing a local competitor have found that they could bolster profits by dropping one of their wire services.[113] UPI has lost more clients than the AP, and in an effort to economize, has cut staff. The newly apparent graphics needs of the medium-size newspaper—with a small art staff, if any—and the long-standing, widespread dissatisfaction with AP graphics presented UPI with the opportunity for a highly visible enhancement to its service. Although the UPI's strategy of hiring artists while sacking reporters might seem shortsighted, even a reduced UPI staff provided more news and features than a small or medium-size paper could accommodate. With its graphics strategy, UPI hoped not only to retain larger clients receiving both wire services but also to capture some smaller papers seeking better maps and diagrams.

Designed for photowire transmission, the new UPI information graphics took full advantage of the superiority of Univax II over Laserphoto in transmitting line drawings.[114] The stippled symbol used to mark the shoreline in Figure 3.17 is an example of UPI's perceptive compromise between aesthetics and the realities of facsimile transmission. In contrast, AP graphics in the early 1980s employed a harsh parallel-line symbol for water bodies, as Figure 3.18 shows. Also, by coordinating its graphics with its written material, UPI promoted both its news service and its package of special-section features.[115] In mid-1985, when further staff cutbacks helped avoid bankruptcy, UPI's editor-in-chief promised to maintain, and possibly even expand, his photo and graphics service.[116] Whether the financially strapped wire service can survive remains questionable.[117] In 1986 several prominent newspapers, including the *New York Times*, dropped their UPI wire. Moreover, as examined in Chapter 4, in 1987 both the AP and a Knight-Ridder syndicate offered improved microcomputer-based graphics services meeting the high aesthetic standard set by UPI in the early 1980s.

Figure 3.17. A UPI locator map, with stippling used as a symbol for coastal waters. [Source: Telephoto map published in the Syracuse *Post-Standard* and other newspapers, 27 May 1985. Reprinted with permission of United Press International, Copyright 1985.]

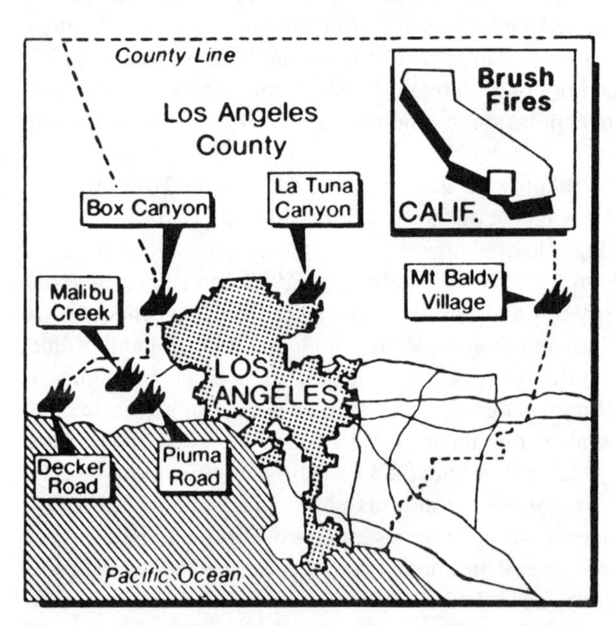

Figure 3.18. An AP locator map, with parallel-line patterns used to represent water bodies. [Source: Laserphoto map published in the Syracuse *Post-Standard* and other newspapers, 16 October 1985.]

Twelve Dailies in Central New York: A Case Study

Examination of historical patterns of map use by 12 daily newspapers in central New York reveals the strong influence of wire services on the cartographic content of small newspapers as well as the effects of advances in printing and engraving technology. The newspapers in this case study are daily newspapers operating in 1985 within a nine-county area centered on Syracuse, New York, the largest city in the region (Figure 3.19). The sample covers 10 cities. Two cities, Syracuse and Utica, have jointly-owned morning and afternoon newspapers, whereas the other eight have only an afternoon daily. Regional editions of Syracuse's two dailies compete with each of the other 10 newspapers, in its city of publication and throughout much of its circulation territory. Newspaper groups, or chains, own all but three of the 12 papers, but no single group controls more than three of the region's newspapers.[118] In 1985 only three

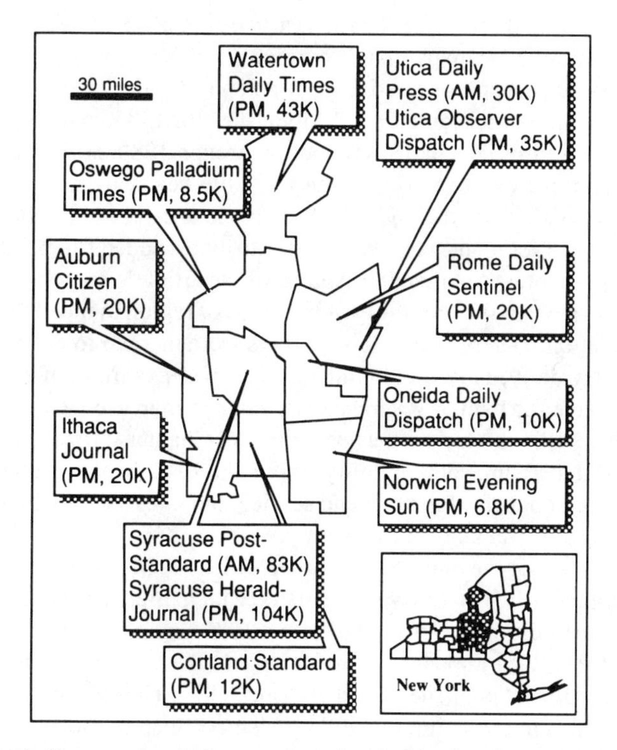

Figure 3.19. The sample of 12 newspapers in 10 cities in a 9-county area in central New York. Newspaper names are followed by publication schedule (A.M. or P.M.) and weekday circulation, in thousands, as reported in the 1985 *Editor and Publisher Yearbook*.

of the cities—Auburn, Syracuse, and Utica—had a Sunday paper, and the evening papers in Auburn, Norwich, and Utica did not print on Saturdays.

As Figure 3.19 indicates, average weekday circulations in 1985 varied from just under 7,000 for the Norwich Even*ing Sun* to over 104,000 for the Syracuse *Herald-Journal.* All but the two Syracuse papers had circulations under 50,000 and would be considered small or medium-size newspapers by most standards.[119] In 1985, 1,435 of the 1,688 daily newspapers in the United States were in the under-50,000 category, and like the papers in the sample, most had circulations under 25,000.

Photowire and syndicate affiliations varied. In 1985 the two Syracuse papers received news photos from both the AP and UPI, seven other papers received only AP Laserphotos, the Auburn *Citizen* received only UPI Telephotos, and the Norwich and Oneida papers had no photowire service. In addition, several papers had contracts with a syndicate supplying graphics by mail: the Syracuse, Utica, and Watertown papers with Tribune Media Services, and the Norwich, Oneida, and Oswego papers with NEA. Only three papers had staff artists: the *Watertown Daily Times* had a single artist for maps, charts, cartoons, and occasional page layouts, and the Syracuse papers shared a seven-person editorial art department.[120]

A sample of all January and July issues for 1985, and for all years ending in zero from the paper's founding through 1980, provided historical data comparable to those for the five elite newspapers examined in Chapter 2. As before, the coding unit was the article, not the individual map, and weather maps were excluded. Attributes coded for examination here or in Chapter 4 included the source of the article's map as well as its general theme and the proximity of the place or region mapped to the city of publication. Where two or more papers were merged to form a single present-day daily, only the dominant paper was examined if microfilm was available for both.[121] Average daily rates of map use were computed for each year for each paper, as described in Chapter 2 for the five elite papers. But to assure comparability among the 12 daily newspapers, most of which did not offer full weekend service, the rates used in this chapter are based upon weekday editions only.

For most of the nineteenth century, cities in central New York were served principally by weekly newspapers, which rarely used maps. When more dailies were started in the 1870s and 1880s, maps still were rare. The oldest of the 12 newspapers in the sample is the *Ithaca Journal*, established in 1815 and published as a daily since 1872. The coarse, two-month sample with a ten-year interval did not detect a map in the *Journal* until 1890, when a map of Henry Stanley's route accompanied an article on the search two decades earlier for African explorer David Livingstone. Like most illustrations in American daily newspapers of the era, this map was a stereotype from an unidentified feature syndicate, possibly Kellogg's. Four other maps appeared in the *Journal*'s sample for 1890, and three of the other papers used maps in January or July of that year. Occasionally, though, one of the larger newspapers prepared its own map, as for Sunday, January 28, 1900, when the Syracuse *Herald* used a seven-column map to examine the potential effect on local business of a state-operated barge canal. None of the samples for 1880 and earlier years contained maps. The most recent of the 12 dailies, the *Oneida Daily Dispatch*, was founded as a weekly in 1851 and converted to a daily in 1926; its first maps in the sample, from the Central Press syndicate, appeared in 1930. The sampling strategy detected maps for only six of the papers by 1900, and identified no maps for three of the papers before 1930. But with one exception, from 1930 onward each newspaper used at least one map during January and July of the sample years.

Like the five elite newspapers, these 12 central New York papers did not use maps with even a modest regularity until well after advances in photoengraving made possible the timely use of line art late in the nineteenth century. Figure 3.20, a time-series graph of average weekday rates of mapped articles, reveals trends faintly similar to those for the elite papers, but showing a much lower frequency of map use. Cartographic artwork was rare before 1930. For some papers a notable peak in 1940, when World War II was underway in Europe and parts of Asia, preceded lower rates of map use in 1950. But for eight of the 12 dailies, the rate of

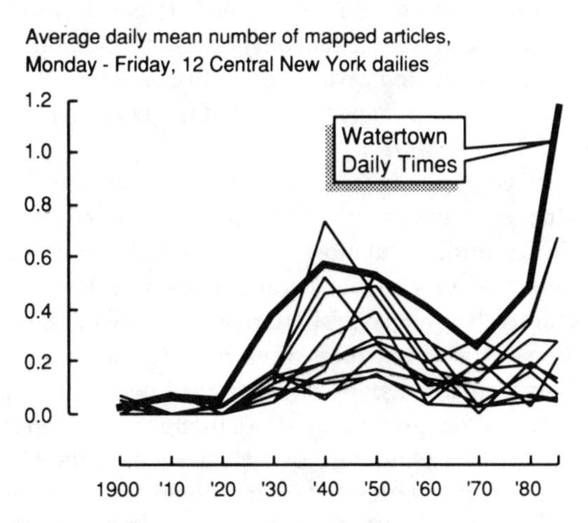

Figure 3.20. Average daily mean numbers of mapped articles, for Monday-Friday editions only, 1900-1985, for 12 daily newspapers in central New York.

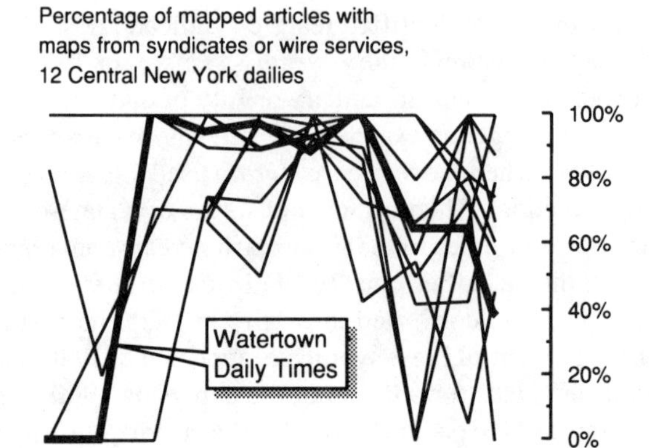

Figure 3.21. Percentages of mapped articles with maps from a syndicate or wire service, 1900-1985, for 12 daily newspapers in central New York.

map use for 1950 exceeded that for 1940. Yet rates for 1960 generally were lower than for 1950, when the Korean War was an important cartographic focus. Modest rises between 1970 and 1980 and a few comparatively spectacular increases between 1980 and 1985 reflect conversion to photocomposition and an increased appreciation of graphics. For half of the 12 newspapers, however, fewer mapped articles appeared in the 1985 sample than in the sample for January and July of 1980. Moreover, only the *Watertown Daily Times* attained a rate of more than one mapped article per day, and then only for the 1985 sample. In general, these smaller dailies adopted cartographic illustration well after the elite papers, and used far fewer maps than all but the *Wall Street Journal* and the Toronto *Globe and Mail*.

Wire services and, to a lesser extent, syndicates have been a dominant influence in the use of cartographic art by these 12 dailies. As Figure 3.21 shows, throughout most of the twentieth century these smaller papers have relied heavily on maps drawn elsewhere. Before 1930, when small dailies rarely used maps, the pattern was less distinct; for a few two-month samples several newspapers offered readers just a single map that might have been prepared locally and engraved by the newspaper's own staff or a nearby photoengraver. By 1950, though, most papers in central New York either had a photowire or were served by the Central Press, which distributed news photos for International Soundphoto. Although more papers acquired direct photowire service in the 1950s, acquisition of an electronic scanner-engraver enabled newspapers to include not only local photos but also portions of planning, topographic, and street maps, annotated to accompany stories about local redevelopment, crime, and recreation. Improved printing technology encouraged news editors to illustrate locally important events with maps, however crude, and after 1960 the trend toward offset printing and photocomposition further eroded the ascendancy of syndicates and wire services as sources of cartographic art. Indeed, much of the increase in mapped articles in recent decades can be attributed to local coverage. Yet even in 1985, only two of the 12 dailies received less than half of their cartographic illustrations from news agencies—only the Watertown paper and the Syracuse morning newspaper used maps by staff artists more frequently than maps drawn elsewhere.

A plausible hypothesis is that map use is more frequent at newspapers with larger circulations, and less frequent at smaller papers with limited resources and smaller "news holes." Yet Figure 3.22, a scatter diagram comparing late-1984 circulation with average daily rates of map use for the 1985 sample, suggests only a tenuous correlation between circulation and map use.[122] To be sure, with larger circulations and at least one staff artist, the two Syracuse papers and the *Watertown Daily Times* used maps more frequently than any of the other nine dailies. And the smallest paper, the Norwich *Evening Sun*, had the second lowest rate of

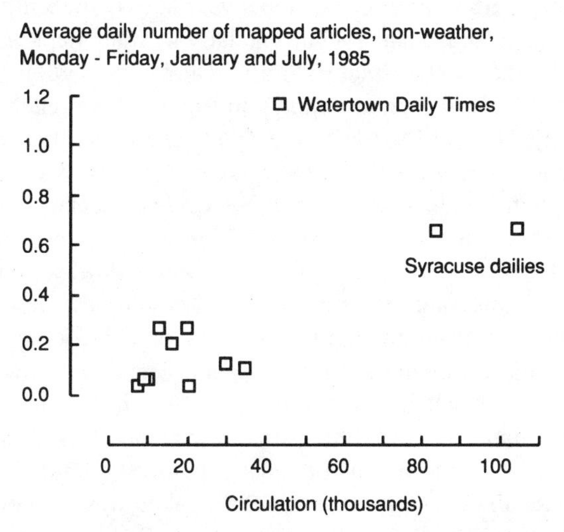

Figure 3.22. Scatter diagram showing the modest positive relationship between late-1984 circulation and frequency of mapped articles for January and July, 1985.

map use. But the two Utica dailies, fourth and fifth in circulation, employed maps less frequently than three much smaller papers, and the *Ithaca Journal*, sixth in circulation, had the lowest frequency. Editorial attitudes toward graphics, examined in Chapter 4, would seem a more important influence on the use of maps by smaller papers than market size and financial resources. Each paper, after all, even those without a photowire and receiving only the NEA graphics mailing, had substantially more maps available than its editors chose to use.

Weather Maps and Electronic Communications

Perhaps the quintessential news map is the weather map: no other type of map enjoys as consistent a presence in so many daily newspapers, and no map has been more dependent upon telecommunications technology for both its construction and distribution. Although meteorologists attribute the first weather map to Heinrich Brandes, a mathematics professor at the University of Breslau, the maps of barometric pressure that Brandes published in 1820 were based upon 1783 data.[123] Maps of current weather, which in the late nineteenth century became the meteorologist's principal forecasting tool, depended upon synchronous weather observations delivered by the electric telegraph. Extensive worldwide development of the electric telegraph after 1844 was followed by the telegraphic collection of weather data, in 1848 in England, and a year later in the United States.[124] The telegraph provided data for current daily weather maps issued briefly in 1851 in London at the Great Exhibition, and more regularly beginning in 1863 in the *Bulletin International*, published by the French astronomer Le Verrier.[125] On April 1, 1875, the *Times* of London became the first newspaper to publish a daily weather map.[126] Figure 3.23, an example from later that month, shows how these first newspaper weather maps employed isobars to show the pattern of barometric pressure, arrows to show wind direction, encircled dots to show calm seas, and type to indicate spot temperatures and sky and sea conditions. Based on 8 A.M. data from 50 stations in the British Isles and on the Contintent, the map appeared in the following morning's paper. The government Meteorological Office, which had prepared daily weather maps since 1872, collected the data and drafted this smaller version of its regular weather chart. The *Times* supported the full cost of publication, including making both the plate and the copy from which the plate was prepared.

Lacking the convenience with which its British counterpart could deal with comparatively few London-based national newspapers, the U. S. Weather Service faced a more formidable challenge in disseminating

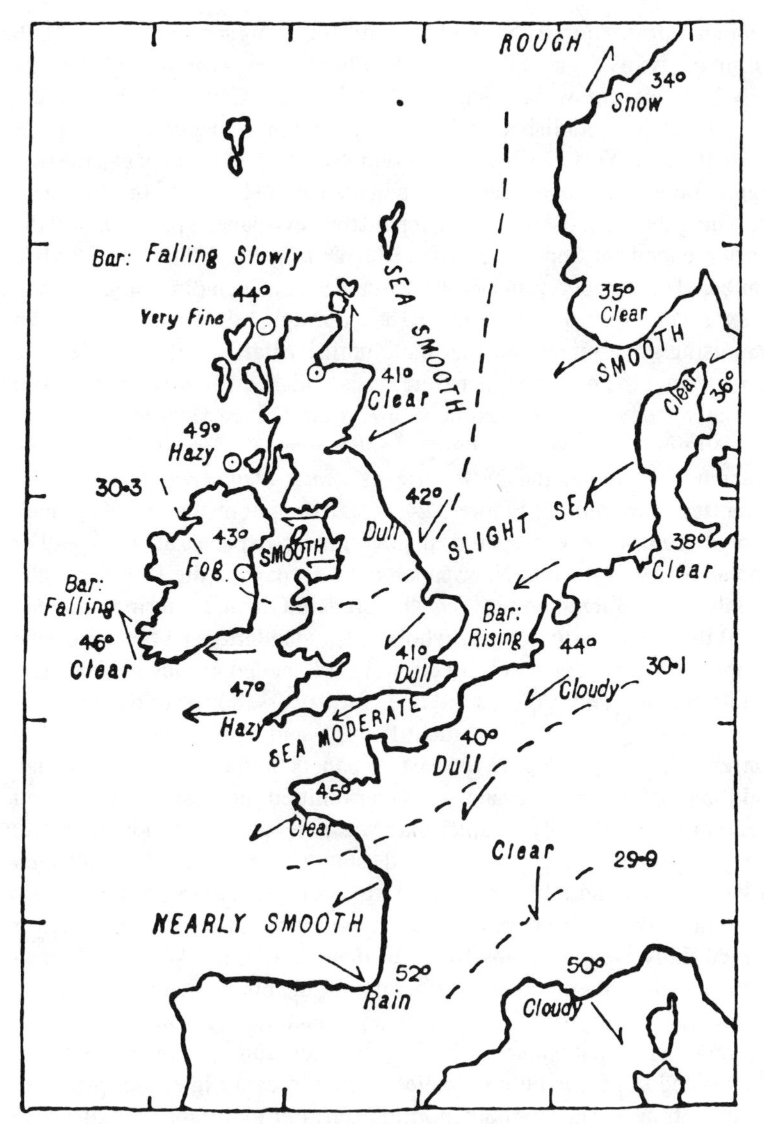

Figure 3.23. Weather map from the *Times* of London of 14 April 1875. Caption beneath the map notes: "The dotted [pecked] lines indicate the gradations of barometric pressure, the figures at the end showing the height, with the words 'Rising', 'Falling', etc., as required. The temperature at the principal stations is marked by figures, the state of the sea and sky by words. The direction and force of the wind are shown by arrows, barbed and feathered according to its force. [A small circle with a dot at its center] denotes calm." [From "The 'Times' Weather Chart," *Nature* 11, no. 285 (15 April 1875), 473-74.]

weather information to a country many times bigger. Formed in 1870 within the Army Signal Office, the Weather Service provided maps to the New York *Herald*, which on May 12, 1876, at the Centennial Exposition in Philadelphia, published the first American newspaper weather map, and to the New York *Daily Graphic*, which on May 9, 1879, began the first regular publication of a weather map in the United States.[127] In addition to providing the map, the government paid the newspaper $10 a day until the service ended on September 14, 1882. During the 1880s and 1890s a number of newspapers printed a daily weather map, including the Cincinnati *Commercial Gazette*, which from 1881 through 1892 produced its own map using government weather data, until its staff weatherman left the paper to join the new Weather Bureau, established two years earlier as a civilian agency in the Department of Agriculture. In 1894 four newspapers offered their readers a daily weather map: the Boston *Herald*, the Cincinnati *Tribune*, the New Orleans *Times-Democrat*, and the San Francisco *Examiner*.[128] Figure 3.24 is an example of the *Herald*'s map, printed from a plate prepared at government expense at the Weather Bureau office in Boston. Newspaper weather maps of the 1890s bore the initials of the Bureau employees who prepared them. Lettering reflected the skill of the cartographer, who used a standard set of hand-drawn symbols—arrows for wind direction, small shaded circles to show sky conditions for major cities, solid lines for isobars, dotted or dashed lines for isotherms, and words to identify high and low pressure cells. To conserve space, the maps in the eastern papers omitted the Pacific Coast, and the map in the San Francisco paper omitted the East. Decentralized preparation of the daily weather map was repeated in the more than 100 Weather Bureau offices that prepared "station maps" from data collected in Washington and telegraphed to stations throughout the country.

In 1910, in an effort to reduce the number of stations at which it printed daily weather maps for local distribution, the Weather Bureau introduced the "commercial weather map," an abbreviated version of the station map designed for newspaper reproduction and made available to any newspaper that wanted it.[129] The Weather Bureau furnished either a clear inked copy for photographic engraving or a plate, and provided separate, timely maps for both morning and afternoon papers. Figure 3.25 shows a typical commercial weather map, with its standard border, title, and state outlines. The campaign was a rapid success: by the end of 1910 the map appeared in 65 papers in 45 cities, by mid-1911 adoption had spread to 132 papers in 74 cities, and by mid-1912 the map was reaching the combined 3,036,000 circulation of 147 daily newspapers in 91 cities. Newspaper distribution of the commercial map was almost 150 times the daily circulation of its printed station maps, estimated at 20,550 in 1912. The Weather Bureau had succeeded not only in widening the distribution

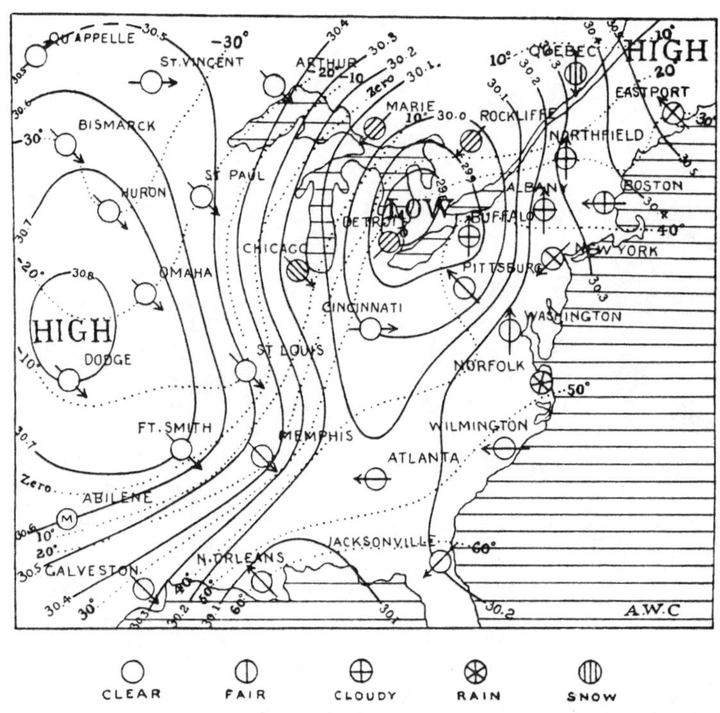

Figure 3.24. Weather map for 8 A.M., 24 January 1894, printed in that day's afternoon issue of the Boston *Herald*. [Source: Robert De Courcy Ward, "The Newspaper Weather Maps of the United States," *American Meteorological Journal* 11, no. 3 (July 1894), 96-107; plate following p. 100.]

of the daily weather map but also in reducing from 110 to 59 the number of cities where it printed a station map.[130]

The Weather Bureau's success in disseminating weather data through the American press suffered the fate of many fads that catch the media's fancy. Subsequent annual reports of the Weather Bureau provide no data on attrition, but 1912 probably was the peak year. No weather maps appeared in the 1920 samples for the three elite United States newspapers and the 12 central New York dailies, although the Syracuse *Herald* carried a daily weather map Monday through Saturday, from May 15, 1910 until July 13, 1919. The *New York Times*, which used a long article on progress in weather forecasting to mark the initiation of its daily weather map on August 4, 1934, apparently never succumbed to the enthusiasm of the 1910s for the commercial weather map.[131]

In early 1935 the AP Wirephoto network began the electronic dissemination of a centrally prepared newspaper weather map, and many

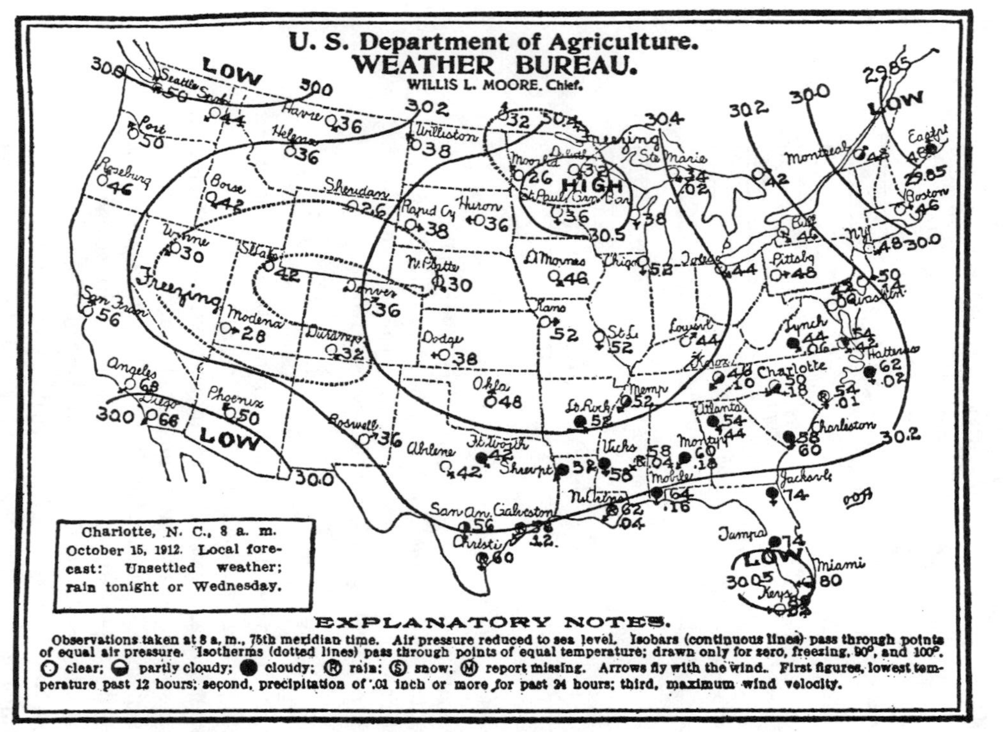

Figure 3.25. An example of the commercial weather map, prepared for a Charlotte, North Carolina, newspaper, with the local forecast in a box at the lower left. Explanatory note at bottom of the map reads "Observations taken at 8 A.M., 75th meridian time. Air pressure reduced to sea level. Isobars (continuous lines) pass through points of equal air pressure. Isotherms (dotted lines) pass through points of equal temperature; drawn only for zero, freezing, 90°, and 100°. [open circle] clear; [half-darkened circle] partly cloudy; [black circle] cloudy; [circle enclosing the letter R] rain; [circle enclosing the letter S] snow; [circle enclosing the letter M] report missing. Arrows fly with the wind. First figures, lowest temperature past 12 hours; second, precipitation of .01 inch or more for past 24 hours; third, maximum wind velocity." [Source: Henry L. Heiskell, "The Commercial Weather Map of the United States Weather Bureau," in *Yearbook of Agriculture, 1912*, pp. 537-39; map on p. 538.]

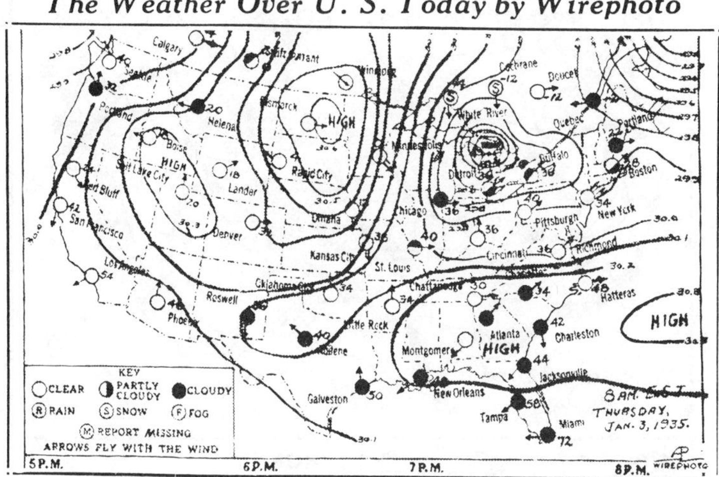

Figure 3.26. Early AP Wirephoto weather maps emphasized the national pattern of atmospheric pressure. [Source: *Syracuse Herald*, 3 January 1935, 6. Courtesy of the Onondaga Historical Association, Syracuse, N.Y.]

newspapers large enough to afford a direct Wirephoto connection began to print the map for their readers. The Syracuse *Herald*, one of the original Wirephoto customers, carried the weather map from January 3, 1935, the third day of Wirephoto transmission, through December 15, 1941, a week after the Pearl Harbor attack, and resumed the service on February 4, 1946, shortly after the end of World War II. Figure 3.26, an early Wirephoto weather map, shows only isobars, pressure cells, and general weather conditions for selected cities—the Weather Bureau didn't add lines showing warm fronts and cold fronts to its maps until 1936, after the air mass theory of Vilhelm and Jacob Bjerknes revolutionized forecasting.[132] Other Upstate dailies, which acquired photowire service somewhat later, followed in two spurts: the *Ithaca Journal*, the Rome *Daily Sentinel*, the Syracuse *Post-Standard*, and the *Watertown Daily Times* between 1957 and 1959, and the Auburn *Citizen*, the *Cortland Standard*, and the Utica *Daily Press* between 1969 and 1971. Figures 3.27 and 3.28 are examples of the forecast maps offered twice daily by the AP and UPI. Yet not all papers with a photowire used the weather map; for example, the Oswego *Palladium Times*, which has had Wirephoto service since the 1950s, apparently has never used the weather map, and the Utica *Observer Dispatch*, which joined the Wirephoto network in the 1940s, carried no weather map in 1950 and used it irregularly in 1960 and only on Sundays in 1970 and 1980. Without a photowire, the Norwich *Evening Sun* never

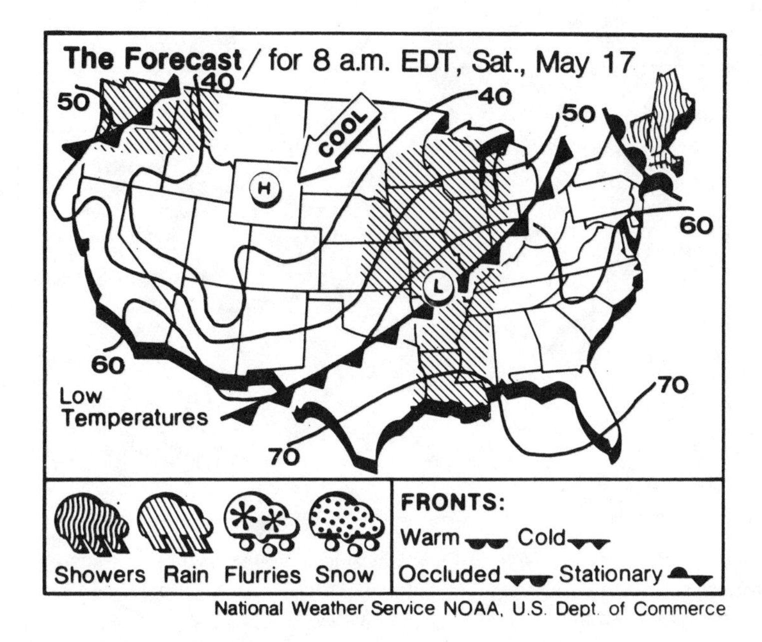

Figure 3.27. Typical mid-1980s AP "LaserGraphic" forecast weather map. Map with a forecast for 8 A.M. was intended for afternoon newspapers published the previous day, and was transmitted in the morning, together with a past-conditions weather map for 2 A.M. that day. [Source: Courtesy of the Associated Press.]

used a daily weather map, and the *Oneida Daily Dispatch*, which started a daily weather map in late 1975, dropped it at the end of 1980, when it cancelled its photowire. The advent of geostationary meteorological satellites in the 1960s brought another type of photowire weather map, the satellite photo showing cloud cover over North America.

The early 1980s witnessed a renewed interest in newspaper weather maps. In 1982 the Gannett chain's national newspaper, *USA Today*, delighted weather fans and caught the attention of news publishers with its full-page, full-color weather package, featuring the large national temperature-band/precipitation map shown in Figure 3.29.[133] Printed in numerous cities throughout the country from pages made up outside Washington, D.C., and delivered by satellite, *USA Today* immediately won high praise from news publishers and the public for its consistently high-quality color printing and its lively information graphics. Although *USA Today* no doubt spurred some newspapers to redraft the wire-service weather map or to add color, these changes predated the Gannett daily at many newspapers, and reflected an industrywide concern with newspaper design and graphics that began in the early 1970s.[134] Yet, even though *USA Today* might be seen more as a product of the revolution in newspaper design than as one of its instigators, many newspapers redesigned their weather packages in the mid-1980s to emulate the Gannett paper or to compete better with it in their own area. A number of papers dropped the informative but drab satellite photograph and introduced a regional temperature map, such as Figure 3.30 shows for the *Ithaca Journal.*

Following the lead of *USA Today*, which received its weather forecast and rough map by facsimile from a private meteorological forecasting service—rather than from the National Weather Service, the AP, or UPI—a number of newspapers arranged to receive their weather maps from one of several private firms providing reproducible artwork by dial-up telephone facsimile. This service was particularly attractive to smaller papers wanting to exploit a new color offset press but lacking a full-time staff artist. Figure 3.31 shows the black-and-white United States outline map used by Air Science Consultants, Inc., of Bridgeville, Pennsylvania, which employed telephone lines to transmit its national temperature-band/precipitation/sunshine map in four color-separated panels. For newspapers with an Apple Macintosh computer and high-resolution

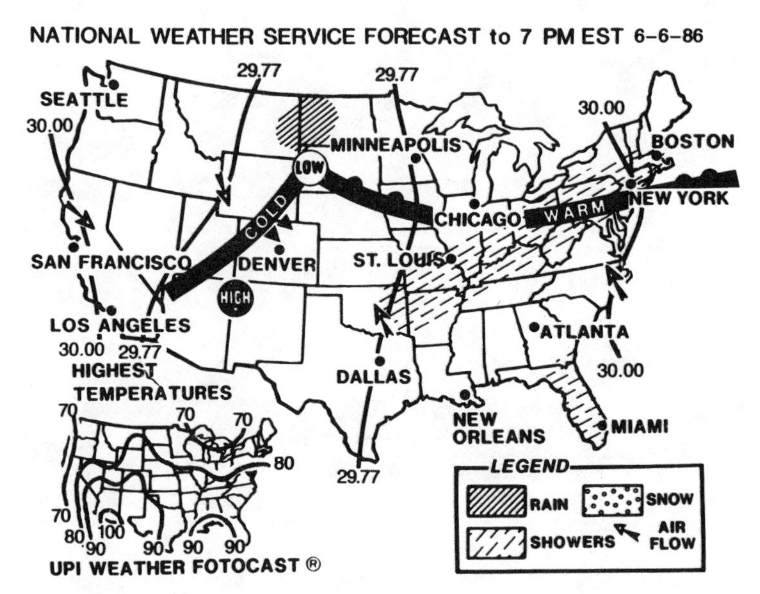

Figure 3.28. Typical mid-1980s UPI "Fotocast" forecast weather map. Map with a forecast for 7 p.m. was intended for newspapers published that morning. [Source: Reprinted with permission of United Press International, Copyright 1986.]

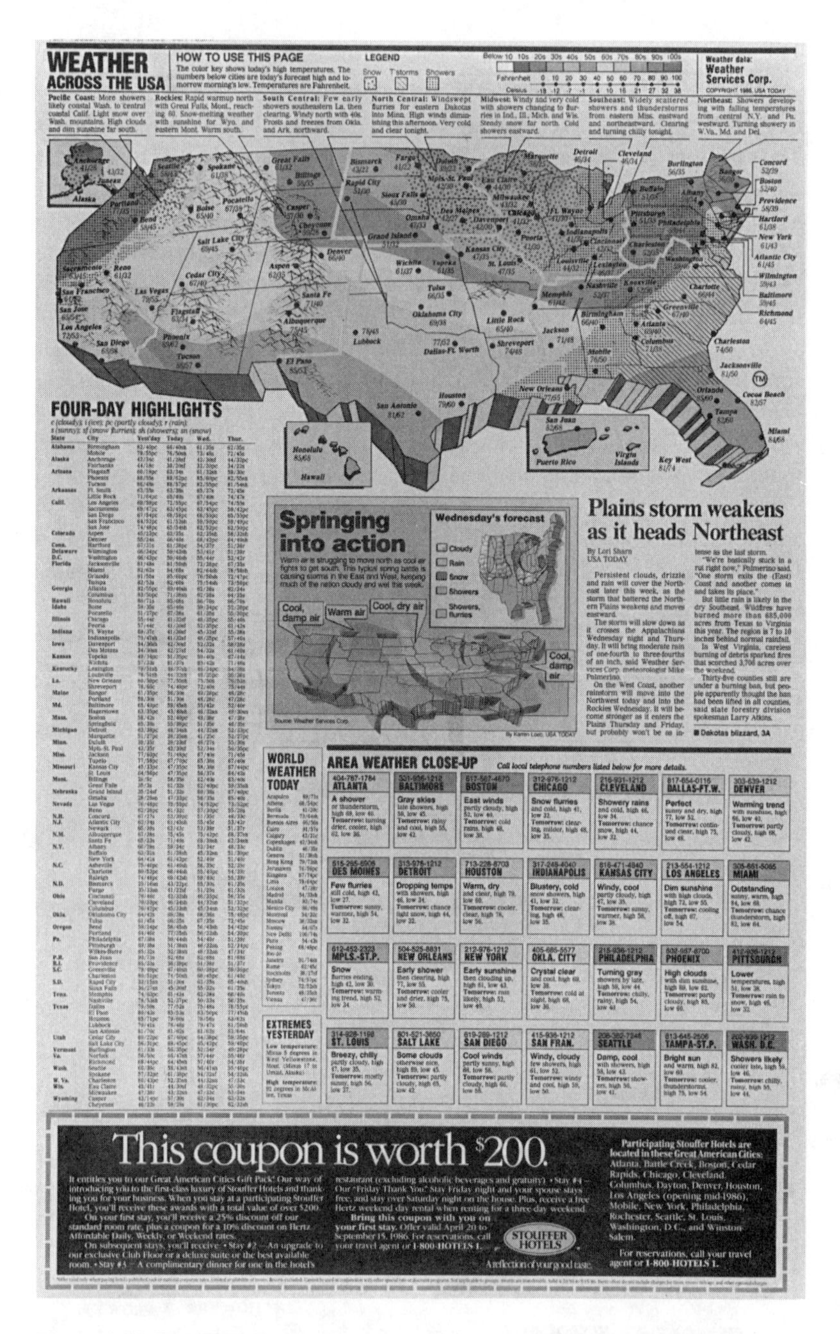

WEATHER ACROSS THE USA

HOW TO USE THIS PAGE

The color key shows today's high temperatures. The numbers below cities are today's forecast high and tomorrow morning's low. Temperatures are Fahrenheit.

Pacific Coast: More showers likely coastal Wash. to central coastal Calif. Light snow over Wash. mountains. High clouds and dim sunshine far south.

Rockies: Rapid warmup north with Great Falls, Mont., reaching 60. Snow-melting weather with sunshine for Wyo. and eastern Mont. Warm south.

South Central: Few early showers southeastern La. then clearing. Windy north with 40s. Frosts and freezes from Okla. and Ark. northward.

North Central: Windswept flurries for eastern Dakotas into Minn. High winds diminishing this afternoon. Very cold and clear tonight.

Midwest: Windy and very cold with showers changing to flurries in Ind., Ill., Mich. and Wis. Steady snow far north. Cold showers eastward.

Southeast: Widely scattered showers and thunderstorms from eastern Miss. eastward and northeastward. Clearing and turning chilly tonight.

Northeast: Showers developing with falling temperatures from central N.Y. and Pa. westward. Turning showery in W.Va., Md. and Del.

FOUR-DAY HIGHLIGHTS

Plains storm weakens as it heads Northeast

By Lori Sharn
USA TODAY

Persistent clouds, drizzle and rain will cover the Northeast later this week, as the storm that battered the Northern Plains weakens and moves eastward.

The storm will slow down as it crosses the Appalachians Wednesday night and Thursday. It will bring moderate rain of one-fourth to three-fourths of an inch, said Weather Services Corp. meteorologist Mike Palmerino.

On the West Coast, another rainstorm will move into the Northwest today and into the Rockies Wednesday. It will become stronger as it enters the Plains Thursday and Friday, but probably won't be as intense as the last storm.

"We're basically stuck in a rut right now," Palmerino said. "One storm exits the (East) Coast and another comes in and takes its place."

But little rain is likely in the dry Southeast. Wildfires have burned more than 685,000 acres from Texas to Virginia this year. The region is 7 to 10 inches behind normal rainfall.

In West Virginia, careless burning of debris sparked fires that scorched 3,700 acres over the weekend.

Thirty-five counties still are under a burning ban, but people apparently thought the ban had been lifted in all counties, said state forestry division spokesman Larry Atkins.

■ Dakotas blizzard, 3A

WORLD WEATHER TODAY

AREA WEATHER CLOSE-UP

Call local telephone numbers listed below for more details.

404-787-1784 **ATLANTA** — **A shower**

BALTIMORE — **Gray skies**

BOSTON — **East winds**

312-976-1212 **CHICAGO** — **Snow flurries**

216-931-1212 **CLEVELAND** — **Showery rains**

817-654-0116 **DALLAS-FT.W.** — **Perfect**

303-639-1212 **DENVER** — **Warming trend**

DES MOINES — **Few flurries**

313-976-1212 **DETROIT** — **Dropping temps**

713-228-8703 **HOUSTON** — **Warm, dry**

INDIANAPOLIS — **Blustery, cold**

KANSAS CITY — **Windy, cool**

213-554-1212 **LOS ANGELES** — **Dim sunshine**

MIAMI — **Outstanding**

612-452-2323 **MPLS.-ST.P.** — **Snow**

504-525-8831 **NEW ORLEANS** — **Early shower**

212-976-1212 **NEW YORK** — **Early sunshine**

405-685-5577 **OKLA. CITY** — **Crystal clear**

PHILADELPHIA — **Turning gray**

PHOENIX — **High clouds**

PITTSBURGH — **Lower**

ST. LOUIS — **Breezy, chilly**

801-521-3650 **SALT LAKE** — **Some clouds**

619-289-1212 **SAN DIEGO** — **Cool winds**

415-936-1212 **SAN FRAN.** — **Windy, cloudy**

SEATTLE — **Damp, cool**

TAMPA-ST.P. — **Bright sun**

WASH. D.C. — **Showers likely**

EXTREMES YESTERDAY

Figure 3.29. The *USA Today* weather page features a large map showing temperature zones in color and one or more maps or diagrams explaining an important facet of the day's weather. [Source: *USA Today*, 15 April 1986. © 1986, USA TODAY, All rights reserved, Reprinted with permission.]

LaserWriter printer, Accu-Weather, Inc., which once supplied the forecast and rough map for USA *Today*, in late 1986 introduced their MacWeather service, providing by telephone facsimile not only national forecast maps for morning and afternoon newspapers but regional weather maps and explanatory weather graphics as well.[135] Even the venerable

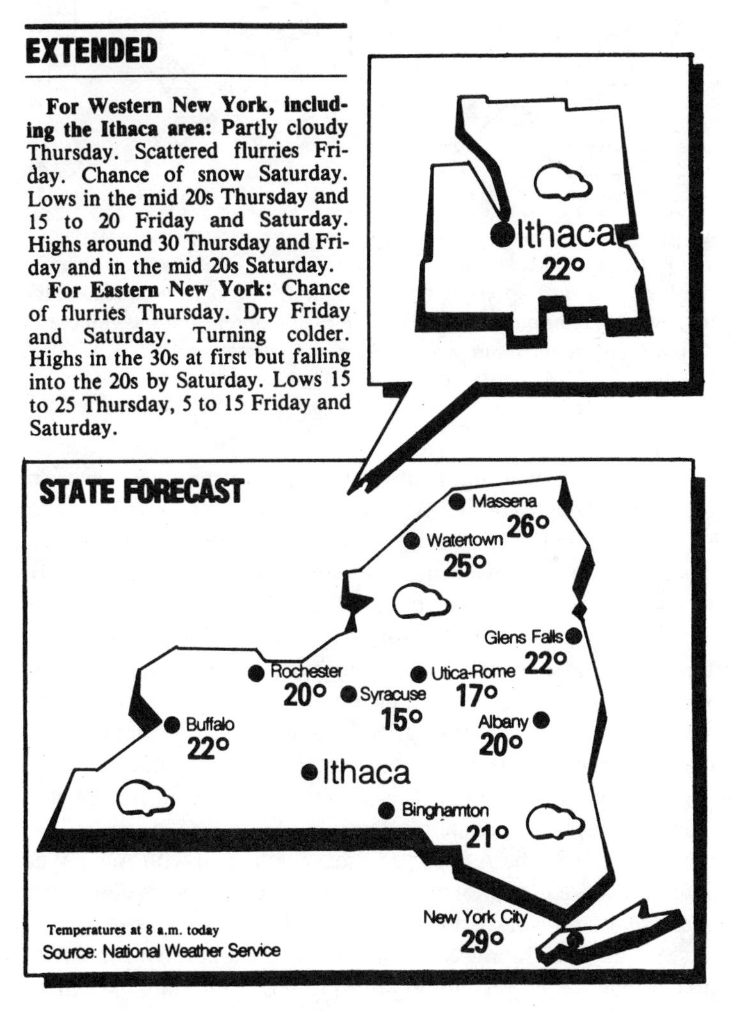

Figure 3.30. Regional temperature map introduced in early 1985 by the *Ithaca Journal* appeared to the right of the AP national weather forecast map and included an inset for Tompkins County and the city of Ithaca. [Source: *Ithaca Journal*, 6 January 1987.]

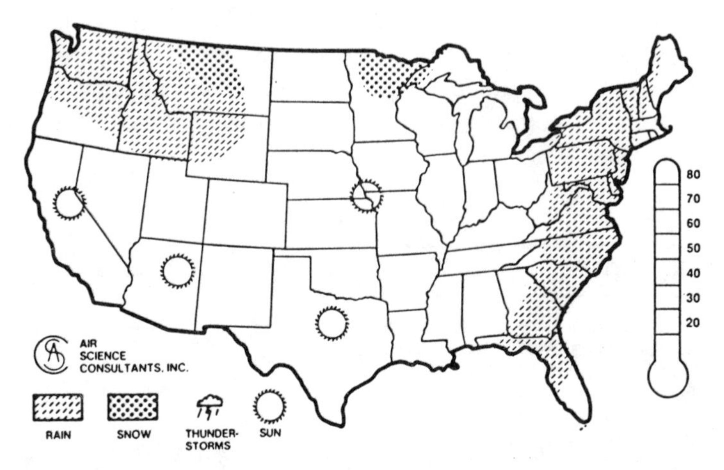

Figure 3.31. National basemap used by Air Science Consultants, which transmitted color separations by telephone for a color-band temperature map. Client newspapers printed the base map in black, in registration with color bands portraying temperature zones. [Source: Courtesy of Air Science Consultants, Inc., Bridgeville, Pennsylvania.]

New York Times, which could not handle color illustrations because of the large number of copies produced at its few printing plants, in 1986 adopted a new black-and-white temperature-band map, with artwork delivered by telephone facsimile from the meteorology department at the Pennsylvania State University. In contrast, the *St. Petersburg Times*, a metropolitan paper with a strong reputation for its graphics and color reproduction, drafted its own color weather map using as a source the National Weather Service's latest forecast map, transmitted late in the afternoon by telephone from the local NWS office.[136] Electronic telecommunications, which made the weather map possible, enabled its fullest distribution through a variety of routes—leased satellite and land-line photowire, telephone, and full-page satellite facsimile.[137]

In early 1987 the Associated Press contracted with Accu-Weather for two new services: national and regional weather maps delivered over the Laserphoto network, and a telephone dial-in graphics service called AP Access, designed for use with a modem, the Apple Macintosh microcomputer, and a high-resolution laser printer.[138] Providing AP members with news graphics—including nonweather maps as well as weather graphics—the AP Access service finally liberated the cooperative's electronically distributed news maps from the continuous-tone bias of Wirephoto and Laserphoto. As an example, Figure 3.32 illustrates the fine graytones that replaced the comparatively harsh parallel-line water pat-

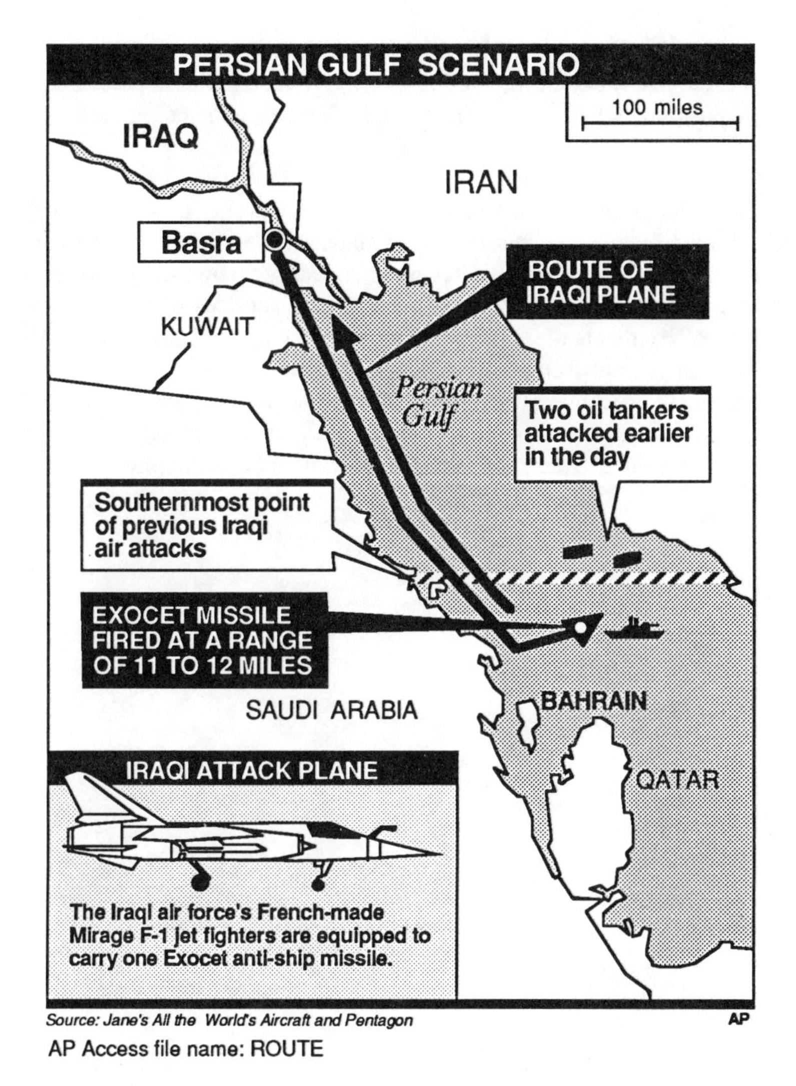

Figure 3.32. Associated Press artists used an Apple Macintosh minicomputer to compose this AP Access map of the May 1987 Iraqi attack on the U.S.S. Stark. Using the AP Access file name "ROUTE," member newspapers could retrieve it by telephone, customize it on their own Macintosh, and print a clean copy on their LaserWriter. The AP also delivered its Macintosh graphics over the Laserphoto network, but newspapers with both kinds of copy preferred the LaserWriter version for image clarity. [Source: Courtesy of the Associated Press.]

tern. Moreover, not only was transmission time reduced from eight minutes to about two or three minutes, but AP member newspapers now could customize their news maps by changing type and adding their own cities.

While technological factors placed limits on the ability of journalists to share and disseminate maps, the use of maps was not shaped primarily by these forces. After all, equally severe technical problems are involved in the transmission of photographs, yet the wire services made a greater effort to disseminate these pictures, rather than maps, because of journalistic practices, not technical problems. Chapter 4 examines the pivotal role of the microcomputer in promoting the rise of both locally produced and syndicate news maps and explores the equally important role of editorial attitudes toward the value of journalistic cartography.

4 Map Use in Print Journalism

Although technological development may be seen historically as the prime prerequisite for the rise of journalistic cartography, two additional factors requiring careful examination are the organizational structure of the newspaper firm and the attitudes toward graphics of its artists, editors, and management. Newspapers vary widely in their use of cartographic art, and perhaps the single most important impetus for using maps is having an art department in which to produce them. Maps offered by wire services and syndicates focus largely on prominent national and foreign events, and only rarely on occurrences within the paper's circulation territory. Yet a sustained, comprehensive cartographic treatment of the "where" aspect of local news calls for more than the occasional photographic "pick-up" of a zoning map to illustrate a development project, or of part of a street map embellished with a Maltese cross to mark an accident or crime site. Whether a map accompanies a story with a spatial component depends not only upon having artists available, but also upon having editors and writers willing to ask, and argue, for an artist's time. In making the modern newspaper both attractive and informative, the art department produces several types of graphics, among which the map must compete—with the editorial cartoon, the business graph, the front-page color photo, and the illustrated layout for the weekly food section. Particularly important, then, are the wider attitudes toward map graphics of the writer's peer group, the art director and staff artists, the paper's management, and the national communities of news publishers and professional journalists. A large "quality" newspaper, with its own reporters and correspondents covering world and national news, might well use the art department to generate its own maps for nonlocal news and features, and even to redraft wire-service maps to match its own style conventions. This strong commitment to journalistic cartography usually is rooted in an appreciation by top management of the map's intrinsic value for reporting facts and explaining ideas as well as for catching the eye.

This chapter examines the institutions and personnel responsible for the use of maps in the American newspaper, and for their design and production. Of particular importance is the art department, its personnel and director, and the editorial budget meeting, at which the paper's editors discuss the news value of recent and imminent events, allocate space in the next issue's news hole, and plan features and special projects. Another significant institution within the newsroom is the style book used by some newspapers to promote a cohesive appearance. Journalistic structures outside the newsroom also influence the use of maps. These external institutions include the networks developed by a few newspaper groups, or chains, to exchange computer-designed graphic artwork among their properties; the Society for Newspaper Design, an influential professional organization of layout editors, artists, and graphic designers; and the graphics departments at the major wire services, news syndicates, and weekly news magazines. Concerned also with the attitudes of editors and artists, the chapter explores the effects of cartographic ignorance, anti-illustration biases, and pictorial hype on the design and use of news maps. An examination of variations among newspapers in their cartographic responses to the same set of national and foreign news events reveals not only a wider potential for the use of maps but also a bias toward sensational stories and little agreement about the significance of location as an element of news reporting. This examination is based on the map-content data discussed in Chapters 2 and 3. The discussion of art departments, graphics editors, and editorial budget meetings reflects the author's visits to numerous nationally prominent and local newspapers.

Cartographic Art and the Structure of the Newsroom

The increased use of news maps reflects the rise of the editorial art department. Most of the commitment of specialized personnel, working space, and equipment to the production of information graphics has occurred since 1945, as stronger papers have added staff and pages, and weaker ones have merged or disappeared altogether. The trend toward information graphics has experienced a marked acceleration since 1975, as discussed in Chapter 2. Yet, cartographic efforts remain subordinate to a wider range of graphic endeavors, and occasional embarrassments result because few newspaper artists have formal training in cartographic principles. In the newspaper's internal organization, the unit producing maps and graphics is under the control of the paper's editorial managers—not its business managers. The editorial art department has virtually no involvement with advertising illustration, most of which is produced by advertising agencies at the request and approval of advertisers. But although the art department clearly is a service unit, subordinate to the needs of editors and reporters, a variety of strategies exist for identifying and meeting those needs.

Cartography and the Newspaper Art Department

Editorial art departments vary widely in size, range of responsibilities, and commitment to cartography. Among the newspapers visited by the author or examined in Chapters 2 and 3, the *New York Times* has the largest art department; a 10-person "map group" draws maps and non-cartographic "charts" within a department of over 60 people, over a third of whom are art directors individually responsible for the paper's numerous weekday and Sunday special sections. Each month the *New York Times* map group produces approximately 300 maps and 350 charts, often by altering or bringing up-to-date an existing drawing on file in the paper's collection of approximately 50,000 hand-drawn maps. The smallest art department was found at the *Watertown Daily Times*, in upstate New York, where a single staff artist produces approximately 20 to 30 maps and several non-cartographic illustrations a month, in addition to laying out pages and drawing an occasional (noneditorial) cartoon. At Watertown the normal workday is from 7 A.M. to 3 P.M., Monday through Friday—the paper is an afternoon daily that only recently added a modest Sunday edition. In contrast, the *New York Times* map group uses staggered workdays to provide seven-day coverage for the morning daily paper and its massive Sunday edition: a single artist is on duty for eight hours on Saturday and Sunday, but on weekdays at least one artist is present from 10:30 A.M. until midnight. Artists work overtime to meet tight deadlines, and on occasion an artist is called in after midnight or on a weekend because of breaking news with a spatial component. According to Andy Sabbatini, who joined the staff in 1943 and has managed the map group since 1960, the *New York Times* has used maps and charts regularly since 1925, when the paper hired artist Russell Walrath. By comparison, the *Watertown Daily Times* hired its first full-time artist in 1983.

Surprising perhaps, given these differences in size and history, is the similarity of these two papers in their reliance in the mid-1980s upon pen-and-ink drafting. At the time of the author's visit, the Watertown paper had no computer graphics and the New York City paper used only a system, linked to its typesetting computer, for making bar charts from business data. At the *Watertown Daily Times* artist Mark Dietterich employed technical fountain pens to draw on tracing paper. He drew at an "upsize," a common practice to minimize flaws in inking, and used an electrostatic copier, downstairs in the paper's business-advertising-circulation offices, to enlarge and reduce drawings and source materials. Yet Dietterich employed a computer terminal to order type from the paper's typesetting system, and often carefully cut labels to follow a curving feature, as shown in Figure 4.1. Maps at the *New York Times* also were drawn at an upsize—usually at "half up" so that an illustration to be printed in a space 4 inches wide, for example, was drawn with a width of 6 inches. Type ordered through a computer terminal connected to the typesetting system was delivered by messenger to the map group, and a copy camera located nearby in the art department provided enlargements, reductions, and prints for editorial approval and page makeup. Tint area symbols, to appear on maps as graytones indicating land, water, or

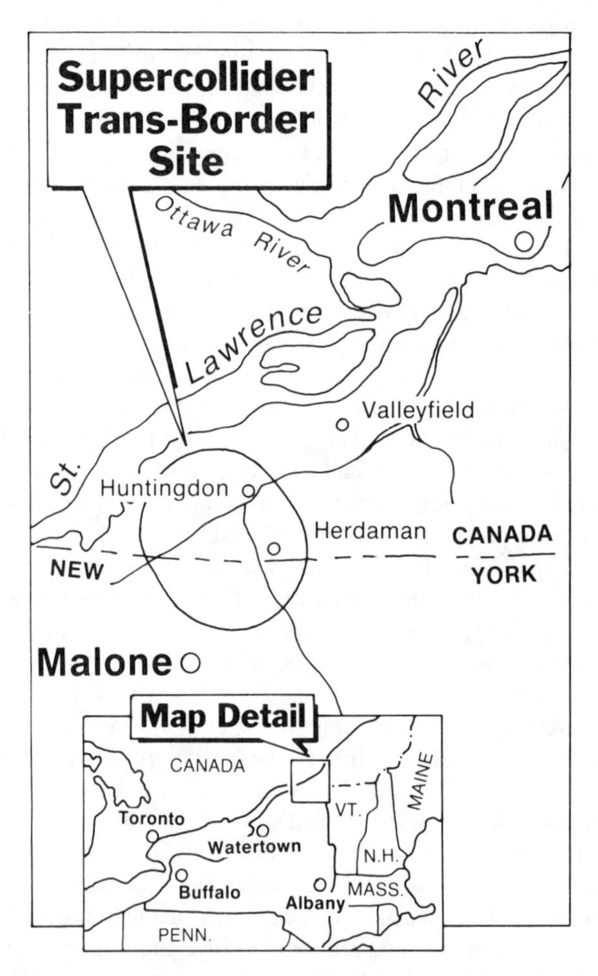

Figure 4.1. On map showing the proposed site of a high-energy physics laboratory in northern New York, artist Mark Dietterich used curved labels for major rivers. [Source: Courtesy of the *Watertown Daily Times*.]

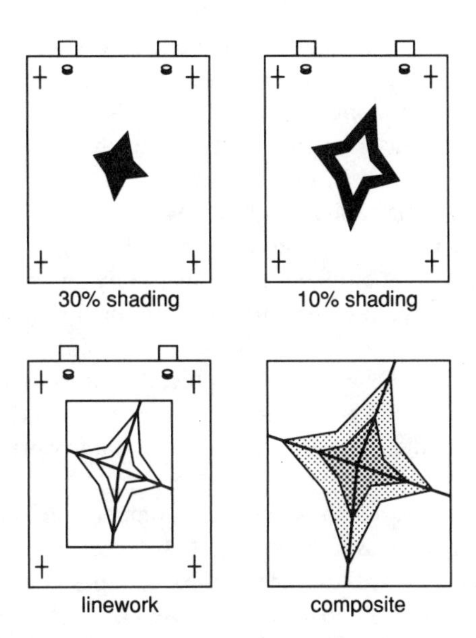

Figure 4.2. Linework flap and two shading-symbol flaps with registration pins (studs) and targets (plus-signs) can be used to produce the composite drawing with graytone area symbols (enlarged at lower right). Areas to be shaded are opaque on their respective flaps; label stating desired percentage of black identifies each flap.

distribution patterns, were produced as overlays called "flaps". As shown in Figure 4.2, flaps are taped to the board on which the inked drawing is mounted. Each flap is a sheet of clear acetate on which the area to be shaded is covered with an orange tint.[1] A separate flap is used for each graytone, and is labeled to indicate the percentage of ink to be applied by the press. Darkroom technicians in the engraving department combine the inked linework with appropriate tint screens for the graytones. At the *Watertown Daily Times* flaps were used to add a spot-color overprint to highlight areas on the map as well as to produce the occasional full-color map for a special section. Using color only for the Sunday magazine section and an occasional magazine supplement with a specific theme, the *New York Times* made frequent use of screened graytones on its maps to differentiate various area, line, and point symbols.

In the mid-1980s the typical editorial art department shared many of the practices of the two papers discussed here: pen-and-ink linework, screening produced from opaque-orange areas on transparent flaps, type ordered through a computer terminal from a centralized typesetting system, enlargements or reductions made optically with a copy camera or high-quality electrostatic copier, and little or no use of computer graphics.

Among those newspapers visited, the most unusual practice was the *Boston Globe*'s restriction against artists "touching type"—the result of an archaic contract with an engravers union.[2] *Globe* artists could order type and specify its size and style, but they received only a Xerox copy of the type, not the photocomposed type provided artists at other newspapers. Labels had to be cut from this paper copy and pasted onto a thin, transparent tissue overlay. The artwork was then sent to the engraving department, where no one but a union engraver could secure to the inked artwork the labels that were photographically copied with the rest of the image onto the printing plate. In marked contrast to the contemporary interior design of the *Globe*'s offices, this quaint work rule in no way diminished the quality of the paper's cartographic artwork, which has received several awards in a national competition sponsored by the American Congress on Surveying and Mapping.[3]

A 1984 survey by the Society of Newspaper Design examined the size and functions of art departments.[4] Based on responses from 62 daily and weekly newspapers in the United States and Canada, the SND survey noted that the most commonly reported responsibility was the design and layout of feature and special-section pages. Additional primary duties were the production of information graphics and pictorial illustrations, and "numerous small jobs," such as retouching photographs and designing logos. Art departments varied in size from 58, reported by the *New York Times*, to one—or none, for a few of the small-circulation papers included in the survey. The *New York Times* clearly is exceptional because the second-largest art staff, reported by three newspapers, was 15. About half of the papers listed five or fewer artists, and forty percent indicated an art staff between six and ten. Among daily papers with circulations under 100,000, the art department generally consisted of one to three artists. The survey report also noted that art departments tended to be smaller than photo departments, which often were twice as large. A few smaller newspapers reported that staff artists were responsible also for advertising and promotional work—duties not required of any of the art departments visited by the author.

Making no mention of maps or map making, the SND report treated maps under the broader category of information graphics. A third of the newspapers listed one or two artists specializing in information graphics, but page designers apparently were more numerous, with a quarter of the art departments reporting three or four layout specialists. Most common, though, were generalists; three-quarters of the papers indicated that members of their art department routinely handled the full range of newspaper artwork. Specialists seem to have evolved from generalists because of talent, interest, the needs of the employer, and the demands of the marketplace, rather than because of formal instruction in an academic or vocational program addressing maps and charts. Indeed, very few newspaper artists have any formal background in cartography.

Among the art departments visited by the author, only the *Boston Globe* and the *Washington Post* had staff artists with extensive experience in designing or making maps prior to taking a newspaper job. Deborah Perugi, who had been drawing maps for the *Globe* for six years at the time of my visit, has a bachelor's degree in fine arts and printmaking but had worked for five years at Cities Corporation, a street-map firm in nearby Cambridge, Massachusetts. Similarly, a job at the *Post* was not what attracted two cartographers to Washington. Brad Wye, who received an undergraduate geography degree at the University of Massachusetts, had worked in the Washington area for several years for the Defense Mapping Agency and the National Oceanic and Atmospheric Administration. Dick Furno, the *Post*'s assistant art director, had worked in Washington for 13 years at the National Geographic Society. His work at the NGS included making mathematical calculations for map projections, a skill used later at the newspaper in developing a microcomputer system for plotting meridians, parallels, coastlines, and political boundaries. Furno's computer plots frequently were incorporated in impressive, specially tailored oblique views of the globe, as illustrated in Figure 4.3.[5] Themselves fully competent as graphic artists, Furno and Wye were outnumbered at the *Post* by generalists trained as graphic or commercial artists.

That lack of formal training in the use of map projections, cartographic symbols, and map design might occasionally lead to blatant blunders is not surprising, particularly under the pressure to meet deadlines for breaking news. In a master's thesis at the University of Wisconsin, Judith Leimer explored the effects upon the quality of newspaper maps of both the limited training of art staff and the technical constraints of newspaper production.[6] In her evaluation of the structure, content, design, and form of newspaper maps, Leimer noted that such specifically cartographic errors as an inappropriate projection or an inaccurate or missing scale were far less frequent than sloppy artwork and labeling, inaccurate map content, and poor graphic design. Large papers, with "a strong graphic commitment" and a large art staff skilled in the techniques of illustration and experienced in newspaper work, produced maps of markedly higher quality than did professional graphics services such as the NEA, the wire services (AP and UPI), and most small and medium-size newspaper art departments. Among those papers Leimer cited for their generally higher-quality cartography were the *Los Angeles Times*, the *New York Times*, the *St. Petersburg Times*, and the *Washington Post*, all of which have artists specializing in information graphics. Many of the

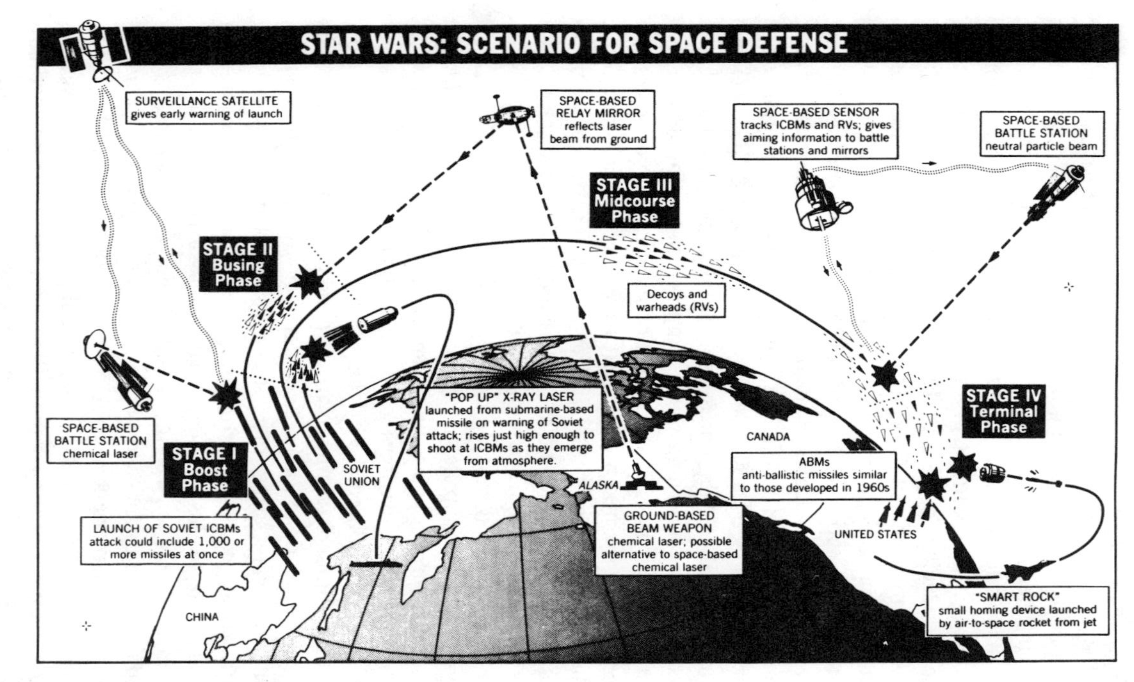

Figure 4.3. *Washington Post*'s "Star Wars" graphic, designed in early 1985 to explain the Reagan administration's "Strategic Defense Initiative" (SDI) technology, was produced by staff artists Dave Cook and Larry Fogel using a computer-generated plot of coastlines, national boundaries, and grid of meridians and parallels. [Source: *Washington Post National Weekly Edition* 2, no. 20 (18 March 1985), 6.]

poor examples used by Leimer to illustrate her thesis were poorly designed or sloppily printed offerings from wire services or syndicates, or graphics hastily or crudely drawn by staff artists at small- or medium-circulation newspapers. But even the bigger, more graphically committed papers occasionally produce flawed maps, as demonstrated by examples from the aforementioned newspapers as well as from the *Chicago Tribune* and the *Christian Science Monitor*. Common deficiencies were maps larger than required for the level of detail provided, incomplete or confusing information, poor use of type, and sloppy execution. Leimer ascribed most of these defects not to cartographic ignorance but to poor printing (or failure to compensate for the pressroom limitations), inadequate research or checking (as when place symbols were not labeled and names were misspelled or mislocated), and poor graphic design or technique. More serious than the limited cartographic knowledge of artroom personnel, Leimer concluded, were the adverse effects of deadline pressure on research and manual drawing, the small or overly large spaces allotted as integer multiples of a standard column width, poor presswork, and inexpensive newsprint unable to record unblemished the intricate details of small symbols and type. Most of these deficiencies can be avoided, even with cartographically ignorant personnel, by hiring good graphic artists, making them aware of source materials and technical limitations, setting and enforcing high standards of accuracy and clarity, and anticipating needs by acquiring current reference materials and promoting communication with reporters and editors—in short, by sound management.

Two promising points must be mentioned. First, Leimer noted that a small amount of cartographic training might well obviate most of the defects associated with map scale, projection, content, and labels. Second, her research was based on maps published in Spring 1982, when awareness of cartographic principles and appreciation of graphic quality were less widespread than later in the decade, particularly among the wire services that supply most of the cartographic art used by smaller newspapers. Larger newspapers, moreover, have given greater attention to both the management of the art department and the role of information graphics in journalism.

Managing the art department is usually the job of the art director, sometimes called a "design director" to emphasize the page-makeup function. Yet a small, one- or two-person art "department" hardly requires a director, for in the open, unpartitioned environment of most newsrooms, the managing editor or news editor can readily see whether the artist is present and working, and is conveniently available to settle disputes. At the *Watertown Daily Times*, for instance, the single staff artist occupies an area toward one corner of the newsroom, near the doorway leading to the page-makeup area and in full view of editors and reporters. He receives assignments for breaking news from the assistant managing editor or the state editor, and for projects with more lead time, from the features editor and the editorial page editor. In contrast, at the *New York Times* the map group occupies an elongated room, with files for artwork and reference materials along one long wall across a narrow corridor from a row of artist workstations, separated by partitions. The chief cartographer has an office at one end, near the entrance and the computer terminal for ordering type. Reflecting none of the openness of most newsrooms, including that of the *New York Times*, the map group's quarters have been designed to minimize distractions and eye contact while the artists are drawing. Walls and movable partitions, of course, are also convenient for posting reminders, research notes, and recent or notable drawings. The *Boston Globe* and *USA Today*, among others, also partition their drawing areas. Although the artists work with the appropriate writer, editor, or special-section art director, they are assigned projects and supervised by the chief cartographer, who reports to the managing art director responsible for the entire art department. In the *New York Times* hierarchy, the managing art director reports in turn to the corporate art director, who is responsible directly to the publisher and also oversees the art directors at the 25 other daily newspapers owned by the New York Times Company. The director of a four- or five-person art department generally must divide his or her time between supervisory and production duties, whereas the person responsible for a larger group of artists is more likely to occupy a desk than a drafting table.

Recruiting freelance artists for feature-page illustration is an important responsibility of many art directors. According to the 1984 survey of art departments by the Society of Newspaper Design, freelances are less expensive than full-time art staff and allow a newspaper to meet temporary needs for increased artwork. Two-thirds of the newspapers surveyed purchased art from freelances, and 60 percent planned to increase their expenditures for freelance work.[7] Many staff artists were hired after serving as freelances, and this system affords art directors an opportunity to observe a potential employee's skill and commitment. But freelances usually contribute little to a paper's cartographic endeavors, or to information graphics in general, because maps and charts commonly refer to breaking news and require consultation with an editor or reporter.

A new art director can be important as an agent of change in promoting a paper's new image or in improving communication between artists and journalists. The *Christian Science Monitor*, for instance, hired Robin Jareaux as design director to oversee its five-person art department and promote the integration of the paper's art with its overall design. As part of a marketing effort, the *Monitor* was redesigned in 1983 by noted design consultant Robert Lockwood, who in his own prospectus described this redesign as a strategy in which "Typography becomes the ART of the paper [while] CONTENT dictates the design." Jareaux's role was to serve as a "design filter" between the editors and the page-makeup function and, in her words, "to get the department working at all levels." Previously the art department had been run like a "filling station, to which an editor or reporter might come with a request of the gimme-a-quart-of-art sort." Jareaux, who had worked at the *Washington Post* and the Dallas *Herald*, takes a holistic approach to newspaper design, page layout, and information graphics. Without the day-to-day in-house support of the art director and other news executives, a bold yet functional redesign easily collapses into little more than a forced, easily tarnished newness. The publisher's or managing editor's call for a redesign or an increased use of information graphics often exposes the need for a new art director.

Art directors come from a variety of backgrounds. A 1971 article on the "explosion in graphics," published in the *Bulletin of the American Society of Newspaper Editors*, lists five routes to newspaper art management: through the production department, right out of school, through the art department itself, through the city (news) room, and from an advertising agency.[8] Yet the talent pool is broader still, and includes art education and television, the previous occupations of two prominent art directors in the vanguard of organized, outgoing, creative people skilled in artistic technique, able to design for both function and aesthetics, and competent in dealing with publishers and managing subordinates. Robert Lockwood, who in the late 1970s became a design executive at the Allentown, Pennsylvania, *Morning Call*, had taught art in the Philadelphia area, and Ron Couture, managing art director at the *New York Times*, had worked in television. Couture, who designed the widely familiar title sequence for the Public Broadcasting System's *Masterpiece Theatre*, readily adapted to print journalism, first as a page designer and then as a manager. Lockwood attained fame as a redesign consultant when he left the *Morning Call*, and for a time produced information graphics syndicated by the Field Newspaper Syndicate, before forming a partnership to provide information graphics and build a cartographic database for the Associated Press. Among the newspapers visited by the author, art directors with a background in production and news reporting were rare. Many art directors had moved to their present jobs from similar positions at other newspapers. Like other news executives and journalists, art directors seem willing to move frequently, after only a year or two, to better paying, more challenging positions. Many a large newspaper has gone through three or four generations of art directors, reflecting an escalating concern with the look and organization of the newspaper as well as organizational tensions and conflicting personalities. Success as an art director seems tied to a holistic view of the newspaper, a flair for page layout, and the ability to work on and carry through a redesign; but equally important is managerial skill.

In an art department committed to cartography, management must be concerned with an organized collection of existing artwork and reference materials. Maps drawn for previous stories, even if years old, can often be resurrected and used with little or no alteration. If the collection is large and its use frequent, the savings in labor and time can be substantial. Old maps, of course, must be checked to confirm that the information shown is both current and appropriate to the new use. But even when symbols and type must be modified to accommodate geographic change and a different news event, existing artwork can have two serious drawbacks: an inappropriate centering of new symbols when the principal location has shifted toward one side of the map frame, and the vestigial remnants of previous style conventions, or lack thereof. Maps that *look* out-of-date can be almost as troublesome as maps that *are* out-of-date, although sometimes whether a cartographic style is quaint or merely conventional depends upon individual taste. In 1985, when Michael Keegan joined the *Washington Post* as art director, tension developed between cartographic specialists proud of the department's large, well-organized collection of map drawings, and their new boss, who considered these existing maps "technically correct but design deficient." Keegan wanted maps having "a fresher look, with more impact, more pizazz." But even though more than half of the *Post*'s 1985 map graphics relied partly upon file maps, this collection was at least equally important as a resource for research and reference, especially for local stories. Because the *Post* has made a strong commitment to covering Washington and its suburbs and to providing maps for regional editions distributed in Maryland and Virginia, two-thirds of the 12 lateral file drawers contained metropolitan area maps and only one-third addressed the rest of the nation and the world.[9]

At the *New York Times*, which has had a commitment to cartography for over 60 years, and for even longer has assumed the role of the nation's newspaper of record, the map files are larger and more geographically diverse than those at the *Washington Post*. In 1963 chief cartographer Andy Sabbatini devised a four-digit geographic code coupled to a sequence number to indicate, for example, that map 7217-13 is the 13th map of Thailand (code 7217). The scheme is hierarchical, with 7200 identifying maps of Southeast Asia, Australia, and Oceania, and 7000 indicating maps showing all of Asia. But separate major categories are reserved for the world as a whole (1000), the United States (2000), and New York City (3000), which is subdivided further by borough, with letters used for a still-finer disaggregation of the borough of Manhattan (3108) into Uptown (3108-U), Downtown (3108-D), and Midtown (3108-M). Existing maps are infrequently reused; the file is organized to promote browsing, with copies of the maps cut from the newspaper and pasted, several to the page, on large pages of black construction-grade paper, and mounted in a book resembling a large photo album. Index tabs identify areal subdivisions. Over 20 of these large books, grouped according to area, line a shelf behind a waist-high counter. Code numbers also refer to the original artwork, which is stored in drawers below the counter. Before Sabbatini started the present system, maps were merely arranged in chronological order in a large book, which became awkward and less useful as the number of maps increased. The map group has a clerk, one of whose duties is to file reference copies, original artwork, and source materials.

Cartographic reference materials come from a variety of sources—public, commercial, and private. Cartographically committed papers require readily available reference maps—in the building and easy to locate. Although most newspapers have their own reference libraries, cartographic materials, including atlases, almost universally are kept in the art department. Even papers close to such outstanding cartographic collections as the Library of Congress and the New York Public Library cannot afford the delays of city traffic and restrictive hours. Andy Sabbatini at the *New York Times* for years has urged journalists on assignment or vacation to "bring back maps." Highly valued are maps of war zones and of cities prone to terrorist attack. Reporters covering stories that might require maps are encouraged especially to collect and annotate maps. In addition to its collection of atlases and of numerous commercial and administrative sheet maps, the *New York Times*'s map group keeps back issues of *National Geographic Magazine* and, to verify the spelling of domestic place names, the reports of the U.S. Board of Geographic Names. The *Boston Globe*'s collection of cartographic reference materials includes the common and well-regarded atlases published by National Geographic, Rand McNally, and The Times of London, as well as aeronautical chart books and rolled maps in tubes from the Boston Development Authority and the Massachusetts Department of Public Works. Graphics specialists Deborah Perugi and Jane Simon have used slow workdays to acquire new sources. At the *Christian Science Monitor* artist Joan Forbes keeps near her desk personal copies of two recent thematic atlases acclaimed for their fresh and visually striking graphics—the *State of the World Atlas* and *GAIA: An Atlas of Planet Management.*[10] In addition, the *Monitor* has the standard commercial atlases and the *National Atlas of the United States*, published in 1970.[11] Reference collections at smaller papers tend to be primarily regional, particularly when wire services and graphics syndicates provide most, if not all, of the nonlocal cartographic artwork. For several counties within its circulation area, the *Watertown Daily Times*, for instance, collects the 1:24,000-scale planimetric quadrangle maps and 1:9600-scale town and village maps of the New York State Department of Transportation.

Another source of information, particularly for maps designed to illustrate breaking news, is the paper's "editorial front-end system," the word processing computer system linked to its typesetting computer. Most journalists prepare their stories on the editorial system, the wire services transmit their offerings directly into the same system, and the paper's editors use the same word processor to edit their writers and to select and rewrite wire copy. Access to the editorial system is particularly important for the map maker, who must be certain that map and text agree in spelling and that the map shows the locations and spatial relationships relevant to the story. All art departments visited in 1985 had their own terminal or used one nearby, and artists could print copies of stories for use at their workstations. But because of cost and limited use, no paper provided individual terminals at each artist's workstation.

Editorial Budgets and Graphics Editors

Senior editors budget the news hole. Newspapers have far more words, from their own staff as well as from their wire services, than they can print. And many stories without maps might benefit from having one. Hence, like a family stretching a modest income to cover necessities and a few luxuries, a newspaper must budget its news hole and its personnel, including the worktime of its art staff. In the news business, though, budgeting is a daily affair, and an even more frequent task when several editions are published. Most large newspapers hold several regularly scheduled editorial budget meetings throughout the day, to select content for the front page, discuss imminent news events, initiate feature stories, and assign personnel. Editors must consider the current and potential significance of breaking news, recognize the diverse interests of readers, and attempt to tailor a balanced presentation of local, regional, national, and foreign news. Their finished product must also succeed in competing for the buyer's attention, on a daily basis to boost or sustain street sales as well as throughout the year to retain subscribers. Within such constraints as the need for at least one local story on the front page and for at least one picture on most news pages, the editors often must weigh the relative merits of, say, a story about a local robbery against an article on drought in Africa.

A large newspaper committed to graphics must guarantee that the art department is represented at editorial budget meetings. The representative might, but need not, be the art director; indeed, more than one person should be qualified to attend, voice suggestions, carry back requests, and advise about conflicting commitments. On slow news days, the art department might recommend initiating a map for a forthcoming feature or a stand-alone graphic for the business section. When breaking news has a strong spatial component but the news value of the story is as yet uncertain, the art department might be advised to collect reference materials, to retrieve a similar recent map, or even to start drafting while standing by for changes. The editors need to know early in the development of a news story what information and drawings the art department has available, and what staff time might be available. A good art representative can identify possible uses of graphics, particularly for stories in which words alone might prove cumbersome or inadequate. A good news story, after all, not only tells but explains. The publisher who wants a visually effective newspaper will encourage members of the art staff to be entrepreneurs, and to see themselves as journalists as well as illustrators.

Because a skilled artist might not be as adept as a graphically aware journalist in recognizing the need for graphics, and because the responsibilities of directing a large art department preclude extensive work with individual journalists, some newspapers have hired a graphics editor or graphics coordinator to improve communications between artists and writers and to make the paper's information graphics more relevant to its text.[12] An important distinction between the art director and the graphics editor or coordinator is that the latter need not have the artist's technical skill; indeed, because of the need to deal with a variety of editors and reporters, the graphics editor is more likely to have a background in journalism than in art. Equally important is the emphasis placed by the job's title and duties on coordination of art and words, with the emphasis on information graphics rather than decorative illustration. In a sense, the graphics coordinator plays a role similar to that of the page designer who coordinates illustrations and articles to create a cohesive front page for the food or travel section. More concerned with explanation than with aesthetics or flavor, the graphics coordinator works with a smaller unit—the story—rather than with an entire page or section.

A graphics editor or coordinator might work at one of a variety of levels in the newspaper's organizational structure—within the art department, above it, or outside but on a roughly equal footing. At the *Washington Post*, for instance, art director Michael Keegan has deployed three "advanced graphics planners" throughout the paper, to work on assignment to the various "desks," or sections, such as metro, national, foreign, financial, or sports. Their duties include ferreting out graphics needs, hounding reporters for graphics possibilities and source materials, checking statistics and finished artwork, editing captions, and assuring compatibility between graphics and story. At the Los Angeles *Herald*, where Keegan worked before going to the *Post*, a journalist had been assigned this role, but the job was too big for one person. With greater resources and a commitment to enhancing further its reputation for insightful, thorough reporting, the *Washington Post* willingly made the commitment to new staff positions to enhance the relevance and informativeness of its graphics.

The *Chicago Tribune*, another large, graphically committed paper, has both a graphics desk and an art department, organized as separate units under its graphics director. The graphics editor supervises the graphics desk, and the editorial art director oversees the art department, which has separate divisions for information graphics and page illustration and design. The graphics desk—almost literally a large desk, or actually several desks pushed together to form a U, in newsroom fashion—is staffed by specialists who identify graphics needs and recommend designs to the art department. They also decide on Friday which of the week's graphics are to be syndicated by Tribune Media Services. Each desk on the paper (local, foreign, national, and so forth) each day prepares a list of stories, illustrated by the example in Figure 4.4, that it "offers to the paper." Graphics personnel have computer terminals linked to the editorial system, and can "pull up" stories to check for facts or concepts that might require a map, statistical chart, or explanatory illustration. Upon identifying a need, the graphics coordinator talks first with the editor or reporter and then with an artist. The coordinator fills out the "Art room work request" shown in Figure 4.5, and often orders type for the artist as well. When I visited the *Tribune* in 1985, the graphics desk had, in addition to the graphics editor, four coordinators with individual assignments to: financial and travel stories; special projects and the Green Streak edition, sold downtown and in city neighborhoods in the afternoon; the Midwest and home-delivered Five Star editions, both "locked up" in the later afternoon and evening for distribution early the next morning; and the science and news sections for the Sunday edition. Work hours vary but overlap, and are adjusted to the deadlines of the editions for which each coordinator is responsible. For instance, Marty Fischer, responsible for special projects and the afternoon city edition, worked the "early shift," starting at 8:30 A.M., whereas his coworkers began at 11 A.M., noon, or 2 P.M. Like four of the five people on the *Tribune* graphics desk, Fischer has a background in journalism, not art. Three of the coordinators, in fact, came from the paper's copy desk. The sole person with a background in art was "hired as an experiment."

Chicago Tribune

foreign
Daily schedule for
wednesday, 15 may 1985-

Slug [zone] Status		Description [Source]	5 Star/5* Final C	D	N	S	GS GSR	MW1 MW3
EMBARGO managua schodolski	30½	won't have much effect on already devasted economy					om	✓
NICAR local worthington	10	culture minister says embargo will fail					om	✓
NAM ho chi yates	32	3d in series: nation still divided mwl pork					✓	off
DISAPPEAR baries beard	26	assignment: dirty war trial not mending wounds					om	✓
SHULTZ vienna atlas	20	meets with gromyko					nu	
POPE	15	big, warm welcome for pontiff					nu	

Figure 4.4. Part of the foreign daily schedule of the *Chicago Tribune* for Wednesday, 15 May 1985. The six columns at the right represent the various editions, for example, "GS" for the Green Streak edition and "MW" for the two Midwest editions. "om" refers to old material, and "nm" to new material. A one-line "slug" identifies each story, for example, "EMBARGO" for a story from Managua in which correspondent Schodolski reported that the U.S. trade embargo "won't have much effect on [Nicaragua's] already devast[at]ed economy." Occupying 30 1/2 column inches, this story is longer than most offered by the foreign desk that day. By early afternoon, when this schedule was developed, the "EMBARGO" story had been used in the paper's earlier editions, and was thus considered "old material" for the Green Streak, a street edition sold downtown in the late afternoon. [Source: Courtesy of the *Chicago Tribune*.]

Art room work request

☐ Map ☐ Chart ☐ Graphic ☐ Photo/Map combo ☐ Other________

Size________ **Section**________ **Edition**________ **Story slug**________
(Width & depth) (When needed) (If available)

Details (List cities if request is for a map; for other materials attach copy)

Headline:________
(if any)

Sub headline________
(if any)

Source________
(if any)

Requested by________ **Date**________

Figure 4.5. *Chicago Tribune* work request form includes "map" as the first of four common options. [Source: Courtesy of the *Chicago Tribune*.]

The Baltimore *Sun* and *Evening Sun* offer an interesting contrast in their approaches to graphics coordination. Jointly owned and sharing the same building, the two newspapers have separate editorial staffs and separate art departments. The *Evening Sun*'s art department, which is responsible for the single, joint Saturday morning edition as well as for the weekday afternoon editions, is a five-person staff, headed by artist Chuck Lankford, who at the time of my visit had worked for the Sunpapers for 18 years. The *Evening Sun* also has a graphics coordinator, hired in 1984 by managing editor Jack Lemmon, who created the position to provide a "cushion between the art department and the managing editor." Jim Day, the graphics coordinator, is a former city editor, not an artist. Lankford reports to Day, who reports to Lemmon. Day's job includes planning, attending the daily editorial meeting, supervising quality control, and working closely with the production departments.

At the *Sun*, where a seven-person art staff is responsible for the weekday morning editions and the *Sunday Sun*, a graphics editor works within the art department, under the design director, who fills the role of art director. At the time of my visit in 1985, design director Dick D'Agostino had joined the morning *Sun* three years earlier from the Baltimore *News-American*, a less-successful competing afternoon daily, which closed in 1986. D'Agostino has experience in news art and page design, and had been actively involved in the redesign initiated by the *Sun*'s new managing editor, Jim Houck, and its new publisher, Reg Murphy. Houck joined the paper three months after D'Agostino and a year after Murphy. The redesign was implemented in early April 1984, after 18 months of study and planning with the assistance of consultant Robert Lockwood. Tony DeFeria, the graphics editor, joined the *Sun* in mid-1984 from the *Washington Times*, a colorful, graphically supportive daily owned by the Rev. Sun Myung Moon's Unification Church. DeFeria is an information graphics specialist, skilled in designing and drawing maps and charts. Although he was spending about 70 percent of his time "on the board," more often drawing charts than maps, he devoted the other 30 percent to negotiating with editors on the use of graphics. In contrast to the *Evening Sun*, where redesign is a continual process, the morning *Sun* made an abrupt, carefully orchestrated change in its appearance and organization. Its management considered a graphics editor specializing in information graphics important to the success of the redesign.

The Style Guide

A redesign often introduces greater coherence to a newspaper's information graphics—not the "foolish consistency" decried by Emerson but a uniformity in style that makes the graphics look as if they were designed *for* the paper, not merely acquired and inserted. The goal is not to restrain creativity but to direct it, by simplifying decisions about type and line symbols so that the artist's creative expression can concentrate upon explanation, communication, visual balance, and the integration of the graphic with the written part of the story. This uniformity is not unreasonable, for without guidelines, decisions on type and symbols often reflect haste and the materials at hand, rather than careful thought or a flash of brilliance. Moreover, because a successful redesign requires staff participation, useful suggestions often come from the paper's own artists and page designers, once they understand the goals and concepts of the redesign. But whatever design specifications and standards are agreed upon need to be documented in a style guide that can be distributed to employees for reference and used to train new workers. By providing both a sense of direction and specific strictures, the style guide serves as bible for the redesign.

Style guides vary in specificity, and reflect the extent to which the appearance of maps and other information graphics is considered a vital part of the paper's design. But the inherent focus is typography and page layout. The *Baltimore Sun Typographic Design Stylebook*, for instance, a

45-page guide developed by Dick D'Agostino and designer Michael Dresser, devotes only two pages to graphics standards. Some of these specifications are general, but many refer only to maps. One highly exact section tells the artist which pens to use:

> *PROPER INKING OF ROAD MAPS*
>
> The inking on road maps should be handled in a categorical fashion. That is, line weights should vary. The thickest line for the most important roads on down. The following inking rules should be followed:
>
> 1. Use a No. 3 pen for all major highways and interstates.
> 2. Use a No. 2 pen for all secondary roads (i.e., routes).
> 3. Use a No. 1 pen for secondary byways such as streets.
> 4. Lastly, use a No. 000 for rivers, county boundaries, state lines and inset boxes.
> 5. Do not use the dot-dash method for boundaries. Always use a continuous line.[13]

Other prescriptions call for a neat line (frame, or box) with a drop shadow on the bottom and righthand sides, a scale showing both miles and kilometers for "every map that shows a large land area," insets on locator maps to "show the relationship of the map area to a wider geographic area," and the use of size and style of type to indicate geographic hierarchies and relative importance. Conscious of possible criticism for constraining creativity, the *Stylebook*'s authors appended a caveat to the end of their graphics standards section:

> *SPECIAL NOTE*
>
> Please keep in mind that the above guidelines are designed for individual development. These guidelines should only lay the groundwork for creativity. Invariably, every assignment will be different, with a different set of challenges.[14]

A 16-page supplement, for zoned editions carrying local news for nearby counties, includes a half-page statement on graphics but no further mention of maps.

Cartographic standards and specifications can be far more complete and detailed, as demonstrated by the 75-page *Chicago Tribune Graphics Stylebook*, which devotes nine pages to news maps and the *Tribune*'s syndicated Base Map System, and five pages to weather maps.[15] Prepared in the early 1980s by the graphics desk, the *Stylebook* also outlines in detail the operations of the desk, sets forth specifications of charts and the use of color, and contains numerous examples of graphics guidelines. Not part of a major redesign, this style guide serves primarily as a training document for employees new to the graphics desk. Another set of cartographic rules is set forth in the *Washington Post*'s 14-page *Style Book for Maps and Charts*, which includes a five-page section entitled "General Guidelines for Creating Maps," by staff cartographers Dick Furno and Dave Cook. The mapping section contains examples of maps as well as practical, how-to-do-it instructions, and is followed by an even more technical section, "How to Create Tab Charts on the Ratheon," describing in detail the generation of data tables on the paper's typesetting system. That much smaller papers occasionally have graphics style guides is demonstrated by the Syracuse *Post-Standard*, which in the early 1980s had a two-page style sheet with specifications for type, highway symbols, a north pointer, and other graphic elements. This guide's illustrations, as sampled in Figure 4.6, reflect the graphic style and wit of Tim Atseff, an editorial cartoonist who served as art director before his promotion to

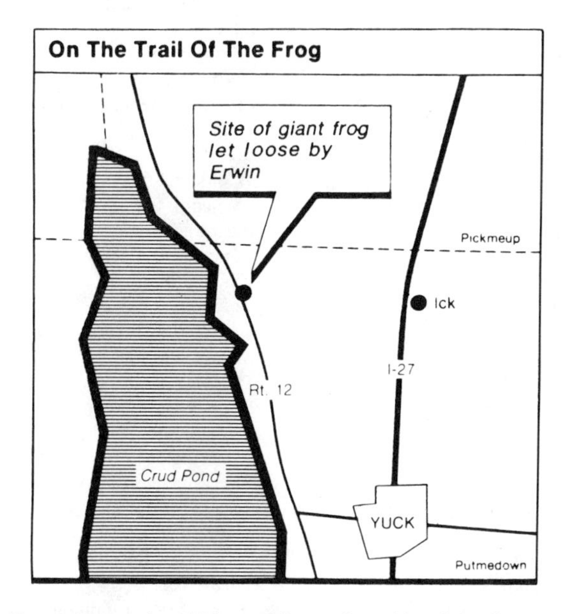

Figure 4.6. Example map in 1984 graphics style guide of the Syracuse *Post-Standard* shows the art director's sense of humor and a preference for approximating curved lines with straight-line sections of border tape, as for the shore of fictitious "Crud Pond." [Source: Courtesy of the Syracuse *Post-Standard*.]

deputy managing editor for the *Post-Standard*'s sister paper, the afternoon *Herald-Journal*. Graphics style sheets for newspaper art departments might contain any of the 54 items listed in Table 4.1, based in part on the style guides of several newspapers.

Table 4.1 Items That Might Be Addressed in a Cartographic Style Guide for Newspapers.

- Area symbols
  - Land-cover patterns (for park, orchard, forest, and so forth)
  - Standard series of graytones for choropleth maps
  - Water bodies (lakes, oceans)
- Associated text within map border to be coordinated with associated story
- Basemaps (supply of, at standard sizes, for frequently mapped areas)
- Border (map frame) design
- Color symbols; special guidelines for artwork
- Credit lines
  - For the artist
  - For the newspaper
- Drop shadows
  - Position
  - Solid or shaded
  - When to use
- Filing system for artwork and source materials
- Inset maps
  - Design of (examples)
  - Placement
  - When to use
- Keys (legends)
  - Design of (examples)
  - Preferred location
  - When needed; content
- Line symbols
  - Boundaries
  - Highways, streets
  - Railways
  - Standard widths
- Logos and standing sigs (use of maps in; examples)
- Maps and atlases for reference
  - Acquisition policy
  - Classification, filing
- Modifications to wire-service and syndicate artwork
  - Caveats
  - Color (converting monochrome to color)
  - "Localizing"
  - Weather map
- North arrow
  - Design (examples)
  - When to use
- Point symbols (standard symbols for bomb blasts, capital cities, highway shields, and so forth)
- Scale bars
  - Design, number of divisions
  - When to use
- Spelling (references, coordinate with story)
- Source notes
  - Content, style
  - When to use
- Standard sizes
  - Standard reductions
  - Standard widths
- Typography
  - Areas (position of labels within)
  - Capitalization
  - Classes of Feature (streets, water, landmarks, and so forth)
  - Credit line
  - Curving features
  - Focal places (and use of balloons)
  - Fonts (uniformity, standard fonts)
  - Key (legend)
  - Key-lines from label to feature
  - Scale bar
  - Size of type (minimum, standard)
  - Source notes, explanatory notes
  - Subtitle (use, position, style)
  - Titles (position, style)

Yet some art departments work well without a graphics style guide. Papers with a single artist for their maps and charts reflect the style of only one person. With two chartmakers, the senior artist might train a more junior assistant, and impose some uniformity. Standing artwork, such as an in-house weather map, periodic business and economic charts, and weekly maps with fishing and traffic information, have an established style that an artist is unlikely to change without approval from management. Papers with one or more long-time employees, particularly an entrenched art supervisor, tend not to have style guides for their graphics. Although maps in the *New York Times* appear to have a unique style, there is no written list of do's and don't's. According to head cartographer Andy Sabbatini, the maps now are far less standardized than before the "sectional revolution," and reflect variations in preference among the 22 feature and special-section art directors. But conventions need not be in writing—as any anthropologist will attest. Moreover, Sabbatini's staff understands the limitations of the presses and would seldom use type smaller than 7 points or a graytone screen finer than 65 lines per inch.[16] The Baltimore *Evening Sun* also works without a graphic or typographic style guide, in part because of the continuity provided by veteran art director Chuck Lankford and in part because the afternoon paper has a slower, more casual and incremental approach to redesign than its its morning counterpart, which changed its appearance overnight.

External Journalistic Structures Affecting the Newsroom

Several institutions outside the newsroom operate to promote the use of news maps. These external structures include computer graphics networks, which foster both sharing and central distribution; a new but rapidly developing professional organization, which is establishing standards of quality through the sharing of information about design and techniques; and the art departments at wire services and national weekly news magazines, whose highly visible, often innovative information graphics also set and reinforce standards. These outside institutions promote communication and a sense of community, and the contacts they foster often lead the industry's more talented art directors and information graphics specialists to change jobs. External journalistic structures are particularly useful in promoting the use of new technology for the design and preparation of news maps.

Computer Graphics and Graphics Networks

In the mid-1980s computer graphics had just begun to make significant contributions to maps and charts in newspapers. But almost all news graphics were still being drawn with pen and ink, although most art directors and news executives believed that electronic workstations would eventually replace their drafting tables. With a heavy use of computers for accounting, classified advertising, typesetting, and editorial word processing, the newspaper industry was well aware of the power and cost-effectiveness of electronic computing. Yet the industry also was wary of the rapid obsolescence and embarrassing failure likely in a fledgling technology, partly because of mixed experiences in the 1960s and early 1970s with computerized typesetting and editorial systems that either were less reliable than promised or were "orphaned" when the manufacturer went out of business or dropped the model, and parts and service were no longer available.[17] In the mid-1980s computer graphics, like the computer industry at large, was threatened by an imminent "shakeout," in which many companies were thought likely to fail.[18] Moreover, recent events in the computer marketplace also suggested that by waiting one more year a newspaper might for less money buy a faster, more versatile microcomputer system with more memory. Being a trend-setter could clearly have its drawbacks, for the few newspapers that had invested a bit too early were almost universally disappointed. The *New York Times*, for example, acquired the SAS/GRAPH software, for making maps and charts on its mainframe computer, as well as IBM's Interactive Chart Utility system.[19] Although both were somewhat useful as design tools once the artist mastered the command and coding conventions, the plots had to be redrafted for use in the newspaper. By late 1984, in fact, SAS/GRAPH was used more frequently for marketing research, by the paper's advertising and circulation departments, than for designing news graphics in the art department.

Another deterrent to an earlier adoption of computer graphics was uncertainty about electronic pagination systems, widely discussed as the ultimate tool of the newspaper designer.[20] A pagination system links the newspaper's editorial and typesetting systems with a CRT screen of sufficient size and resolution to display a readable image of an entire page. An editor or page designer laying out the page electronically can change headlines, move stories onto or off the page, and reposition other graphic elements (Figure 4.7). A high-resolution electro-optical film writer plots the page on photosensitive paper, to yield a positive image of the page, or on transparent film, to produce a negative image from which a press plate can be made directly. Because of the high cost of providing the resolution and memory needed to handle news photos, early pagination systems

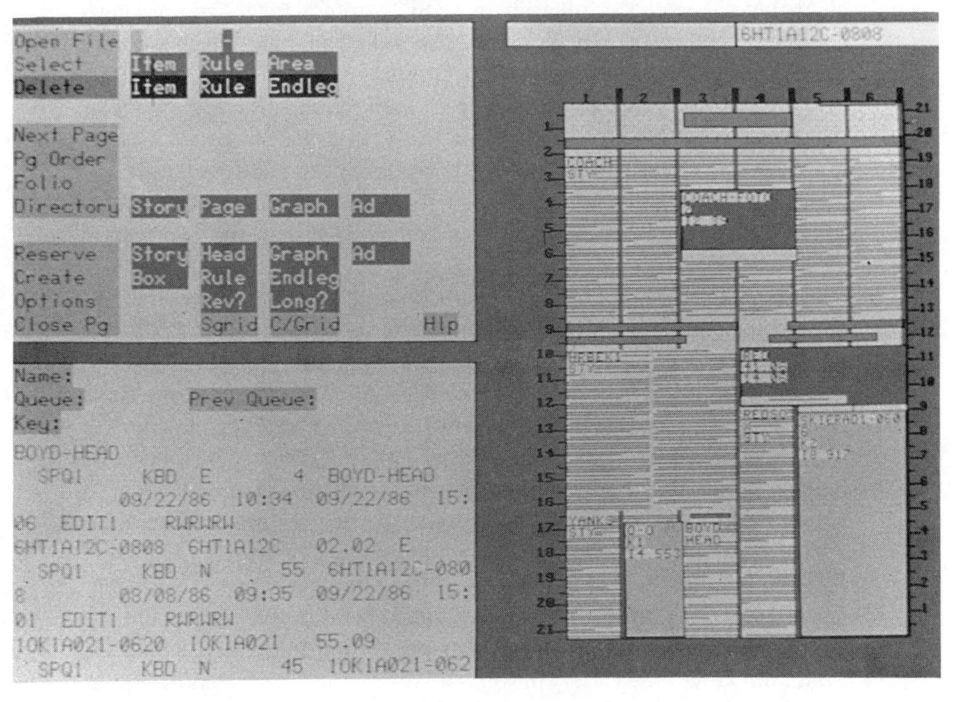

Figure 4.7. Integrated pagination systems serve a variety of news publishing needs, incuding news pagination, editorial layout, interactive graphics, display-ad make-up, and classified-ad pagination. These systems require large-screen, high-resolution interactive display terminals (above) and screen menus designed for creating page layouts (below). [Source: Courtesy of Crosfield Data Systems Inc.]

marked the position of a photo with a box, and the printers had to "strip in" a halftone negative of the photo onto the page negative before making the plate.[21] The trade literature of the early and mid-1980s chronicled the uneven progress of pagination systems in demonstration trials at a handful of newspapers, including the Utica *Daily Press* and its afternoon counterpart, the *Observer Dispatch*.[22] The Utica experiment, which lasted from January 1983 through April 1985, had no discernible effect upon these papers' use of maps. Although line art such as maps and charts required less resolution and less memory than photographs, pagination systems available in the mid-1980s provided little software support for cartography and statistical graphics.

But the potential of pagination was clear, and further advances were imminent. Only marginally impressed by SAS/GRAPH and the IBM charting system, the *New York Times*'s managing art director foresaw pagination systems able to deal not only with photographs, maps, and charts but also with the full array of 12 to 24 or more pages comprising a complete section of a newspaper. The newspaper designer, Couture maintained, needs to work with the section as a whole, to assess the effect of "jumping" (continuing) a story from the section front to a particular inside page, and to experiment holistically with the placement of text, graphics, and advertising. The likelihood of widespread future development of truly versatile electronic pagination systems was a further reason for publishers to defer purchase of a computer system for information graphics.

Nonetheless, inexpensive microcomputers offered a low-risk opportunity for an early adoption of computer graphics. The early 1980s saw impressive reductions in cost and gains in computing power. By 1984, anyone with a thousand dollars could walk into the computer store at a local shopping mall and buy a microprocessor comparable in memory to a full-size mainframe computer that during the 1960s might have cost a million dollars.[23] Although earlier forecasts predicting computers in most homes by the end of the decade were proving overly optimistic by the mid-1980s, sales of microcomputers were impressive, as shown in Figure 4.8. In 1981, introduction of the IBM Personal Computer, or PC, for short, assured wavering buyers of the interest and commitment of the country's largest manufacturer of electronic data processing equipment, and of the availability of a wide variety of software for business and scientific applications and for personal amusement. The electronic architecture of the IBM PC quickly became the industry standard—for microcomputers serving business and scientific research, at least—and domestic and foreign manufacturers drove prices still lower with a spate of "IBM-compatible" clones. Staying ahead, IBM introduced faster PCs with bigger memories: the XT in 1983, the AT in 1984, the RT in 1985, and a new series, the PC/2, in 1987.[24] For less than $5,000, a newspaper or news graphics service could purchase a computer system to help it meet deadlines and increase production of maps and charts.

Whereas PC mapping software developed for use in businesses and universities was limited largely to choropleth and other statistical maps, two prominent cartographic systems developed for newspapers focused on geographic databases and the timely, visually effective display of location. Both were developed with IBM microcomputers, and both involved an entrepreneur and the support of a potential user in the news business.

Dick Furno, cartographer and assistant art director at the *Washington Post*, developed the simpler and earlier of these two mapping systems.

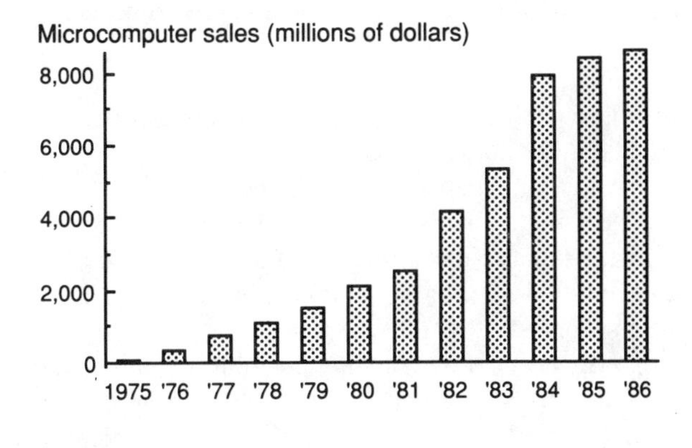

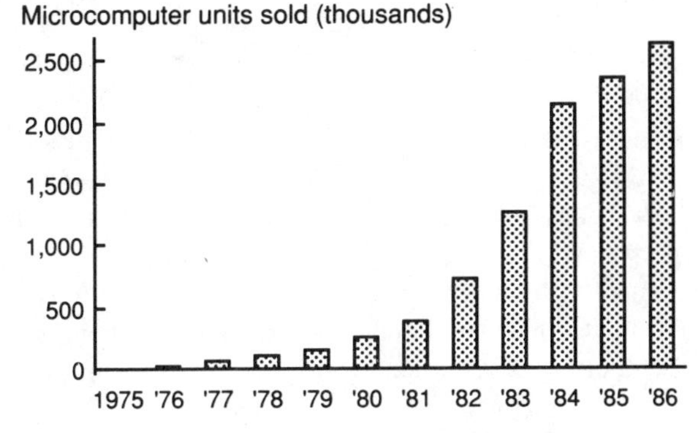

Figure 4.8. Time series showing spectacular growth in the early 1980s in annual domestic consumption of microcomputers, in millions of dollars (above) and thousands of units (below), for the United States, by year from 1975 to 1986. A microcomputer is defined here as a computer with a list price between $1,000 and $15,000. [Source: Data courtesy of the Computer and Business Equipment Manufacturers Association.]

As discussed earlier, with a map example shown in Figure 4.3, Furno's system reflects his background at the National Geographic Society in the mathematics of map projections, as well as his several years as a computer hobbyist. In 1979, he bought a Tandy TRS-80 computer for home use, and two years later began experimenting with maps. In 1982, he intensified his efforts, and with the support of the newspaper started to work at home and in the office, as time permitted, on a system that in 1985 included an enhanced 640K IBM PC, a plotter, a hard-disk peripheral memory unit, two monitors, and a tape back-up system. At the time of my visit, Furno had developed a world database, digitized from published maps, of coastlines and national boundaries, with additional detail for the United States, Mexico, Central America, and the Washington metropolitan area. His menu-driven software, called Azimuth, allows an artist to compose a dramatic view of part of the earth by selecting, in sequence:

1. an aiming point, say, 52°N, 8°W;
2. the distance from the globe of a viewing position, for example, 23,200 miles above the surface;
3. the scale of the map at the aiming point;
4. the view angle, that is, the angle between a line from the viewing point to the aiming point and the horizontal plane at the aiming point; for an oblique view, a viewing azimuth is also specified;
5. whatever shifts of the viewing position (up, down, left, right, and so forth) are needed to include all relevant areas; and
6. the grid interval, for a graticule of meridians and parallel (5°, 10°, 15°, 30°, 45°, or 90°).

Although the computation of new, transformed coordinates for points representing boundaries slows down the drawing of a map on its screen, the system is unique in providing dramatic, specially tailored oblique views of any chosen part of the Earth. The *Post* has published numerous plots drawn on this system using disposable liquid-ink pens and enhanced manually through the addition of type, area shading, and other symbols. Furno also has generated experimental film images on a high-resolution typesetter, as illustrated in Figure 4.9. Having invested substantial amounts of his own time on the system, Furno holds the rights to the software. In mid-1985 he and a partner experienced in marketing computer technology offered buyers a complete, "turn-key" system. In 1987 he rewrote Azimuth for use with the Apple Macintosh and LaserWriter.[25]

Another approach traded versatility in viewport and projection for a richer database and more rapid display. Artist and redesign consultant

Figure 4.9. Map generated on an IBM Personal Computer using Azimuth 87 software developed by Dick Furno of the *Washington Post* editorial art department and plotted in 1985 as a high-resolution image on a Camex Supersetter typesetter. [Source: Courtesy Dick Furno and the *Washington Post*.]

Robert Lockwood developed the Computer News Graphics system with three partners and the additional financial support of the Associated Press, for whom Lockwood's other venture (News Graphics, discussed in Chapter 3) each week produces five manually drawn information graphics.[26] The partnership evolved from Lockwood's investigations of computer graphics with Craig Ammerman in 1982. Two years earlier Ammerman had been executive editor of the Philadelphia *Bulletin*, which Lockwood helped redesign. Other principals were Ed Miller, editor of the Allentown *Morning Call* in the late 1970s, when Lockwood served as the paper's art director, and George Wieland, a computer consultant with wide experience in software development. Their system's database included meridians and parallels, place names, waterways and lakes, mountains, highways, and boundaries of individual counties, states, and countries, as well as a variety of corporate logos, flavor symbols, missile and aircraft drawings, and other graphics elements of use for business charts and explanatory graphics. Geographic data were digitized manually from a single small-scale world map and several separate, overlapping map sheets covering portions of the world in more detail at a somewhat larger scale. Coordinates and projection were fixed, but the user could select an area for display by entering either latitude and longitude or a place name. According to Lockwood, "newspapers use mostly locator maps," and speed, accurate information, and useful geographic reference features are fundamental. Programmer and artist collaborated in designing the system's command menu and display functions, and Lockwood is particularly proud of the "snap function," used to snap into alignment linear features digitized (recorded) from different source maps. The system also produced simple business charts and allowed users to digitize their own local-area databases as well as to receive graphics from the AP, over telephone lines, and to exchange graphics with each other. Its zoom feature, according to the firm's advertising, enabled the artist to "window in and work in detail on any part of the composition." In mid-1985, workstations were in operation at Lockwood's News Graphics studio (in the basement of his home in eastern Pennsylvania), at the AP graphics department in New York, and at the Dallas *Morning News*, which was producing 20 graphics a day with the system. Lockwood and his partners were marketing complete systems, with software, microcomputer, monochrome and color monitors, plotter and digitizing tablet, for $80,000 to $100,000, and had held a number of demonstrations at trade shows as well as at Lockwood's home. The software was designed for use on a variety of computers, and they planned to link it to an electronic scanner, a high-resolution electrostatic laser printer, and a pagination system.

A prime deterrent to the wider use of Lockwood's sophisticated but "artist-friendly" system has been the versatile, low-cost Apple Macintosh microcomputer, introduced in early 1984.[27] Unless spread widely over many users, the staggering costs of developing a system such as Lockwood's precludes its use by small and medium-size newspapers. According to a question-and-answer sheet prepared for potential buyers, "As of May 1985, the investment in Computer News Graphics exceed[ed] $800,000, not counting time invested by the principals." Rapid adoption by newspapers of the $3000 Macintosh diminished Lockwood's market and raised the unit price. In 1986 Lockwood altered his strategy, and with the continued support of the AP set to work on a system based on the IBM AT microcomputer and marketable at between $15,000 and $20,000, or about $30,000 with the database. Although his system is designed to be hardware-independent, Lockwood has viewed the Macintosh as too confining for the artist as well as too slow for such features as progressive zoom enlargement.

In 1986, though, newspaper art departments everywhere were buying from Apple, not IBM. The Macintosh was Apple Computer's

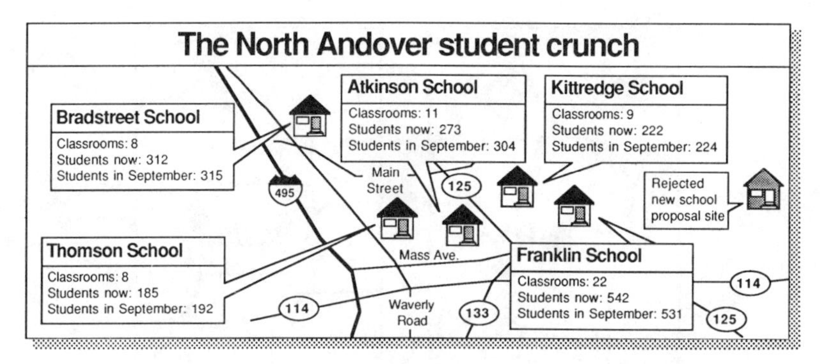

Figure 4.10. Journalist Steve Segal used MacDraw to produce this map for a story on the need in North Andover, Massachusetts, for additional elementary school classrooms. [Source: *Lawrence Eagle-Tribune*, 16 June 1987, B3.]

attempt to repeat the success in the late 1970s of its Apple II, and to market more widely at a lower price the "easy to run" design of its slow-selling Lisa computer. An innovative Lisa feature transferred to the Macintosh was a tracking device called a *mouse*, which controls the movement on the screen of an arrow. The mouse sits on the table beside the computer, to which it is attached by a wire. By moving the mouse to the right and back of the table, for instance, the user can shift the arrow toward the right and top of the screen. Commands can be entered through the keyboard, or chosen with the arrow from the selection on a "pull-down menu," lowered by mouse and arrow from the top of the screen like an electronic windowshade. With the mouse-controlled arrow, an artist can sketch and brush lines, indicate areas to be filled with a pattern, choose the pattern from a menu, select parts of the drawing to be moved or deleted, and indicate where text is to be inserted.

Particularly useful is the MacDraw software, selling for less than $200.[28] MacDraw works with points and lines—what the cartographer calls *vector data*—and with it the artist can prepare technical drawings, graphs, charts, and maps, such as Figure 4.10. Software packages available for information graphics include Microsoft Chart, for making histograms, time-series graphs, scatterplots, and other data graphics, and MacAtlas, a collection of "clip-art" templates for use with MacDraw.[29] MacAtlas templates available in 1985 included the United States by state, various regions of the U.S. by state, major world regions by country, and the states of the U.S. by county. With MacAtlas a newspaper artist might quickly compose a locator map similar to Figure 4.11. Numerous other graphics programs and databases available from a variety of software publishers add to the flexibility and cost-effectiveness of the Macintosh.

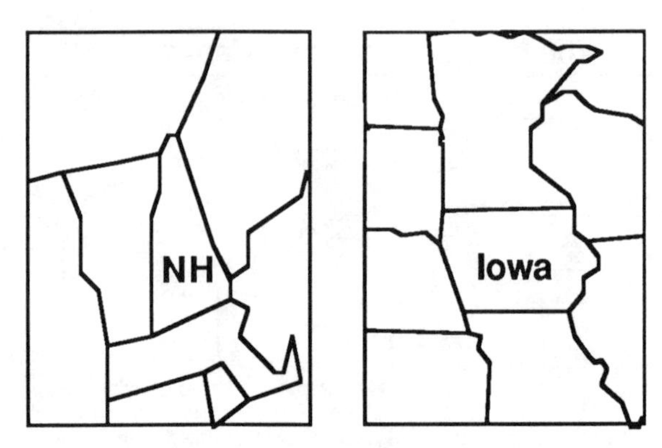

Figure 4.11. MacAtlas templates yielded these linework separations used in producing color maps to decorate a story on the New Hampshire and Iowa presidential contests. [Source: *Lawrence Eagle-Tribune*, 15 June 1987, A1.]

Like the IBM PC, the Macintosh spawned enhanced models with more memory: the 512K "Fat Mac" in 1985, the more versatile Macintosh Plus, with a built-in 800K diskette unit, in 1986, and the even more powerful Macintosh SE and Macintosh II in 1987.

Low-cost, high-resolution plots are also important, particularly if the artist is to add type electronically rather than paste on labels by hand. The LaserWriter marketed by Apple Computer provided the perfect solution for Macintosh buyers interested in graphics or electronic publishing. With a "laser engine" manufactured in Japan by Canon, the Apple LaserWriter, introduced in early 1985, produced a fine-grained raster image with 118 dots per cm (300 dots per in.), very close to the 157 dots per cm (400 dots per in.) commonly regarded as necessary for aesthetically pleasing type.[30] Apple Computer was only one of several firms to market a low-cost laser printer, but its LaserWriter included a sophisticated microprocessor for the efficient generation of a variety of typefaces, lines, and area-fill patterns. A marriage of xerography and the laser beam, and an outgrowth of advances in electronic typesetting, the laser printer uses a finely focused beam of coherent light to draw the image on the surface of a rotating photoconductive drum. Where light strikes the drum, an electrostatic charge remains to attract an oppositely-charged black powder, or *toner*. As shown in Figure 4.12, the laser printer's microcomputer fires the laser at times carefully chosen to direct the beam through a system of lasers and mirrors onto selected spots on the rotating drum. Before applying the powder, the laser printer generates the complete image, including typographic characters and other symbolic elements. After toner is allowed to adhere to the charged parts of the drum, the image is transferred to a sheet of paper, which is heated briefly to fuse the powder to the paper fibers. In 1985, when a LaserWriter sold for less than $7,000, waiting for improved pagination systems no longer was a valid excuse for deferring purchase of a computer graphics system. Macintoshes and LaserWriters began appearing in newspaper art departments, large and small, and MacDraw maps began to illustrate a variety of local, domestic, and foreign news stories.[31]

For two large newspaper groups, Gannett Newspapers and Knight-Ridder Newspapers, the Macintosh presented further opportunities for low-cost information graphics. Both groups already operated telecommunications networks for moving feature and news stories among their

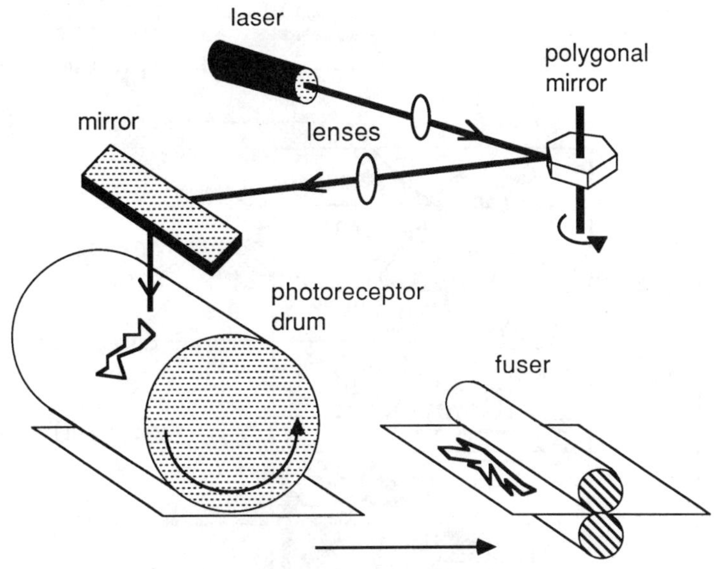

Figure 4.12. Principles of the Apple LaserWriter. Microcomputer synchronizes narrow pulses of coherent light from the laser to assure their arrival at intended locations on the surface of a rotating drum, as directed by a mirror above the drum, various lenses, and a rotating polygonal mirror. Timing of the light pulse determines the direction at which it reflects from the polygonal mirror, and hence where it strikes the top of the drum. The LaserWriter first composes the entire image on the surface of the photoreceptor drum, and then applies a toner, or black printing powder, which adheres only to the image areas on the drum. Next, the toner is transferred to a sheet of paper, and heat is applied to fuse the toner into the paper.

newspapers, and the ability to transfer artwork by wire persuaded Knight-Ridder to install a Macintosh and laser printer at each of its 27 dailies. Roger Fidler, the group's director of graphics and newsroom technology, supervised the move of the central graphics office from Miami to the National Press Building in Washington, D.C., and the initiation in April 1986 of graphics-sharing among 21 Knight-Ridder papers. Fidler saw the

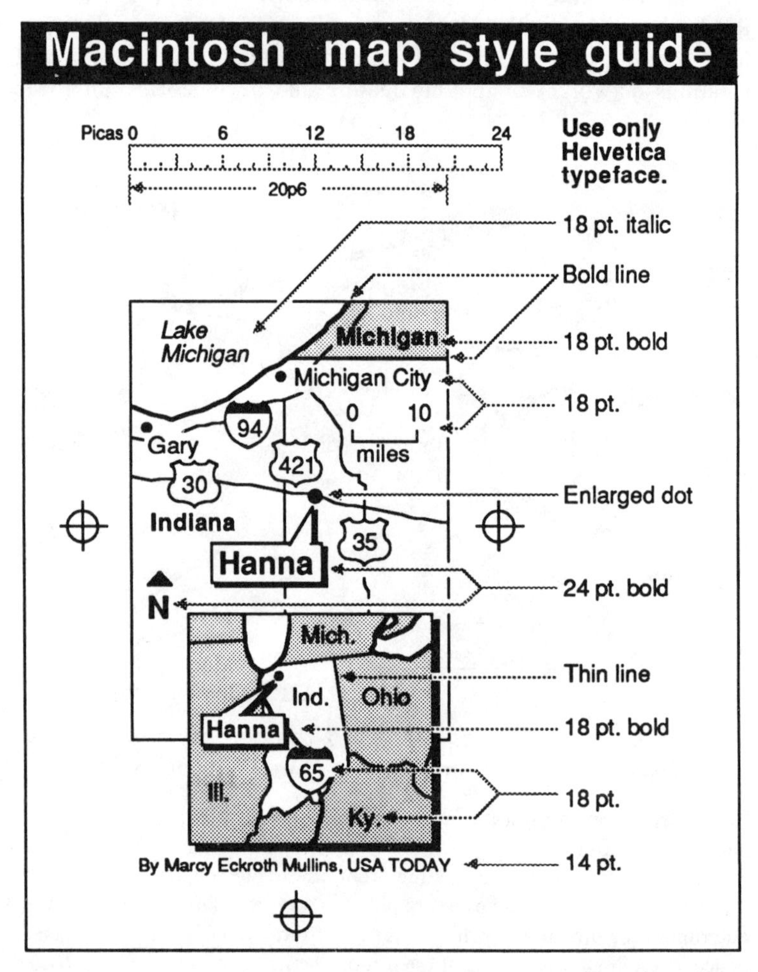

Figure 4.13. Example locator map used in *USA Today*'s Macintosh map style guide. [Source: Marcy Eckroth Mullins and Buzzy Albert, "USA TODAY: How to Draw Maps," *Deadline Mac* 1, no. 2 (May 1986), 1. © 1986, USA TODAY, All rights reserved, Reprinted with permission.]

potential of a wider clientele, including papers outside the group who subscribed to the KNT (Knight-Ridder-Tribune) News Wire or Knight-Ridder's PressLink network.[32] Graphics posted on one of PressLink's electronic bulletin boards could be "downloaded" into a Macintosh equipped with a modem and customized to fit the space available and the graphic style of the receiving newspaper. Tribune Media Services, which markets the Knight-Ridder Graphics Network, also markets the Chicago Tribune Graphics Service's Tribnet.

Equally enthusiastic about the Macintosh, the Gannett chain even started a monthly "MacGraphics" newspaper, produced for its member papers by the group's flagship, *USA Today*. After distributing 80 "Fat Macs" among 38 sites in September 1985, and 41 MacPlus computers to another list of 38 in February 1986, Gannett found the equipment reliable and widely used for "not only charts, maps, tables and illustrations, but promotional materials and even advertisements as well."[33] As shown in Figure 4.13, *USA Today* even developed a "Macintosh map style guide" for use by its 21 artists.[34] Gannett's network, similar in principle to Knight-Ridder's electronic bulletin board, uses a central computer to store illustrations, which can be retrieved over 800-number WATS (Wide Area Telephone Service) telephone lines. In 1987 the AP initiated a similar service, called AP Access. Telecommunications software and modems, which convert between purely electronic impulses and audio signals carried through the telephone system, have given the newspaper industry a further incentive to adopt the Macintosh for information graphics.

Unlike electronic bulletin boards, the satellite communications networks used by several national newspapers promote the uniformity of information graphics by moving entire pages. Indirectly, though, they have encouraged a fuller use of maps and charts and of fresher designs. Satellite communications have not only made possible the timely distribution throughout the nation of the well-established *Wall Street Journal* and *Christian Science Monitor* but also encouraged the Gannett group to launch in 1982 its colorful and highly graphic *USA Today*. Facsimile scanners able to convert entire pages into electronic code permit the distribution of a single national edition of *USA Today*, made up in a high-rise building in Alexandria, Virginia, and delivered by satellite to more than 30 printing sites throughout the United States. Maps and charts have been an important part of the *USA Today* concept, particularly the large, full-color daily weather map, its accompanying explanatory weather graphics, the frequent full-color information graphics on the front page, and such standing artwork as the daily "USA Snapshot" graphic, with a colorful illustration enhancing an otherwise-mundane chart or table, and the black-and-white locator map accompanying the daily "USA Journal" feature, which commonly focuses on a unique event in an obscure yet interesting small town. Richard Curtis, managing editor of graphics and photography, acknowledged that "one of our goals [was] to be different from other newspapers." Curtis promoted the use of stand-alone graphics because "we feel that there are some news developments we can report better in a graphic than in story form."[35] Maps routinely drawing attention to human-interest stories in various parts of the country fit well the notion of *USA Today* as a traveler's paper, and stand-alone explanatory graphics were a natural addition to the paper's strategy of relatively short but numerous stories.

Satellite transmission has had a far more subtle effect upon the design of the *Wall Street Journal*, which in late 1983 hired design director Jerry Litofsky to coordinate its redesign. One concern of the *Journal* was to attain a greater consistency among its regional editions for the East, Southwest, Pacific Coast, and Midwest, which are made up at regional "originating plants" using material assembled in the East under direction of its New York headquarters. The *Journal* was transmitting full-page facsimiles among five originating plants and 12 additional printing sites, and management was planning a fuller use of pagination and computer graphics. When I visited in late 1984, Litofsky was preparing to standardize the folio heads (standard rule and text at the top of the page), to facilitate the eventual adoption of a pagination system. Despite its conservative image and its traditional—some might say "drab"—appearance, the paper has long been active in exploring newspaper technology, and has provided its art department with a variety of microcomputers and graphics hardware, including a Macintosh system. A phototypesetting system identical to those in its originating plants had been installed to help with the redesign, and Litofsky and the editors with whom he worked had produced a number of experimental pages to study various designs and test the equipment. But the redesign reached beyond standard rules and page numbers to a new emphasis on art and design.[36] Acknowledging the visual blandness of the mass of type on a typical *Journal* page, Litofsky considered illustrations, charts, maps, and even editorial cartoons important as a source of "white space" for more attractive page layouts. "Because the page designer can't find it in the *Journal*'s text," he noted, "the white space must be built into the graphics." Figure 4.14, which illustrates the contribution of maps and other information graphics to the overall appearance of a page lacking advertising or photographs, demonstrates that the reasons for adopting journalistic cartography are aesthetic as well as informational.

PAGE 33
WEDNESDAY, APRIL 8, 1987

THE WALL STREET JOURNAL.

© 1987 Dow Jones & Company, Inc. All Rights Reserved

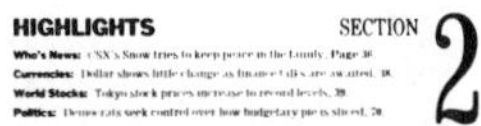

HIGHLIGHTS

SECTION 2

REAL ESTATE

Troubled Properties for Some Are Opportunities for Others

Lawyers: Lessons in Appraising Worth

If That Leash Ever Breaks, Nancy And Sluggo Had Better Run for It

Auto Prices

Phone Owners Face Dilemma Of 'Inside Wire'

YOUR
MONEY
MATTERS

Baseball Teams Field Lavish Offers For Spring-Training Sites in Florida

Naming Terms

'Little Benefit'

State-of-the-Art

Few Problems

Figure 4.14. Front page of second section of the *Wall Street Journal* illustrates the use of map, chart, and headlines to add white space to an otherwise gray page. Offshore stippling to differentiate land and water on the map obviates the need for a notably darker graytone screen or line pattern to represent water. [Source: *Wall Street Journal*, 8 April 1987, 33. Reprinted by permission of the Wall Street Jounral, © Dow Jones & Company, Inc. 1987. All Rights Reserved.]

The Society of Newspaper Design

Growing technical specialties, like successful academic disciplines, tend first to develop a cognitive identity, which matures later into a professional identity. Newspaper design seems to have attained a cognitive identity at least by the late 1970s, and to have made many advances during the 1980s toward a more formal professional identity. Indeed, the field has acquired not only journals, reference works, and teaching texts, identified as necessary infrastructure by the disciplinary sociologists Arnold Thackray and Robert Merton, but the equally important "sense of common orientation and purpose" derived from a central problem and focusing on appropriate techniques and concepts.[37] And a professional identity is emerging from a new professional society, the recruitment of "followers and students," and the usual propaganda that any profession uses to promote its contributions and distinguish adherents from outsiders. Despite decades of textbooks and the articulate writings of British designer-scholars Allan Hutt and Harold Evans, most of these developments are recent and most of them are American. Indeed, the single catalytic event was the formation in 1979 of the Society of Newspaper Design, which produces the field's only journal, holds workshops and an annual convention, sponsors an annual design competition, supports student chapters, and promotes internships. Before the SND, organizational support for newspaper design was limited largely to college courses and textbooks and the efforts of a few individuals working through the American Society of Newspaper Editors, the Associated Press Managing Editors, and the American Newspaper Publishers Association.

The roots of the SND might be traced to a newspaper design seminar held in July 1978 by the American Press Institute.[38] Among those attending were Robert Lockwood and Ed Miller, at that time art director and executive editor, respectively, of the Allentown *Morning Call*. Lockwood and Miller agreed to make arrangements for a meeting to be held 19-21 January 1979, at a country inn in Buck Hill Falls, Pennsylvania, near Allentown. Twenty-one graphics experts and newspaper editors attended, including Roger Fidler, of Knight-Ridder Newspapers, Robin Jareaux, then at the Dallas *Morning News*, and Richard Curtis, Baltimore *News-American* art director, who later became executive art director at *USA Today*. Lockwood, respected and well liked, was named chairman of a seven-person steering committee. In the late 1970s and early 1980s, the SND resembled a large, far-flung fraternity, with many familiar names decorating its mailings. But competent newcomers were welcomed eagerly, and many professionally active designers rose rapidly in SND ranks, as founding members such as Lockwood and Curtis became occupied largely with their own business ventures and executive duties.

The society grew rapidly and steadily, to over 1,100 members by 1984 and to a total of 1,483 for 1986. Despite an understandable concentration in the United States, SND members resided in 33 countries, with strong contingents in Canada (60), Norway (43), and Sweden (39).[39] Annual dues in the mid-1980s were a modest $45.

Particularly significant is the diversity of SND membership, which includes editors and news executives in addition to artists, art directors, and page designers. A 1984 survey examined the job titles listed in SND's first membership directory, for 82 percent of its members.[40] The survey recognized three major job categories: newsroom "nonvisual" (42 percent), newsroom "visual" (34 percent), and non-newsroom (24 percent). The 388 job titles in the "nonvisual" newsroom category were skewed toward top management: members working as editors, managing editors, and executive editors (183) outnumbered news editors and assistant managing editors (100), section editors (65), and reporters and copy editors (40). The 311 "visual" job titles were dominated by managers, with graphics editors (89) and art directors (71) more numerous than photographers (69), artists (47), and designers (36). The 224 non-newsroom job titles were almost evenly divided among owners and general managers (75), educators and consultants (73), and others (76). Upper-level managers and owners tended to be from smaller papers, which are more numerous, whereas "visual" members commonly worked at larger papers, which are more likely to have larger, more specialized art and design staffs. The survey not only suggests a concentration of talent at larger newspapers but also reveals a broad interest among less prominent publications, with a strong interest overall among persons well positioned in their firms to implement innovative design concepts. Equally apparent is the limited involvement as SND members of nonmanagerial artists and illustrators. Nonetheless, people working full time "on the board" are reached through workshops and regional meetings sponsored by SND, as well as through workshops at the Poynter Institute for Media Studies, in St. Petersburg, Florida, and the Rhode Island School of Design, in Providence.[41]

Since its inception, SND has been important as an instrument of both collegiality and continuing education. Among its early activities were the annual design workshop and convention, started in 1979, and the annual design competition, for which the number of entries had risen from 2,474 in 1980 to over 8,000 by the middle of the decade. Each year's entries are judged by a new panel of noted designers, artists, and news executives, and pictures of winning entries are published in the annual *The Best of Newspaper Design*. Another SND publication, its quarterly magazine *Design*, begun in 1980, offers short, well-illustrated articles addressing a variety of practical concerns in page design, news illustration, and new

DeadlineMac

From concept to copy on the Macintosh computer

SND

Sponsored by the Society of Newspaper Design

Vol. 1, No. 6 ©1986 DeadlineMac **September 1986**

In Detroit, the weather is a breeze

By George Rorick
Assistant Art Director, USA Today and Acting Graphics Director, The Detroit News

The Detroit News' production plant is located almost 28 miles from the downtown art department where we produce the new weather package. Through the combined use of a Macintosh computer and a modem, we have cut hours off the daily production time. The art is now completed four hours earlier than before and travel time is eliminated. Sending the graphics by phone rather than courier saves as much as an additional hour an a half. And because the production department starts stripping color much earlier, we've eliminated the overtime created by the previous weather page.

The printing plant keeps a copy of the Macintosh-generated base art. After we produce the overlays at the downtown office, we transmit the precipitation patterns, color separations, weather fronts and symbols to the production plant.

Making the map

The finished art was executed on the Macintosh using the MacDraw application and a Summagraphics MacTablet. I wanted the maps to be drawn in perspective and in the same style that I would have drawn them with a pen. I drew them using the polygon tool (Option key down), configured for a 25% reduction. That means I drew everything big!

Drawing with the polygon tool allows me to later fine-tune the elements by choosing Reshape Polygon in the Edit menu. Holding down the Option key keeps the polygons from closing prematurely as they are drawn. Without this technique it would be impossible for me to do professional artwork using a variety of line styles.

The map of the USA was drawn state-by-state then grouped. An outline was then drawn around the entire map. This outline was then duplicated, filled with black and sent to the back to create the shadow.

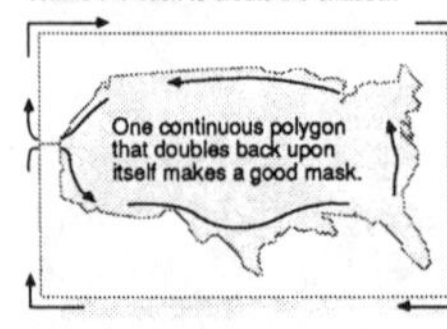

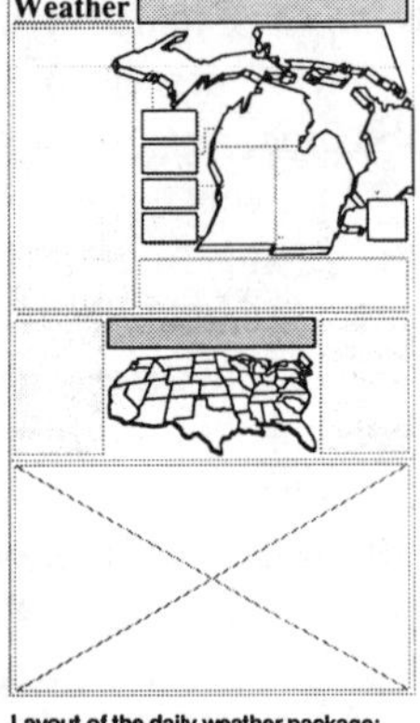

Layout of the daily weather package: Macintosh portions are outlined in black.

Using the polygon tool, a mask was drawn around the map. The mask and its border are both filled with white. This is used to trim any overlapping elements that fall over the edge of the maps. No one has to physically trim anything from around the complex shape of the USA map.

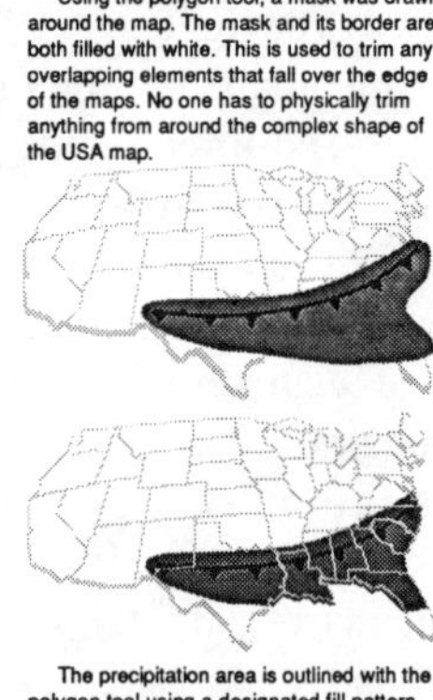

The precipitation area is outlined with the polygon tool using a designated fill pattern. The polygon line is then filled with the same pattern in the Pen menu. This shape is smoothed for printing. No effort is made to keep the precipitation bands precisely within the borders of the map. The overlap is eliminated by bringing the white mask and a None-filled copy of the state borders to the front. (The full-size printed version is in four-color with the state lines solid black, not toned as shown above.)

Continued...

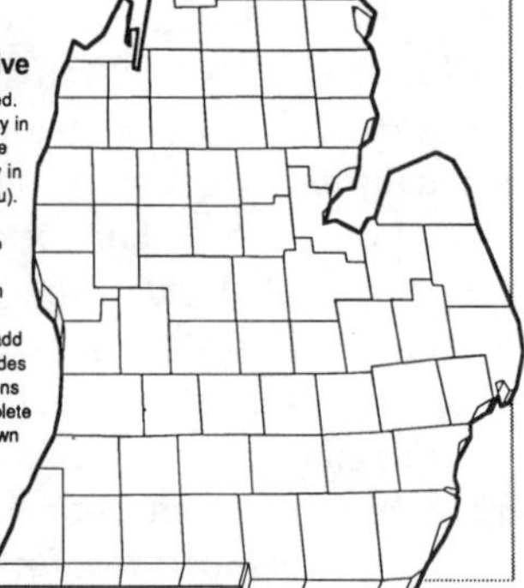

Three-point perspective

Here is another technique I used. As in the USA map, each county in Michigan is drawn as a separate polygon with the screen display in the Reduce mode (Layout menu). The grid is kept off to allow the shapes to accurately follow the actual county boundaries. Reshaping the polygons is then done at Normal Size. Vertical slash marks are positioned to add dimension. These serve as guides for the placement of the polygons which form the ridges. To complete the map, a heavy outline is drawn around everything. This map is not a part of The Detroit News weather package.

Figure 4.15. *Deadline Mac*, a two-page newsletter for Macintosh users, was edited by John Monahan, of the Quincy, Massachusetts, *Patriot-Ledger*, and published by the Society of Newspaper Design. [Source: *Deadline Mac* 1, no. 6 (September 1986).]

technology. Not an academic journal, *Design* devotes considerable space to examples of creative page layouts and effective information graphics. *Deadline Mac*, a newsletter started in 1986, welcomed numerous contributions from members and promoted the exchange of helpful hints on the use of the Macintosh and opinions about software (Figure 4.15). Many volunteers contribute their own time to SND (and in some cases, employers contribute some of their employees' paid time as well). The society also enjoys the support of the American Newspaper Publishers Association (which shares its headquarters complex—the Newspaper Center, on an attractive campus in Reston, Virginia—with SND), the American Society of Newspaper Editors, and several other organizations serving the journalism profession. Clerical assistance is available for SND's full-time executive secretary.

Journalistic cartography has been a minor but recurrent topic in SND's publications and workshops, but maps are often combined with other varieties of information graphics. The newsletter *Deadline Mac* mentions maps more frequently than other SND publications, but most issues of *Design* have addressed maps in at least one article. The Spring 1984 issue, for example, was organized around the theme "Information Graphics," and included a comparative evaluation of maps of the Grenada invasion; the Winter 1985 issue included a half-page article on mapmaking at *USA Today*.[42] But no more than five percent of *Design*'s content addresses cartography directly, and maps receive even less coverage in SND's annual *The Best of Newspaper Design*. In the design competition, which has divided entries over more than 30 categories in recent years, maps constitute less than half the information graphics winners. In the sixth edition, for instance, information graphics accounts for 15 of the 299 pages of award winners, and only 29 of the 64 winners in this section are wholly or partly cartographic.[43] Indeed, maps are more numerous, if less densely concentrated, in the remainder of the volume, where they frequently decorate a winning page layout. SND's treatment of the map seems an accurate reflection of the status of journalistic cartography as an indispensable but decidedly minor ingredient of news presentation.

News Services and News Magazines

Art departments at wire services, graphics syndicates, and news magazines are similar in many ways to those at large newspapers. For the most part they have been comparable in their drawing media and their limited use of computer graphics, as well as in the number and backgrounds of their staff. Most staff members were trained in art or design schools, and some had worked previously at newspapers, the most promising labor pool for employers seeking artists experienced in information graphics. When I visited its Washington, D.C., offices in early 1985, United Press International had a graphics department of eleven, including its director, Joe Scopin, eight artists, one researcher, and one researcher-photographer. Scopin had worked as art director for the *Washington Star*, which redesigned before it closed in 1981. (A number of prominent newspaper closings have followed a redesign, demonstrating that improved graphics and layout are seldom sufficiently successful to rescue a neglected enterprise facing a strong local competitor.) Although rare in newspaper art departments, researchers have proven useful at wire services and news magazines for collecting and organizing source materials as well as for compiling facts for specific illustrations. A visit in late 1984 to Associated Press headquarters, in New York, found a graphics staff of seven artists working with a graphics editor as a unit within the news photo department under Hal Buell, the assistant general manager in charge of AP photojournalism for several decades. The AP also received artwork weekly from Robert Lockwood's News Graphics studio and used several local freelance artists. When a staff artist was not available, during early morning hours, for instance, an editor in the photo department could prepare a simple locator map using the ready supply of base maps, a stick-up symbol and headline type, and the department's copy camera. At the AP's Canadian affiliate, the Canadian Press, photo editors also make locator maps for breaking news when an artist is not available. (The one or two artists at CP's Toronto headquarters produce maps and charts for Canadian stories, and photo editors select AP maps, sent from New York, for non-Canadian stories.) As with UPI, the AP graphics department is an outgrowth of its news photo operations, with which it shares facsimile transmission facilities. AP's final step toward a specialized graphics department was the hiring in mid-1985 of Don DeMaio as graphics director. DeMaio had served five years as art director at United Media, parent of the Newspaper Enterprise Association (NEA), which syndicates maps and other graphics, and for the previous 20 years had worked at various newspapers as artist, art director, and graphics editor.

Because clients or subscribers are dispersed throughout the nation, maps produced by wire services, syndicates, and news magazines have a much wider geographic focus than the cartographic art of a metropolitan newspaper, which devotes much of its map-making effort to local coverage. Delivery by mail excludes graphics syndicates from making locator or explanatory maps for breaking news, although Tribune Media Services and other syndicates offer basemap packages, with which a news editor equipped with razor blade, rubber cement, and a quarter hour of time can assemble a simple yet highly presentable locator map. In contrast, wire services might be expected to provide locator maps within the hour for some stories, and commonly advise subscribers of the planned transmission time for graphics related to a breaking story, so that interested client papers can reserve space on the page.

Both syndicates and wire services offer anticipatory graphics relating to a variety of forthcoming political and sporting events, and occasionally provide maps and charts to illustrate specific feature articles distributed by another department within the news service. The AP mails its members the information graphics produced at Lockwood's studio. Printed on glossy paper by offset lithography, these AP/News Graphics offerings are markedly higher quality reproductions than images regenerated on a Laserphoto receiver. The AP and UPI must also prepare on a routine schedule such timely, standardized graphics as separate daily weather maps for morning and afternoon papers, and charts illustrating trends in financial markets and government economic indexes. Some news magazines, such as *Business Week* and the *Economist*, include many routine business charts, but most information graphics prepared for weekly news publications are designed to summarize or explain the spatial details of important recent events, rather than present routine quantitative data. News magazines usually have more time available than newspapers for designing and researching explanatory maps, and almost always print them in color.

Because of the need to be current—and to scoop competitors, if possible—weekly news magazines prepare much of the artwork for maps and charts in the two days before the Friday evening deadline. These magazines are printed over the weekend at a number of regional printing sites and delivered early the following week to news agents and subscribers. What advantage might have been conferred by modern telecommunications and printing technology seems to have been lost to the fierce competition for circulation and advertising. In the 1930s and 1940s, according to cartographic artist Richard Edes Harrison, who drew maps for *Fortune*, *Time*, and *Life*, deadlines were set to accommodate railroad schedules as well as to include the very latest stories. In a June 1985 interview, Harrison recalled that it was possible to draw higher quality maps for the monthly *Fortune* than for the weekly *Time*, which he considered little better than a newspaper in available lead time. For five years Harrison worked regularly on *Time*, as a freelance, but had to quit because his workload was too great. *Time* was "always a short, rush job." Often he received an assignment on Saturday for artwork that was to be finished and checked and delivered on Sunday at 5 P.M., in time for the New York Central Railroad's Twentieth Century Limited to carry it to Chicago for Monday morning engraving. Only the part of the week and mode of transport were appreciably different in the mid-1980s, when *Time* employed several artists drawing maps and charts, most notably chief cartographer Paul Pugliesi, whose work has won several awards in the American Congress on Surveying and Mapping (ACSM) map design competition. Although Pugliesi might know of the need for a map as early as Tuesday, half of the maps are not started until Friday morning, and for late breaking news sufficiently important to warrant last-minute changes in content, maps have even been started on Saturday.[44]

Like contemporary newspapers, the wire services, syndicates, and news magazines have developed identifiable graphics styles to promote continuity and avoid indecision on minor specifications. But their styles also reflect technical constraints and client objectives. UPI's crisp, open style, for instance, accommodates not only photowire transmission but also the likelihood that some subscribing newspapers will enlarge the artwork, whereas others will reduce it. UPI's offshore stipple (see Figure 3.17) differentiating land from water is an effective compromise between fine graytone screens, which might be reproduced too faint, too dark, or too blotchy, and the coarse, aesthetically unappealing line patterns that once were a hallmark of AP locator maps. In contrast, the carefully printed maps and charts mailed by graphics syndicates easily accommodate screened graytones, and the information graphics in weekly news magazines are largely in color. Although future networks designed specifically to transfer graphics into the client's Macintosh or computer graphics system might well lead to an increased use of color for wire service graphics, from the perspective of the mid-1980s, the news magazines were the most prominent and consistent users of color.

Magazines have been particularly conscious of the contribution of information graphics to news presentation and reader interest. To differentiate its pages from those of its competitors, *Time* went one step beyond in introducing to its charts the lively, controversial, and widely acclaimed pictorial style of Nigel Holmes, whom it promoted to executive art director. Figure 4.16, a typical Holmes "fever chart," uses an inattentive cat resting atop a television set to reinforce the time-series graph's message about the dwindling audience for network newscasts. Holmes, who had worked in Britain as an artist for the BBC's *Radio Times*, joined *Time* in 1978. An active early member of the Society for Newspaper Design, Holmes is noted also for his book *Designer's Guide to Creating Charts and Diagrams* and his participation as critic and lecturer in workshops at the Rhode Island School of Design.[45] His style has not affected *Time*'s cartography as much as its pie charts, time-series graphs, and explanatory graphics, which can more readily incorporate pictorial illustration.

Among weekly news magazines, Britain's *Economist* merits special mention, not only for the high rate of map use noted in Chapter 2 but

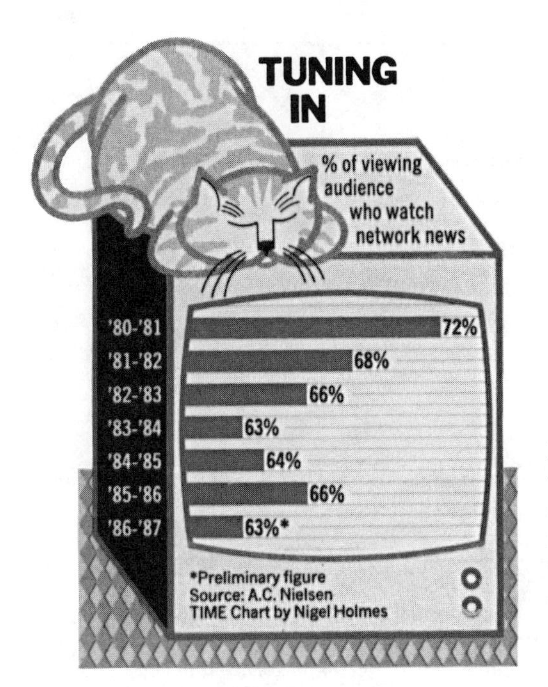

Figure 4.16. "Tuning In," a typical Nigel Holmes fever chart, portrays the decline in viewing of network television news programs during the early 1980s. [Source: *Time* 129, no. 11 (16 March 1987), 63.]

also for the presence in the Economist Building, on London's St. James Place, of two separate art staffs producing maps and charts. On the eleventh floor, a five-person staff operates the official "cartographic department," and on the ground floor, toward the rear of the building, a four-artist staff comprises Richard Natkiel Associates (RNA). Richard Natkiel is the *Economist*'s art director; he supervises the production of maps and charts for "the newspaper," as employees refer to the weekly magazine, as well as for the *Foreign Report*, *Connections*, the *Financial Report*, and other periodic bulletins published for businesses, governments, and investors by the Economist Intelligence Unit. Richard Natkiel is also the principal of RNA, founded several years before my visit in early 1985 at the suggestion of the managing director (president of the firm). The *Economist* bills RNA for space and materials, and takes a share of the profits. RNA, in turn, provides added personnel for Economist publications during times of peak workload, and also serves outside clients, including book publishers, stockbrokers, and other financial magazines. Unlike any American (or other British) art director I met, Natkiel was trained as a cartographer, serving a seven-year apprenticeship at George Philip and Son during his teens, and returning as a journeyman after serving as a cartographer in the British army during World War II. He

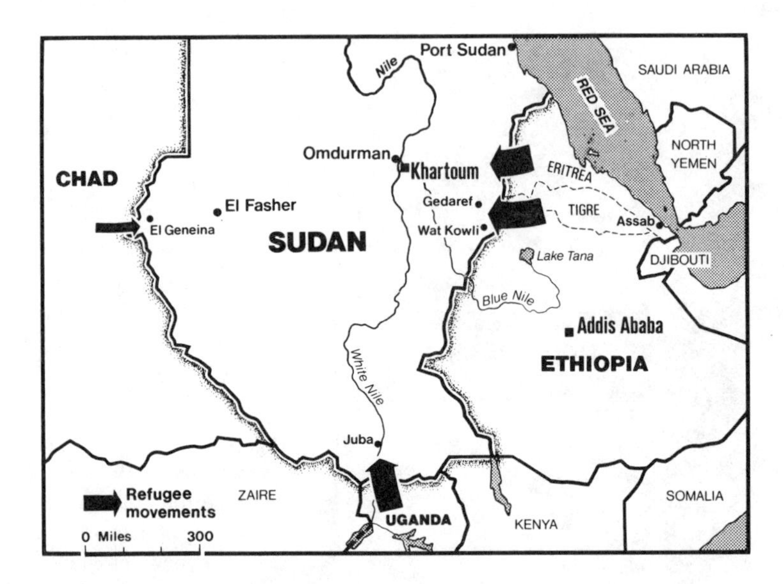

Figure 4.17. Map of the Sudan, Ethiopia, and surrounding areas illustrates the *Economist*'s cartographic style, including the scale bar as a part of the neat line, or map border. [Source: "Famine Watch," *Economist Development Report* 2, no. 2 (February 1985), 6-7.]

joined the *Economist* in 1952, shortly after the magazine hired two or three artists and began to use information graphics with regularity. An accomplished military historian as well as a skilled designer of maps, the genial Natkiel is author of *Decisive Battles of the Twentieth Century* and several other atlases.[46] Unlike other news art staffs, most of Natkiel's workers, in both RNA and the cartographic department, also have backgrounds in cartography. Largely responsible for the *Economist*'s crisp cartographic style, with such features as a scale bar cut into the neat line (Figure 4.17), Natkiel holds that maps and charts must relate to the remainder of the magazine, particularly its typography.

Perceived Role of the Map in News Stories

Elsewhere, as well as at the *Economist*, the attitudes of editors, artists, and publishers affect the frequency, content, and design of news maps. Although the opinions of persuasive or powerful individuals can be important at a particular publication, these opinions often reflect values common to an occupational group. Editors, for instance, generally are more concerned about explanation than are art directors, and thus might be expected to favor locator maps and explanatory maps with abundant type. More sensitive to aesthetics and pleasing contrasts among art and type, art directors are more likely than editors to prefer nonconventional cartographic renderings. Because of their inherent concern with the need to attract and hold readers and advertisers, publishers are more attentive to the map's contribution to the newspaper's image. Conflict is reduced, though, because top management determines that image and tends to hire editors and art directors with supportive views. But the publisher's control is by no means complete, particularly for such a minor journalistic element as the map. Publishers might easily be swayed by the often conflicting attitudes of their editors and artists about the proper role of graphics in news presentation.

By examining attitudes toward graphics, this section complements the foregoing treatment of institutional resources available for journalistic cartography. The focus here shifts from the firm and its structure to the perceptions and opinions of people in the two occupational groups most closely involved with news maps—editors and artists, or as they sometimes see themselves, "word people" and "picture people." What are the dominant attitudes of editors and artists? How deep-seated is the conflict between word people and picture people? Are some picture people more amenable to explanatory maps than other picture people? In addition to addressing these questions, this section also examines attitudes toward maps expressed in the literature on news writing and newspaper design. Textbooks, in particular, reflect group values and serve as instruments for promoting a wider awareness.

Attitudes and Values of Artists and Designers

It would be misleading to suggest a single, simplistic set of values common to all newspaper artists. Specialization based on aptitude and circumstances tends to focus an artist's attention on feature-page illustration, political cartooning, information graphics, or page layout. And page designers need not be skilled in illustration. As a group, artists and designers address the paper's appearance more than editors and reporters, who develop its content. These divergent foci—editors toward facts and opinions, and artists toward decoration and packaging—occasionally generate tension, creative or otherwise, and usually the journalists' view prevails. From an aesthetic standpoint, though, the artist's creativity usually is constrained less by an unappreciative editor than by deadlines, high-speed printing, and cheap newsprint.

Cartographic principles are not intuitive, and traditional maps might easily appear needlessly bland to those conditioned by training and temperament to value aesthetics and creative originality. Thus the skilled illustrator who produces an appealing picture for the weekly food page might inadvertently violate or willfully ignore elementary rules of map design when called upon to draw a world map. Knowing nothing about map projections and little about geography, the artist might look for a "better" projection—one, say, more suited to the rectangular space to be filled—and even transfer lines from the Goode's homolosine grid found in his atlas onto a Mercator basemap found at the bottom of the art cabinet. Or, if particularly "creative," the artist might even improvise by getting rid of the grid, rotating a bit, and adding a "drop shadow" to raise the contents well above the surrounding oceans. The reader thus is treated to a vast exaggeration of Siberia relative to China, or an unwarranted shift of East Africa toward India. The editor or reporter might know little more of the world than the artist, and might even applaud a pointless and misleading distortion of geographic relationships. Fortunately, those art departments regularly producing news maps usually have acquired a cartography savvy, largely though self-study, the hiring of experienced map artists, or the criticism of geographically astute readers.

Cartographers and geographers have been particularly sensitive to map abuse by the media, and this concern has led to a number of recent critiques of news maps. Daniel Beaudat and Serge Bonin, in commenting on maps of Afghanistan appearing in *Le Monde* and other French newspapers after the Soviet invasion, pointed out a number of inaccuracies stemming from obsolete data or a poor understanding of geography.[47] They also deplored missing map scales and a general failure to give readers a firmer grasp of distance relationships by comparing Afghanistan with France. Judith Leimer, who noted a wide range of quality among 207 maps found in 231 issues of 36 U.S. newspapers, observed four common flaws: poor typography, sloppy drafting, a scale too large for the details presented, and incomplete or confusing information.[48] As noted earlier, Leimer attributed these shortcomings largely to deadline pressure and the technical limitations of newspaper printing. Gary Chappell, who looked at news magazine coverage of the 1982 Falkland Islands war between Britain and Argentina, also noted a number of inaccuracies.[49] He concluded that map designers at *Newsweek*, *Time*, and *U. S. News and World Report* had eye appeal as their chief goal, and that this attitude often interfered with a fuller, unambiguous presentation of geographic relationships. Particularly troublesome were overly large pictorial symbols and "dynamic symbols" that attempted to show complex military maneuvers. Chappell attributed a heavy use of maps in Falklands coverage to a dearth of action photos because of tight censorship.

Patricia Gilmartin also focused on a single event with limited photographic coverage—the shooting down in 1983 of a Korean Air Lines passenger jet that had wandered over Soviet territory.[50] Gilmartin, who examined the initial maps carried by 20 North American and European newspapers and weekly news magazines, noted a variety of projections and orientations, including several on which the artist improperly drew a route by connecting the end points with a straight line. She identified five general approaches to portraying KAL 007's intended and actual routes. As shown in Figure 4.18, these include representations that erroneously portray the actual route as a blatant deviation into Soviet airspace (A), as never entering the USSR (B), and as merely a mid-flight short-cut (C). Some designers seemed unaware that a great circle route is a straight line only on a gnomonic projection, whereas others chose a suitable projection and attempted an accurate portrayal of available information (E). Recognizing both the limited cartographic training of many newspaper artists and uncertainty about how much readers rely upon journalistic maps, Gilmartin even suggested that traditional standards of cartographic quality might well be suspended for news maps, as they seem to be waived for subway maps and AAA Triptiks.

A group of British geographers was less tolerant. Irritated for years by blatant expressions of cartographic ignorance, they proposed a "Media Map Watch" to monitor systematically the use of maps on television as well as in the press. The effort was initiated by the Geographical Association, a professional group of secondary and college-level educators, and won the support of the Royal Geographical Society and the British Cartographic Society. During November 1984 several hundred geography teachers and other volunteers scanned newspapers and monitored television broadcasts, and submitted several thousand maps.[51] A principal criticism was the failure of the media to provide maps for many stories with a significant geographic component. Among the flaws noted in those maps presented were the frequent absence of a title, lack of a scale, incorrectly spelled place names, clutter, and excessive simplification. The Map Watch committee identified as an underlying cause the widespread lack of cartographic training among newspaper artists, which others have also noted. As a remedy, a group of British geographers in July 1985 held a one-day seminar on Cartography for the Media. Unfortunately, few media representatives attended besides invited speakers from the BBC and the *Times*.[52]

Like architects, news artists are influenced by trends and sometimes neglect function for form. Many, it seems, were impressed by the often-flamboyant style of chartmaking espoused by Nigel Holmes, who in the mid-1980s conducted several widely acclaimed workshops at the Rhode Island School of Design.[53] Since its appearance in *Time* in the late 1970s,

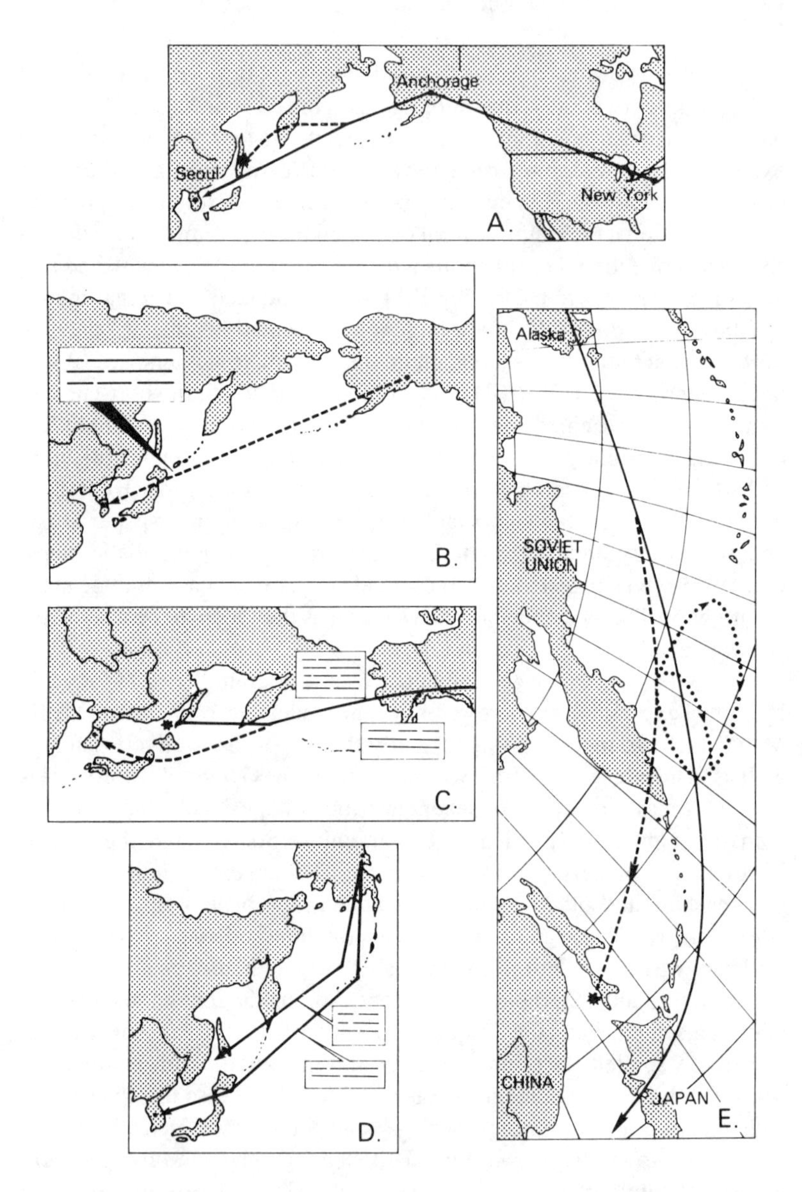

Figure 4.18. Five representative portrayals in newspaper maps of the planned and actual routes of KAL 007's final, fatal flight. Maps D and E are the more accurate representations. [Source: Patricia Gilmartin, "The Design of Journalistic Maps: Purposes, Parameters and Prospects," *Cartographica* 22, no. 4 (Winter 1985), 1-18; Figure 2 on p. 10.]

Holmes' style has been widely imitated, usually for statistical diagrams—"fever charts," he calls them—but occasionally for maps. Obviously one of Holmes' goals is to enhance interest in the news through the lively pictorial portrayal of otherwise bland facts. In the preface of his *Designer's Guide to Creating Charts and Diagrams*, he states the challenge:

> We are deluged by a daily surge of figures. . . . How do we cope with them? Too many of us are scared by figures. We do not understand them. We would rather not look at them.
>
> We would like someone to explain them to us. . . . The chartmaker can unlock the secret of the figures. He or she can make their meaning visible, literally.[54]

Not all chart designers agree with Holmes. Perhaps the most vocal dissent has come from Edward Tufte, a professor of political science and statistics at Yale University. Tufte, who advocates a simplicity in design achieved by avoiding extraneous decoration—what he terms "non-data ink"—attacks Holmsian graphics as needlessly simplistic in content. He is particularly hostile to the assumption of Holmes and other writers that data are boring. Tufte also sees a self-serving motive behind the news graphics revolution: in his view an exaggerated emphasis on lively graphics is little more than an empire-building justification for more artists, a bigger art department, and a more powerful, better-paid art director. More ominous is a possible loss of editorial control to artists. In Tufte's words:

> As the art bureaucracy grows, style replaces content. And the word people, having lost space in the publication to the data decorators, console themselves with thoughts that statistics are really rather tedious anyway.[55]

In his highly regarded book *The Visual Display of Quantitative Information*, Tufte examined the graphic sophistication of *Business Week*, the *New York Times*, *Time*, the *Washington Post*, the *Wall Street Journal*, and ten influential newspapers and magazines from five other countries. Tabulating the proportion of graphics that were "relational," that is, relating two or more factors, he ranked the five U. S. publications near the bottom.[56] Tufte attributed this dearth of sophisticated graphics in the American press not only to the perception that numerical data are boring but also to a lack of quantitative skills among news artists and prejudice among editors and art directors alike that "graphics are only for the unsophisticated reader."

Maps can easily fall in the middle—or to the side—of this debate. Although the map itself is inherently relational, most maps in most newspapers are simplistic locator maps, illustrating a single location relative to several selected geographic reference features. Tufte omitted maps from his tally of "relational graphics," and Holmes has had little to say about them. Although Tufte marvels at the amount of information that a "data map" can store, and Holmes occasionally adds "flavour" to a chart by including map-like shapes, both authors clearly are more fascinated by the challenge of designing charts. For their respective purposes, the data graphic is more amenable than the map to portraying statistical relationships and incorporating pictorial illustration.

Perhaps because of this apparent inflexibility of the map, news artists specializing in cartographic illustration differ from chartmakers and other journalistic designers in both skills and aesthetic values. *Time*'s map makers, for instance, generally have avoided Holmsian embellishments despite an increased use of color, nontraditional perspectives, and pictorial point symbols. Cartographers at the *Washington Post* have had to defend their style as appropriate for "geographic maps" against the preferences of a new art director who viewed maps as but another element of the paper's artwork ripe for a fresh approach. Michael Keegan's call for more "pizazz" might seem a bit surprising, though, given the innovative computer-generated "Earth-from-space" perspectives used by the *Post*'s map staff to focus attention on specific cities, countries, or regions. Although map makers might accept changes in type styles, line weights, graytones, and borders, cartographic license inherently is less flexible than artistic license.

The attitude that the map is but one of several types of newspaper artwork might well underlie the recent abundance of maps in newspapers, such as *USA Today*, that offer readers a profusion of relatively short news stories and feature articles. For these papers, maps—together with photographs, line drawings, data charts, and pictorial illustrations—are a useful tool for the page designer, who uses them not only to inform the reader but also to attract the eye, balance the page layout, and separate headlines between rectangular "modules." This latter role of a buffer is particularly important, as illustrated in Figure 4.19, where a "piece of art" separates headlines that otherwise would run together unless differentiated by contrasting typefaces, a vertical rule or box, or a wider white space between columns. A map or photo can also be useful, as in the lower right of Figure 4.19, for filling out a rectangular module. Appearing below the headline and to the right of two columns of bodytype, the map or photo gives the story a horizontally elongated shape and permits a longer headline than otherwise practicable.

The map's principal competitor for these "art slots" is the photograph. For some stories, a photo clearly is more appropriate—say, when a suitable "head-shot" portrait is available for a heretofore unknown or only moderately well-known person who is the undisputed focus of the story. For accounts of an accident or disaster, many editors would prefer a spectacular action photo or a dramatic shot portraying pain and suffering, yet for breaking stories set in remote locations the map can be a convenient substitute for the photo the editor wants, but the wire service cannot yet deliver. The map is also convenient for the newspaper wanting a controlled splash of color on its front page—say, placid blues for the eastern Mediterranean, a yellowish-tan for the semi-arid Lebanese coast, and drops of brilliant red for bomb blasts. A geographic component to a

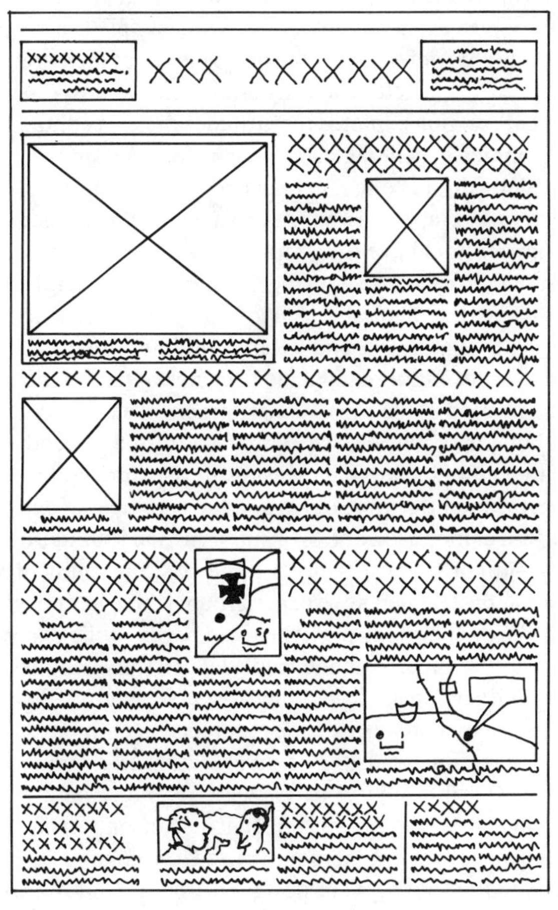

Figure 4.19. Modular layout of front page with maps used as a buffer between headlines (in story at lower left) and as an element below the headline to complete a rectangular module (in story at lower right).

major story can be an invitation to cartographic illustration even when the locational facts are well known to the informed reader. That a concern for packaging and designed pages might promote much of the recent interest in journalistic cartography is not surprising.

Editors and the Decision to Use a Map

Working largely with words and generally starting their careers as reporters, newspaper editors are comfortable with the label "word people." Although most editors have a tolerant view of art in the newspaper and do not see themselves as the inherent opponents of "picture people," others have become highly defensive of the prerogatives of their craft against the depredations of packaged news, decorative art, and even "information graphics." This section explores the attitudes of news editors toward both maps and graphics in general, and searches for consistency in the type of story for which editors believe a map necessary or suitable.

Reaction to graphics is not new: since the emergence of photoengraving, some editors have only grudgingly yielded space in their news hole to photographs and line drawings. In the 1980s, a heightened interest in news graphics led to another round of controversy, with a few vocal proponents and opponents, and a large middle ground for the mildly enthusiastic and the mildly skeptical. Yet much of the reaction might have been less a rejection of the *USA Today* formula of skillful color reproduction, short but numerous news stories, and a full-page weather map, than a reaction to a news business increasingly concentrated into ever fewer, more powerful media empires. Thus, editors critical of what some call "McPaper"—a comparison of *USA Today* with the leading American fast-food chain—might be expressing a more basic resentment against the paper's parent, the Gannett newspaper group, which has bought out many local papers with enormous profits gained partly at the expense of local news coverage.[57] Gannett's announced goal of a truly national newspaper, with home delivery in metropolitan areas, seemed to threaten papers poorly equipped to match *USA Today*'s precision color presswork, not to mention its sports, financial, and lifestyle sections. To some editors, insult accompanied possible injury, as Gannett's venture paid tribute to the newspaper industry's archrival, television. Not only were its graphics intended to appeal to a generation raised on color TV, but even its ubiquitous vending boxes were designed—or so *USA Today* publicists proclaimed—to resemble television sets.[58] Publishers attempting to meet this challenge by adding more graphics and enlarging art staffs, brought the threat of *USA Today* home to traditionalist editors conditioned to value content far more than packaging.

Defensive, anti-graphics attitudes of some editors were spurred years earlier, in 1975, with the appointment of artist Robert Lockwood as art director of the Allentown *Morning Call.* Yet, resentment was directed not so heavily at Lockwood as at Edward Miller, the paper's executive editor, who had hired him. Appointed to the top editorial management post by his father at the age of 27, Miller was an innovator and experimenter—and anything but a traditionalist, despite several years' experience in Paris with the *International Herald Tribune*. He rejected not only the bland, ritualistic formats of most news presentation but also the often-rigid division of labor among reporters, editors, photographers, artists, and page designers. In the lead article of the *Bulletin of the American Society of Newspaper Editors* for March 1979, Miller asserted that "There should be no distinction between content and presentation, that both elements are related through the larger process called design."[59] At the *Morning Call*, he noted, "two people share responsibility for the content of the paper. One is an artist. He doesn't just decorate what the other decrees. He has a hand in the judgments from start to finish." Writing that same year in *Newspaper Design Notebook*, Miller announced that "one of the unique aspects of this newspaper is that unlike most others who have art departments that are clearly auxiliary in secondary functions, the art department at the *Call* is very much in the center. In fact, it is the center."[60] The result was a newspaper far from predictable in layout: large pictures frequently dominated the front page, but occasionally, as on the morning after the November 1978 congressional and state election, a single small photo was dwarfed by an index-menu occupying over half the page. Reader reaction generally was positive, and news publishers elsewhere were impressed. Despite the validity of Miller's claim that his paper's packaging not only served but was inseparable from its content, traditionalist editors were repelled, if not intimidated, by the shock effect of his epistles, the striking redesign of the ASNE *Bulletin* following his ascension to the chairmanship of its editorial board, and the acclaim accorded the *Morning Call* among newspaper designers. Two central New York editors I interviewed mentioned Allentown without a direct prompt, and readily disparaged Miller as a spoiled, inexperienced son of a rich, overly indulgent father, and the *Morning Call* as a paper in which "context dominated content." Several others who admired the design of the *Christian Science Monitor* and the packaged sections of the *New York Times* were less hostile toward the Miller paper but believed that graphics and design were "being oversold." None of these editors had a noticeable enthusiasm for page design, and their papers used maps and charts sparingly.[61]

Particularly relevant to a discussion of editors' attitudes toward graphics is Ed Miller's observation that because the newspaper is inherently a visual medium, "editors who can't visualize how to present information are incomplete craftsmen."[62] Research on cognitive style suggests that this deficiency in visualization might be common among news editors, selected for editorial management positions at least in part because of demonstrated effectiveness as "word people." This conclusion is linked tenuously to studies of brain-damaged patients that made psychologists aware of the distinctly different functions of the two sides of the brain, with the left hemisphere associated with language skills and analytical reasoning, and the right hemisphere supporting visual-spatial and musical skills and holistic thinking.[63] Popularized accounts of brain lateralization commonly exaggerate these effects by asserting as fact unproven claims, such as that artistic, right-brained people tend to be deficient in verbal skills, whereas literary, left-brained people are visually deficient. Yet, more carefully qualified theories of the operation of the human mind appear to have some validity. In short, for reasons that might be more environmental than biological, *some* people have a dominant brain hemisphere that might not only promote one cognitive style over another (say, language over visualization) but also cause them to favor a career in which such skills are particularly helpful. Despite the possibility that left-brained people might be more common in the newsroom than in the general population, though, neuropsychology has yet to demonstrate that editors are inherently left-brained or visually deficient. Indeed, among print journalists, cultural conditioning might well play a more important role than genetic endowment in fostering "word person" attitudes and values.

Whatever their cause, attitudes toward graphics and maps do vary among editors. Although open hostility seems rare, the attitudes of most of the editors to whom I spoke approached a complacent neglect. Yet some editors, clearly not a majority, were enthusiastic supporters of maps. At a few central New York newspapers, for instance, the news editor clearly appreciates the value of maps for reporting spatial facts about crimes, zoning decisions, and other local events. The Norwich *Evening Sun* and the Rome *Daily Sentinel* don't have staff artists, but do maintain collections of area maps from the state transportation department and various local agencies. In a quarter hour or less a news editor can easily copy a portion of one of these maps and add an arrow or Maltese cross to obtain a simple, straightforward, and useful locator map like that in Figure 4.20. In contrast, one editor saw little need for local maps. In his words, "Everybody knows where everything is." Surprisingly, his paper serves a slightly larger city with a major Ivy League university—and a substantial turnover each year in students, faculty, and researchers. That Rome also has considerable residential turnover because of Griffiss Air Force Base thus seems less significant an explanation for the *Sentinel*'s more frequent use of maps than does its having a news editor who recognizes that geographic facts sometimes might be communicated and explained more readily by showing than by telling.

How widely do newspaper editors differ in their use of maps? Are not some stories so intrinsically geographic that all, or almost all, papers with a map available might use one? To examine the consistency with which news editors use maps, I compared the 12 central New York newspapers by listing by event the 138 nonlocal mapped articles they published during January and July 1985. (I omitted travel articles and star charts.) Because some stories had been illustrated cartographically by more than one paper, there were only 96 separate stories, 25 percent of which had been accorded a map by more than one newspaper. Agreement on mapworthiness was not high: over half (13) of these 24 stories were

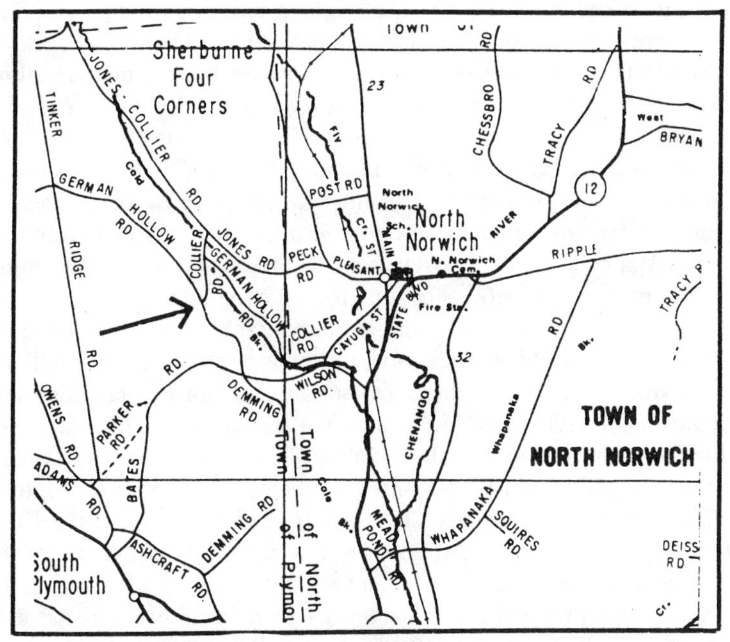

KILLER BROOK?—The State DEC is fining a Plymouth sewage disposal firm for creating water quality problems in Cold Brook, below. Fishermen have complanined a decrease in fish population and an examination of the brook reveals a heavy concentration of apparent pollutants. The area in question is indicated above, by the arrow.

Figure 4.20. Map with title "Killer Brook?" was customized from a public-domain county highway map by the Norwich *Evening Sun*, which added a hand-drawn arrow to show the approximate location of Cold Brook, where discharge from a sewage disposal plant had allegedly killed fish. [Source: Norwich *Evening Sun*, 11 July 1980, p. 1.]

accompanied by a map at only two papers, and another quarter (6) were illustrated cartographically by only three papers.

Agreement was higher for some of the period's more sensational stories. The most frequently mapped story, for which seven newspapers carried maps, reported the crash in Bolivia, on January 2nd, of an Eastern Airlines jetliner. The story was noteworthy because all 29 (later reported as 33) passengers and crew were presumed dead. Because of the remote location, an AP or UPI locator map showing the crash site in the Andes southeast of La Paz was a useful substitute for a photograph of wreckage or victims. Less morbid but no less sensational was the story that broke the following day, about a Soviet cruise missile that a week earlier had gone astray during a test flight, crossed northeastern Norway, and landed in Finland. The missile had not yet been found, and a map was an obvious illustration, at least to the editors of the six newspapers that used a wire-service map to show the proximity of Finland and Norway to the U.S.S.R. and the presumed crash site near Lake Inari. Perhaps the previous day's disaster had made some of these editors more receptive to a map. The third most frequently mapped event was a train wreck that killed an estimated 392 people in Ethiopia, the focus of a locator map carried by five central New York dailies in their editions of January 14th or 15th. Africa was also the locus of the fourth most frequently mapped article—the successful Ugandan army coup reported cartographically by four papers on July 27th or 29th. Again, because of the remoteness of these two African locations, a locator map was a useful substitute for a photograph.

News stories with more limited cartographic consensus included several events involving neither loss of life, armed conflict, nor military threat. Moreover, most of these less sensational stories were domestic, and most addressed economic themes. On January 15th, the *Watertown Daily Times* and, on the 16th, the Syracuse *Herald-Journal* used a Tribune Media Services map of Brazil with a syndicated story on that country's internal politics. On January 20th, the *Herald-Journal* and the Utica *Observer Dispatch* ran a *Chicago Tribune* map showing plans for the presidential inauguration, and on January 26th, the *Herald-Journal*, the Utica *Daily Press*, and the *Watertown Daily Times* used an AP map illustrating the Pope's visit to 17 cities in four South American countries. Two days later the *Herald-Journal* and the *Daily Press* included a map with an AP Newsfeatures story discussing the plight of the aluminum industry in the Pacific Northwest. In July, five domestic stories on economic or regional planning topics were accompanied by maps in two newspapers. On July 2nd, using an AP map, the *Cortland Standard* and the *Watertown Daily Times* reported the redrawing of ward boundaries in Chicago. An NEA map showing nuclear warheads stored in 10 states appeared in the Norwich *Evening Sun* for July 5th and the *Oneida Daily*

Dispatch for July 11th. An Empire Airlines route map accompanied stories in the July 11th Utica *Observer Dispatch* and the July 12th Rome *Daily Sentinel* on the planned relocation of the firm's headquarters to Syracuse. On July 25th, both the Cortland and Rome papers used an AP map with a story about urban growth encroaching on military bases in Arizona. And on July 26th and 28th, respectively, the Syracuse *Herald-Journal* and the Auburn *Citizen* used a locator map to show the heretofore obscure town of Spring Hill, Tennessee, which General Motors had just announced as the lucky recipient of its new Saturn manufacturing plant. That week many other American dailies, including the *Christian Science Monitor*, the *New York Times*, and the *Wall Street Journal*, used a similar simple, single-site locator map. Location, after all, was an important part of the Saturn plant story, and a map was more efficient and interesting than a photograph of auto executives or a disappointed governor.

Other stories accorded maps by two or three central New York newspapers addressed war or disaster. Three papers included maps with stories on a Vietnamese invasion of Cambodia (January 7th or 8th), forest fires in Southern California (July 2nd or 3rd), a train derailment in France (July 8th or 9th), forest fires throughout the western states (July 9th, 10th, or 11th), and the drowning of over 200 persons in a narrow valley in northern Italy when a dam collapsed (July 20th). Only two papers employed maps for a number of other, almost equally sensational stories: the crash of an Air Force plane in Honduras (January 24th), a bomb explosion in an Italian airport (July 2nd), an Amtrak derailment in Idaho (July 8th), an Iraqi missile attack on an oil tanker in the Persian Gulf (July 9th), and a toxic fire in Cedar Rapids, Iowa (July 17th). Most of these were simple locator maps, focusing on an obscure place, possibly substituting for a photograph in some cases but clearly supplementing a photo in others, for example, the Cedar Rapids fire.

A still wider variety of news stories was illustrated cartographically by a single newspaper. Some of these clearly dealt with sensational breaking news, such as the Indian earthquake reported on January 1st by the Syracuse *Post-Standard*, one of the few papers studied publishing that day, and the riots in New Caledonia, reported on January 13th in the Utica *Observer Dispatch*. The *Watertown Daily Times* used maps for a number of less sensational, but still significant breaking stories, including the need for increased police guards around shrines in Jerusalem (January 7th), a bombing in West Germany (January 11th), a kidnapping in Texas (January 14th), plans to withdraw Israeli troops from Lebanon (January 15th), snipers in Jamaica (January 16th), and gang robberies in Nairobi (July 12th). But its regionally unique, strong cartographic coverage also addressed a number of far less sensational, continuing stories, such as the Soviet occupation of Afghanistan (January 3rd), Spain's internal hostili-

Table 4.2 Agreement of the 12 Central New York Newspapers among Themselves and with the *New York Times* on Stories Meriting Cartographic Illustration, January and July, 1985.

			Mapped Articles for Stories Covered with Maps by			
	Mapped Articles		At Least One Other Central New York Daily Newspaper		*New York Times*	
Newspaper	Total	Nonlocal	Number	Percent*	Number	Percent*
Auburn *Citizen* (P.M.)	14	13	6	46	4	31
Cortland Standard (P.M.)	14	12	10	83	7	58
Ithaca Journal (P.M.)	4	4	3	75	3	75
Norwich *Evening Sun* (P.M.)	2	2	1	50	0	0
Oneida Daily Dispatch (P.M.)	3	3	2	67	1	33
Oswego *Palladium-Times* (P.M.)	3	3	1	33	1	33
Rome *Daily Sentinel* (P.M.)	14	10	7	70	4	40
Syracuse						
Post-Standard (A.M.)	37	19	6	32	5	26
Herald-Journal (P.M.)	54	38	8	21	8	21
Utica						
Daily Press (A.M.)	9	6	5	83	3	50
Observer Dispatch (P.M.)	12	10	6	60	6	60
Watertown Daily Times (P.M.)	55	27	12	44	5	19
12-newspaper summary	221	147	67	46	47	32

*Percentages are based on nonlocal maps only. Star charts have been omitted from counts of total and nonlocal maps.

ties (January 16th), salt pollution in California's Central Valley (January 18th), famine in Ethiopia (July 8th), drought in Africa (July 19th), and the plight of women in Africa (July 20th). The Syracuse *Herald-Journal* also provided maps for a number of stories, both breaking and continuing, not covered cartographically by other central New York newspapers. Among the more sensational stories it addressed with maps were allegations of widespread child abuse in Jordan, Minnesota (January 8th), hostages in Lebanon (July 1st), a bombing in Kuwait (July 12th), raids on "cocaine factories" east of Utica (July 12th and 14th), and terrorism in Nepal (July 27th). But, like the Watertown paper, the *Herald-Journal* used maps for a number of comparatively nonsensational stories as well, including reports on New York State's forest industry (January 7th), racial tension in South Africa (January 24th), refugees in Central America (January 29th), international variations in personal income (July 4th), agricultural development in Saudi Arabia (July 13th), the growth of health maintenance organizations (HMOs) in the United States (July 14th), rejection of a stiff antiabortion law by Swiss voters (July 20th), and variations among regions in major industrial closings (July 25th). Clearly many subjects existed that other area newspapers could have treated with maps—far more themes, both sensational and otherwise, than the average small or medium-size daily newspaper seemed equipped to address cartographically. As the Syracuse afternoon and Watertown papers demonstrate, the map-conscious editor has far more opportunities to use a map than the few spectacular front-page stories for which a map might be a photographic stand-in of last resort.

An important determinant of a newspaper's cartographic content seems to be whether the editor likes maps, or at least tolerates them. Table 4.2, which summarizes the agreement on mapworthiness among central New York newspapers, suggests that papers that use more local maps tend also to use more nonlocal maps. Particularly obvious as relatively frequent users of maps for both local and nonlocal stories are the *Watertown Daily Times* and the Syracuse newspapers. As a group, these three papers have the largest circulations and thickest issues in the region, and each has at least one photowire, an art department, and maps from the Chicago Tribune syndicate; they thus might be expected to be more cartographically prolific. But within the group, the *Watertown Daily Times*—smaller than either of the Syracuse dailies and without a Sunday edition in 1985—clearly is a newspaper whose management appreciates maps. In contrast, four of the area newspapers used no local maps during the two months sampled, and these papers also made comparatively little use of maps to illustrate sensational nonlocal stories. This second group includes the Norwich *Evening Sun* and the *Oneida Daily Dispatch*, the two papers with the smallest circulations, and presumably the fewest resources, judging from their use of the NEA graphics service as a substitute for a photowire. Yet this group's other members, in Ithaca and Oswego, serve prosperous communities and receive AP Laserphotos. The other five newspapers, which employed at least one local map during January and July of 1985, also used maps more frequently for nonlocal stories than the four low-map-use papers. This intermediate group includes the three highest percentage rates of cartographic agreement—among the *Cortland Standard*, the Rome *Daily Sentinel*, and the Utica *Daily Press*, all of which rely principally upon the AP photowire for nonlocal maps. Interviews suggest that their editors share a basic appreciation for news maps and a belief that location can be an important element of a news story. At these papers, maps showing distant places tend to address breaking, sensational stories about major disasters or international conflicts. Here the map is preferable for some stories but clearly optional for most.

One final note on the limited consistency among news editors is in order: there is surprisingly little agreement on mapworthiness between the central New York dailies and the elite, and more cartographically rich, *New York Times*. As shown in Table 4.2, the highest degree of overlap with the *New York Times*, 75 percent, was registered by one of the low-map-use papers, the *Ithaca Journal*. Three of the *Journal*'s four nonlocal maps addressed themes also covered cartographically by the *New York Times*: the Bolivian air crash, the errant Soviet missile, and the Italian dam collapse. More puzzling is the second lowest rate of overlap, 17 percent, for the *Watertown Daily Times*, the most cartographically prominent newspaper in the regional sample. Is the Watertown paper running maps for events too esoteric or too wildly sensational for its New York namesake? Not really. If anything, the *Watertown Daily Times* might even be described as the more cartographically comprehensive of the two papers, for the *New York Times* devotes many of its maps to travel stories in its Sunday edition, and during January and July 1985 used maps less frequently for breaking news of disasters or military activities than for reports on topics as diverse as the continuing sagas of Central America and the Middle East, a nuclear fuel plant in France (January 1st), a freak snowstorm in San Antonio, Texas (January 14th), cold-wave damage to the Florida citrus industry (January 23rd and 24th), the reorganization of Western Airlines (July 3rd), warfare in Mozambique (July 9th), a land dispute on a Minnesota Indian reservation (July 15th), and an air safety agreement between the U.S., the U.S.S.R., and Japan (July 31st). Most of the news maps in both papers are simple locator maps, but the *New York Times* often uses its cartography to illustrate—and draw attention to—special reports by its own correspondents. By focusing less on breaking news, its cartography also reflects an audience and a publishing mission different from those of the Watertown paper. The minimal overlap between the *New York Times* and the *Watertown Daily Times* further underscores the myriad news stories with an unfulfilled potential for cartographic portrayal.

Map Consciousness in Journalism Textbooks

Like teachers of most other technical and professional disciplines, teachers of journalism use textbooks as well as lectures and critiques to communicate techniques and values. The significance of textbooks extends well beyond the student cohort, though, for instructors tend to select texts that sustain their own views, and experienced journalists occasionally peruse a text as part of an informal, personal program of continuing education. In the newspaper business, texts would be of particular interest to workers intent on career advancement, as from reporter to editor, and to managers eager for information on new trends, such as newspaper design.

References to maps are rare in textbooks on journalism. Where mentioned, maps usually are dealt with rapidly, as worthy of no more attention than, say, boxed tables containing baseball scores. But a few examples indicate an increased acceptance of the map, albeit only as a minor element in news illustration.

In the early 1900s, when news maps were scarce, textbook authors could easily overlook them. For instance, Edwin Shuman's *Practical Journalism*, published in 1903, provided a concise 265-page overview of the newspaper business without any reference to maps, not even in its eight-page chapter "In the Artist's Room."[64] Later authors were not much more appreciative. George Bastian's 252-page *Editing the Day's News*, published in 1924 with the subtitle *An Introduction to Newspaper Copyediting, Headline Writing, Illustration, Makeup, and General Newspaper Methods*, accorded the map a scant, two-sentence paragraph in a 19-page chapter, "Making the Newspaper's Pages Attractive":

> Maps and diagrams of varying sizes serve a useful part in the scheme of newspaper illustration. Some newspapers make it a rule to use at least one map daily.[65]

Yet Bastian's chapter "Newspaper Illustrations" discussed mostly the selection and cropping of photographs, and failed to mention maps. Moreover, his discussion elsewhere of news values ignored the possibility of illustrating the location or geographic situation of a foreign or local news event, and his description of the duties of the art director mentioned only "photographs, cartoons, etchings, and other art features."[66] John

Floherty's *Your Daily Paper*, an introduction to the newspaper and its production published in 1938 for junior high school students, also reflected the low status of news maps.[67] Floherty, who wrote several other occupation-related works of juvenile nonfiction, discussed the role, selection, and engraving of photographs and cartoons, but ignored maps. The anonymity of the map in this popularized account reflects attitudes in the texts he consulted and the reporters and editors he interviewed.

Ideas commonly gain credence in the professional press well before they appear in textbooks. During the 1930s and 1940s articles in periodicals directed toward the working journalist occasionally cited the usefulness of news maps, especially for reporting the spatial complexities of battles and military strategy. In 1936, for instance, a short article in the *Bulletin* of the American Society of Newspaper Editors quoted a Troy, New York, editor who needed a more convenient method of producing maps for reports on the Spanish civil war.

> The lines of the rebels were being drawn closer and closer around the capital city of Madrid. I should very much like to have a map in my office sufficiently large so that I could draw the line of the rebels and the defending line of the loyalists, to be printed in connection with dispatches from the front. This is only one of such incidents.
>
> It is a cumbersome process to try to reproduce a page out of an atlas and, after all, any ordinary map goes into too much detail.
>
> It ought to be possible to have a morgue of maps in every office.[68]

The same article noted that the Philadelphia *Ledger* developed a collection of news maps by assigning one artist "almost exclusively for the preparation of maps." In the mid-1930s, apparently, syndicate maps distributed by mail arrived too late to satisfy some editors, yet small and medium-size papers were unable to afford either Wirephoto or timely photofacsimile artwork from one of the AP's competitors.

Almost thirty years later the ASNE *Bulletin* again mentioned the relevance of journalistic cartography, this time with a three-page lead article by Theodore M. Bernstein, assistant managing editor of the *New York Times*.[69] Bernstein's title, "Maps: An Important Reader Tool," was followed by the subtitle "Are Maps a Luxury Only Larger Papers Can Afford? Should Editors Be More Map-Conscious?" In addition to touting his own paper's use of maps, Bernstein noted that smaller papers could develop local basemaps to which adhesive symbols and type could rapidly be added, as well as use-and-file-away maps supplied by the AP or UPI news picture services. He concluded by placing part of the blame for a minimal use of news maps upon the design of syndicate and wire-service products:

> I have a feeling that editors are not map-conscious now, partly because the subject of maps has not come to their attention very frequently or with much impact and partly because the maps available to them have not been good enough—they are often "empty" and sometimes not up to the minute. If it were made easier for them to produce what would be at least in part their own maps, they might find this form of news illustration more rewarding. And beyond doubt their readers would be appreciative because they would find the news more intelligible.

Immediately following Bernstein's article, under the title "Sure, We Should Be Using More Maps, But . . .," the *Bulletin* printed responses from four editors at smaller papers. All four supported the use of maps, three defended the quality of wire and syndicate maps, and several were critical of newspaper artists. One noted that, "Unfortunately, it seems the ordinary artist takes a couple of days to draw a street intersection," and another observed that, "Our two artists think we like [maps] too much. They'd much rather spend the time drawing cartoons." All seemed to agree that editors could benefit their readers by becoming more "map-conscious," and one optimistically stated that, "Undoubtedly a concentrated effort to make [desk editors] as conscious of the desirability of maps as well as pictures would bring results."

By the late 1970s journalists writing about journalism continued, albeit intermittently, to press for more and better news maps. In the 1978 *APME Red Book*, for instance, the Photo and Graphics Committee began their annual report with the assertion that "Intelligent use of maps and graphics in newspapers today is no longer a luxury."[70] Placement of "maps" ahead of "graphics" reflected the focus on cartographic illustration in the suggestions provided in the report by art supervisors at the *Chicago Tribune*, the *New York Times*, and the *St. Petersburg Times*, three papers widely regarded for their maps, information graphics in general, and overall layout. Although providing numerous practical hints about sources, design and execution, and the use of AP maps, the report clearly addressed small papers, as reflected by an assertion that "The key to good maps in your newspaper is not a well-funded art department, although that helps considerably. The key is organization, and a resolve to do the job."

Writing in *Nieman Reports* in 1982, Howard Shapiro cited the increased interest in demographic studies among newspaper publishers and business journalists alike as a reason for increased map-consciousness.[71] But his focus on statistical diagrams, as well as his title, "Giving a Graphic Example: The Increased Use of Charts and Maps," accorded the map a secondary position among newspaper art.

Although other writers might have viewed the map merely as a type of information graphic, the author of an article published in 1985 in

Publisher's Auxiliary clearly assigned the map a unique and important role. Its introduction was strong and appreciative:

> Let us now say kind words about maps.
>
> The group of illustrations called expo (expository) art consists of maps, charts, graphs and diagrams. Each is a powerful tool for communication and ornamentation. And the greatest of these is the map.[72]

The article's title, a not-so-subtle pun, was "Map Out Future Use of Expository Art," and its author was the highly regarded Edmund Arnold, who wrote a column on page design, and who was the foremost American newspaper designer during the 1950s and 1960s. This column, which discussed the use of maps to illustrate local news stories, concluded with another bit of word-play: "Arnold's Ancient Axiom: Maps are mapnificant."

Arnold is perhaps best known for his textbooks on newspaper design, writings that have had a wide influence on the look of American newspapers. His first text, *Functional Newspaper Design*, published in 1956, mentioned maps rather briefly, in two paragraphs in a chapter entitled "Pictures."

> *Art*, newspaper parlance for all illustrations, covers more than photographs. Maps, drawings, diagrams, cartoons and headings—called *hand art*—are also included.
>
> Weather maps are regular features in a great many dailies. Long-range predictions, shown in map form, are becoming more popular with weeklies and will become more so when they are easily available at budget prices and on a usable schedule. These usually run on inside pages but move to page one when the weatherman makes the news. War maps are regular front page features when cold wars turn into hot ones and maps are often useful adjuncts as X-marks-the-spot's of everything from murder to train wrecks.[73]

Arnold's short testament to the utility of news maps is particularly significant, given the neglect of maps in two earlier, widely used textbooks on newspaper layout. John Allen's *Newspaper Makeup*, published in 1936, reflected the then-current acceptance of vertical layout with few, if any, illustrations.[74] His chapter "Illustrations" concentrated on photographs and photoengraving, and he mentioned maps only in noting—incorrectly—that Stephen Horgan engraved the first daily newspaper weather maps for the New York *Daily Graphic*.[75] In a 1948 text, *Design and Makeup of the Newspaper*, Albert Sutton abandoned pictureless front pages and included a short discussion of color printing, but ignored the map except to suggest that the daily weather map, like the daily comic strips, might be placed among the classified ads "to brighten the design and to attract readers."[76] Although other educators were unable to look far beyond the daily weather map, Arnold accepted and proclaimed the map's prowess for describing obscure locations and complex military maneuvers. Arnold's enthusiasm for maps increased in subsequent writings. His *Modern Newspaper Design*, published in 1969 as a major revision of his 1956 text, mentioned maps in both a three-page section, "Expository Art," and a separate two-page section, "Maps."[77] The postwar independence of many former colonies in Africa and Asia reinforced his appreciation.

> Maps become more useful by the day. New countries are being created at such a pace that only the professional geographer can be expected to know where they are located. . . . Unless the geography is clear to the reader, he may not be able to comprehend the nuances of the story.

Arnold also noted several reasons for local news maps, most notably a mobile, migratory population and expanding cities and transportation networks. Twelve years later, Arnold was equally supportive in *Designing the Total Newspaper*, with a similar amount of space accorded maps and other expository art. His endorsements stressed explanation and information, rather than mere visual appeal:

> We have long recognized the vital need for maps with international news . . . . Wire editors realize that there are few stories too close to the reader not to make a map useful.

He also advocated maps for local as well as foreign news.

> A map that shows the scene of a gas-main explosion as being just two blocks off the reader's regular way to work brings the event closer and heightens its impact. Almost any story with geographical references will be enhanced by a map.

In praising the *Chicago Tribune*'s use of maps in reporting state news, Arnold noted:

> It's a rare reader who knows even the approximate location of every town in the state, so a *dateline* alone is often inadequate to set the scene.

And he encouraged writers and reporters to draw their own maps with pens, straight-edges, border tape, and press-on symbols.

> Maps, charts and graphs don't require the high skill of an artist. Anyone with fewer than 11 fingers can do a creditable job on any of them.[78]

This assertion is not naively glib, for Arnold had for many years taught a practical graphics arts course for journalists, and also had written a text on graphic arts.[79] In his 1981 newspaper design text, he urged newspapers to prepare outline maps for various parts of the local area, and to keep a supply of duplicates on hand for the rapid addition of appropriate self-adhesive symbols and type. He noted how the style and size of type could differentiate various kinds of features and signify their relative importance. Arnold's guidelines for cartographic design included bar scales for all maps showing an area less than 500 miles wide, and unobtrusive north arrows to show orientation. The scale of the map, in his opinion, was for the editor to specify, not the artist.

In the 1970s and 1980s, texts on newspaper design sustained Arnold's support of news maps. Harold Evans, a distinguished British editor who wrote a five-volume treatise, *Editing and Design*, mentioned maps several times in his fourth volume, *Pictures on a Page*, published in 1978. He noted the value of maps as flavor graphics and as information graphics, and provided several examples of both roles. In addressing the words-only mentality of some journalists, Evans observed that some written accounts were inefficient without a supplementary map:

> The journalist must really recognize that his beloved words are indeed only one way of communicating, that if he wants to tell somebody the way from A to B, it is better to draw a map than to talk.[80]

In a four-page section, "Making the Most of Maps," he indicted small type and graphic clutter as the two most frequent failings of news maps.

In a popular 1981 college text, *Contemporary Newspaper Design*, Mario Garcia also recognized the value of the news map, particularly where the story might need to jump to another page:

> Maps, charts, and instantly recognizable symbols should figure prominently in the graphic designer's strategies . . . [but these] work successfully only when not abused. . . . For example, if a photograph has been used to accompany a news story, perhaps the additional map or chart available to go with it can be eliminated or placed elsewhere—where the story might be continued, for example, or where a sidebar (related story) is carried.[81]

Six years later, though, this cautious treatment of the map had matured into a stronger recognition of the map's contribution to news presentation. In his second edition, published in 1987, Garcia devoted seven pages, accompanied by 12 examples, to a discussion of maps as information graphics. World events, he noted made maps essential:

> In a world that seems to get more complicated and more connected by tensions, the map as a form of informational graphic is no longer a commodity. It has become a necessity that helps to place the reader in the midst of the action being reported.[82]

Another journalism professor, Daryl Moen, included a brief, two-page discussion of maps in his 1984 textbook, *Newspaper Layout and Design*.[83] Moen focused on the *St. Petersburg Times*, the maps of its art director, Frank Peters, and the map-consciousness of its publisher, Nelson Poynter:

> Poynter was also an internationalist who believed that it was the newspaper's fault if foreign stories were foreign to readers. Not content to leave maps to *Time* magazine, Poynter ordered his editors to do whatever was necessary to be sure their readers understood the geography of the world. That commitment has been admired and copied, but not as much as it should be.

Moen also recommended that news art departments develop an archive of reference and base maps—"local, state, national, and international"—and subscribe to the National Geographic News Service.

Despite this strong testimony from prominent journalistic educators specializing in newspaper design, the map has yet to gain the attention of textbook authors addressing news writing and news editing. Inspection of an admittedly unsystematic sample of over 25 such texts published in the 1970s and 1980s found few books with even the barest recognition of how a map can complement a written account. Only three books addressed journalistic cartography: one was a intermediate-level text with a focus on feature writing, and the other two treated news editing. More elementary books, dealing with reporting, generally said little about illustrations of any type, and the two news editing texts included a minimal discussion of maps after a much more detailed examination of news photos.

Louis Alexander's *Beyond the Facts: A Guide to the Art of Feature Writing* viewed the map as a research tool, not as a medium of communication. No maps appeared among the book's illustrations, but toward the end of his chapter entitled "Research," Alexander advised:

> Don't overlook the usefulness of maps. If someone says a street is in the northeast part of town, but in a doubtful voice, look it up for your–self . . . .[84]

His single-paragraph endorsement also mentioned airline route maps, aviation charts, and topographic contour maps. In contrast, a 1935 text,

Modern Feature Writing, by Harry Harrington and Elmo Watson, included three maps, praised the use of maps in *Time*, noted that maps must be tailored for the reader, and demonstrated how a map can clarify geographically complex stories.[85]

More recently, in the third edition of *News Editing*, Bruce Westley also used illustrated examples of expository maps. He viewed location as an important element of some news stories, and mentioned weather maps, maps from the wire services, and locator maps.[86] His attitude was clearly enthusiastic:

> Even more important than maps of faraway places are local and regional maps. Some newspapers routinely use maps of the state to locate small towns unknown to most of their readers. . . It's a minor effort—but a help to the more curious of our readers (and they're the best kind).

The three coauthors of *Electronic Age News Editing* also included several examples, and pointed out two principal uses of news maps: to provide a "frame of reference" and to brighten inside pages, and thereby attract readers.[87] But in both editing texts, the map received less space, and clearly was considered less important, than the news photo. Most textbooks on news editing include a chapter on photographs, yet ignore maps altogether.

Recent trends in newspaper layout and the prominence of crisp, neat, syndicated and locally produced graphics inspired by the microcomputer and laser printer have made editors and publishers more aware than ever of the value of news maps. Although newspapers appreciate maps principally as tools of geographic communication, news maps will continue to be important as page decorations and picture substitutes. In many instances, though, the map must play a dual role of providing visual variety while portraying spatial relationships. As examined in the next chapter, the electronic media demonstrate further the added value to journalism of cartographic synergy fostered by computer graphics.

5 Maps in the Electronic Media

Since the development of the telegraph network in the mid-nineteenth century, the news business has relied heavily upon electronic communications for collecting facts. Development of radio broadcasting in the 1920s demonstrated the further potential of telecommunications for distributing news directly to the public. Newspaper publishers were skeptical at first, and then a bit worried. Some set up their own radio stations, and a few experimented with facsimile broadcast newspapers, received in homes with a printing radio of sort and described later in this chapter. But the biggest assault of the electronic media on the daily paper did not come until after World War II, in the form of television, which offered islands of news in a sea of passive entertainment. Enjoying the benefits of postwar prosperity in the early 1950s, most American families bought small-screen black-and-white TV sets, and spent less time with their daily newspapers. Circulations dropped and advertisers fled—to television in some cases, and to a single, large, reasonably solvent, surviving daily newspaper in others. The threat from television increased in the 1970s, as cable systems brought dozens of channels to their subscribers' homes, and in the 1980s, as programmable video cassette recorders allowed viewers to record TV broadcasts for watching when convenient. Spurred by the threat of a new generation of video viewers displacing an aging generation of newspaper readers, news publishers sought to brighten their daily papers with color, abundant lifestyle features, and short-and-to-the-point news reports. Some also explored yet another electronic medium—*videotex.* Under more intensive development in Europe and Canada than in the United States, videotex had been acclaimed the "newspaper of the future" in the 1970s, although in the mid-1980s its imminent rise, and the consequent demise of the centrally printed daily newspaper, seemed less likely than a decade earlier. Nonetheless, the bigger American newspaper groups hedged their bets and set up videotex affiliates. Although various scenarios seem plausible, and timing is uncertain, electronic communications of some sort appear destined for a fuller, more substantial role in the dissemination of news.

This chapter explores ways in which the electronic media, particularly television broadcasting, have affected map use in news stories.

Special attention is given to the technical constraints video communication imposes on map design and to the widespread use of electronic weather maps in TV news programs. Particularly significant in televised weather maps is the combined use of satellite imagery, computer graphics, color, and dynamic displays—an orchestration that the print media can envy but never fully emulate. Roughly chronological in its organization, the chapter concludes with videotex and begins with the facsimile newspaper of the late 1930s and the postwar 1940s. The facsimile newspaper's early hype, short life, and minimal exploitation of cartography provide a useful background for evaluating the enthusiastic claims made for videotex, and the likely place of news maps in an electronic newspaper.

Facsimile Newspapers and the Almost-Instant News Map

Early use of electronic communications in the news business largely involved transmissions from a single sender to a single receiver, and the possibility of a newspaper image arriving electronically in subscribers' homes was little more than foolish fancy. In the first decade of the twentieth century, when Professor Arthur Korn was forwarding news photos between newspapers in London and Paris, resolution was so poor that only the most technologically optimistic news publisher would have considered transmitting a story's typeset text by electronic facsimile. Nonetheless, the growth of telephone systems suggested the possibility of a single sender at least speaking the news to many listeners. American news publishers, with substantial investments in printing plants and equally substantial revenues from advertisers conditioned to the visual pitch, had little incentive to try such a scheme. Yet by 1895 a telephone newspaper, using its own wires, was operating in Budapest, where the "Telefon-Hirmondo," or "Caller of the News," designed by Hungarian electrician Theodore Puskas, who had worked with Thomas Edison, served 15,000 homes with a staff of 200 and 1,800 km (1,100 mi) of wire. Each receiver had two "ear tubes," and for two cents a day, two members of a household could listen to one of the eight "loud-voiced stentors," who read the news as part of a daily program of news and music that ran from 8 A.M. to 10 P.M. The *Scientific American* correspondent who reported this wire-based forerunner of the radio broadcast predicted similar ventures "before long" in New York, Chicago, Philadelphia, Boston, and San Francisco. Oblivious to the inconvenience of sitting around with a tube in the ear, he thought that news publishers would be persuaded by cost savings alone, for "all kinds of expense [could] be eliminated from the cost of publication, such as paper, ink, typesetting, and a great and expensive staff."[1] Essentially a forerunner of the modern cable system, the "Telefon-Hirmondo" had been in operation for nearly 17 years before a similar service was made available, in 1911, to residents of Newark, New Jersey.[2]

Electronically distributed news never caught on in the United States, until the development of commercial "wireless" radio in the 1920s. Costs were shared by the listeners, who bought their own receivers, and shortly thereafter by advertisers, who supported programs such as "The A & P Gypsies" and "The Palmolive Hour." Free entertainment attracted listeners, and a sizable audience lured advertisers. Accounting for only a small fraction of air time, news became an important element of commercial radio. On November 2, 1920, reporting the results of the Harding-Cox presidential election, Pittsburgh's pioneer radio station KDKA offered the first scheduled news broadcast.[3] By mid-1923, over 300 radio stations were operating in the United States. Sales of radios rose from 100,000 in 1922 to over 2,000,000 in 1925.[4] The first nationwide network, the National Broadcasting Company, formed in 1926, and coast-to-coast broadcasts of music, drama, comedy, news, and sports entered millions of homes nightly.

Newspapers responded first by trying to ignore radio stations, then briefly by trying to beat them, and finally by joining them. Impressed by the huge, comparatively captive audiences attracted to radio, advertisers switched substantial amounts of support from print to the airwaves, and newspaper publishers fought back by dropping listings of radio programs and pressuring the Associated Press and the United Press not to serve radio broadcasters. When this news blockade failed, many publishers were sufficiently impressed or threatened to buy out or set up local radio stations.[5]

Newspapers also used shortwave radio to expedite news gathering. Successful experiments in radio facsimile transmission convinced some publishers that radio might be useful not only in transmitting photographs from the scene to the press room but also in broadcasting readable images of type and graphics to readers' homes. Also interested in the facsimile newspaper were electronics manufacturers, such as the Radio Corporation of America, which in the mid-1930s developed a home facsimile receiver that could be mass-marketed at $100. In 1935, RCA interested four New York City papers in a test, and by the end of 1937 the Federal Communications Commission had approved broadcast facsimile tests in seven cities.[6] On December 7, 1938, the St. Louis *Post-Dispatch* initiated the first regularly published facsimile newspaper, broadcast daily between 2 and 5 P.M., from its experimental ultra-shortwave radio station W9XZY.[7] Received at the homes of 15 newspaper and radio executives, the *Post-Dispatch*'s first Radio Edition consisted of only nine pages, with three or four columns of news printed on one side of paper 22 cm (8.5 in.) wide. It included news, sports, current financial and weather information, editorials, a cartoon, and three picture pages. Typeset pages printed on a proofing press were mounted on the revolving cylinder of a facsimile scanner similar to a Wirephoto transmitter. Equipped with continuous rolls of white paper and carbon paper, the receiver produced a dry image by striking down on the carbon paper. At four cents a day, the cost of receiver paper was slightly less than the price of a full-size daily newspaper, but the transmission rate of 15 minutes per page must have seemed excruciatingly slow to eager readers. On February 1, 1939, the Sacramento *Bee* broadcast its first *Radio Bee* over radio stations in Sacramento and Fresno.[8] During a year-long experiment to measure reader acceptance, fifty persons chosen at random in each city were loaned a receiver for three weeks and asked to record their opinions daily.

Interrupted by World War II, demonstration trials of facsimile newspapers resumed with renewed enthusiasm in the late 1940s. Papers that introduced a facsimile edition included the Chicago *Tribune*, the Philadelphia *Inquirer*, the Miami *Herald*, and the *New York Times*.[9] In mid-1948, the FCC adopted a technical standard for commercial broadcast facsimile—FM transmission at an image resolution of 41 lines per cm (105 lines per in.) on paper 21 cm (8.4 in.) wide—thereby encouraging the mass production of low-cost home receivers.[10] Electronics manufacturers such as Finch Telecommunications and RCA stood to gain more than news publishers, and were from the outset the most enthusiastic supporters of radio facsimile. If a large market could be established, the sale of floor- and table-model facsimile receivers, in wood cabinets like that shown in Figure 5.1, might yield enormous profits.

Belief in the inevitability of the facsimile newspaper escalated through the late 1930s and the postwar 1940s, as the technical problems of static-free transmission and legible, low-cost, moderately rapid printing seemed to have been solved. In 1938, a writer for *The Nation* introduced an essay titled "Next—The Radio Newspaper" by asserting, "There is no longer anything speculative about printing newspapers in the home by radio," and moved quickly to an assessment of the social consequences of radio facsimile.[11] She concluded by warning of both a possible monopoly of this new medium by newspaper publishers and a more vulnerable free press if access to the airwaves was by government license. Addressing fellow journalists and production workers in 1940, the author of a short article published by a regional trade journal asserted in the title that "Facsimile Newspapers Are As Certain As Wire Pictures." The article's conclusion was particularly ominous for backshop workers:

> So peering into the future, it seems that an editor will still be an essential on a newspaper, but with additional abilities as a paperhanger and a draftsman. Operators, pressmen, compositors, circulation men, and extensive printing offices won't be required.[12]

Electronics experts and some scholars were more optimistic. Reporting a variety of technical advances and broadcast trials, a 1938 article in *Electronics* noted the endorsement of a presidential commission:

> The prediction by President Roosevelt's National Resources Committee in their *Technological Trends and Their Social Implications* that home facsimile is one of the thirteen inventions which carry vast potential for changing the economic, social and cultural status of the nation has thus been brought a greater stride toward actual realization.[13]

In the late 1940s trade journals in broadcasting and electronics were equally confident. In 1948, for instance, a story in *Radio and Television News* concluded a description of the *New York Times*'s facsimile experi-

Figure 5.1. RCA table-model radio facsimile receiver printed three facsimile newspaper pages per hour on paper 21.6 cm (8.5 in.) wide. Printed area measured 29.2 by 19.1 cm (11.5 by 7.5 in.). [Source: *Radio Facsimile System* advertising booklet (Camden, N.J.: RCA Manufacturing Co., ca. 1938.]

ment with a confident reference to the new FCC standard for facsimile broadcasting:

> in view of the FCC's green light for commercial faxcasting, [the pages of news and pictures broadcast by the *Times* over WQXR/FM] looked like harbingers of another revolution in radio—publishing via the airwaves.[14]

Later that year, *Newsweek* reported a prediction by RCA president David Sarnoff that in retrospect seems the epitome of optimistic hyperbole:

> Radio delivery of a complete Sunday newspaper, comics and all, to the home in one minute.[15]

As might be expected, the trade press addressing the print media was more cautious and skeptical. For example, *Editor and Publisher*, in 1948 and 1949, respectively, carried articles with the titles "N.Y. Times Starts Fax But Adds to Presses," and "No Immediate Threat Seen in Facsimile."[16] By 1949, when McGraw-Hill published a textbook on the facsimile newspaper, coauthored by the managing editor and the facsimile editor of the Miami *Herald*, most demonstration trials had ended after showing technical progress in image transmission but little commercial potential.[17] Television, also under intensive development during the 1930s and 1940s, had emerged as a commercially feasible medium of mass communications. Offering the public a faster delivery, passive entertainment as well as news, and a more innovative visual experience, television quickly demolished whatever hopes electronic manufacturers and a few news publishers had for commercially successful facsimile newspapers.[18]

Maps had a very small, only marginally significant part in the facsimile newspaper and its promotion. As with traditional, full-size newspapers, maps were less common than halftone photographs—perhaps more so because the smaller page and smaller newshole of the facsimile newspaper left little room for pictures of any type. Several articles discussing the facsimile newspaper carried photos of a typical page, by itself or emerging from a receiver. These sample pages commonly contained a photo, or occasionally a cartoon—but almost never a map. Where the article did mention maps, often the only reference was to weather maps, or maps were but one item in a list that included comic strips and crossword puzzles.[19] The most comprehensive single work on the facsimile newspaper, the textbook published in 1949 by editors at the Miami *Herald*, mentioned maps in only three places, one of which was a general discussion of the wider use of facsimile in meteorology:

> The science of weather forecasting is being improved by up-to-the-minute facsimile maps showing the weather conditions in distant areas.[20]

The *Herald* editors apparently considered weather maps of limited interest, for they mentioned them again only in a discussion of how the facsimile newspaper might be tailored for agricultural areas:

> The weather. Farmers and aviators are the weatherman's best fans. What is the forecast today, tonight, tomorrow, and long range? Are there any frosts moving this way? Use weather maps. Most farmers know how to read them. Run an explanation for those who don't. Do a good detailed job on weather, including [statistics].[21]

Their third reference to maps occurred in a general examination of facsimile technology, outside the context of the facsimile newspaper. In noting uses of facsimile in education, the editors mentioned maps at the end of a list:

> Facsimile can be as valuable in schools as in any other large organization. It can transmit study outlines, assignment sheets, reference lists, tests, pictures, maps, and other material.[22]

Clearly neither the *Herald*'s facsimile editor nor its managing editor was a map-conscious journalist. Their attitudes might have been different had facsimile experiments not been suspended during World War II, when full-size papers often ran maps of an individual battle or an entire theater of war.

Oddly, the two writers most aware of the potential role of journalistic cartography in the facsimile newspaper were electrical engineers—inventors heading small firms that manufactured facsimile equipment. In his paper "Facsimile and Its Future Uses," published in 1941 in the *Annals of the American Academy of Political and Social Science*, facsimile pioneer John Hogan referred to maps in a comparison of television and facsimile:

> Television relates to transient images, whereas facsimile has to do with records, with text, with single maps and pictures which are still and permanent, and not (as in television) transient or fleeting.[23]

Hogan noted the advantage of a hardcopy image for later reference, acknowledged the complementarity of spoken commentary and pictures, and advocated adding sound to facsimile broadcasts:

> In broadcasting, facsimile can stand on its own feet as a purveyor of magazine features and of news. If co-ordinated with sound, however, its possibilities may be even greater. Maps and illustrations would add to the value of many sound programs . . . .
>
> In the field of adult education . . . how impressive geography would be if the text were delivered vocally by a good speaker and the maps presented by facsimile.[24]

Hogan's map-consciousness was reflected in a postwar demonstration of equipment at a meeting of the American Newspaper Publishers Association, as recorded by the journal *Electrical Engineering*:

> The wide diversity of copy presented at the Hogan demonstration ranged from news pictures, matter to accompany household programs such as recipes and sample menus, weather maps, comments on musical numbers and pictures and life stories of composers, cartoons, crossword puzzles, instruction texts and reference reading for educational programs, and special bulletin newspapers, made up in half pages tabloid size.
>
> . . . During the usual 15-minute broadcast, four pages of illustrated reference copy can be transmitted by frequency modulation radio and reproduced in the home, thus permitting a 15-minute newscaster on an amplitude modulation station, for example, to accompany his spoken word on an associated frequency modulation station with 4,000 words of printed text or four pages of maps, diagrams, and photographs.[25]

Milton Alden, whose Alden Products Company provided facsimile equipment for early experiments at the *Milwaukee Journal*, shared some of Hogan's map-consciousness as well as his advocacy of adding sound to facsimile:

> Illustrated sound broadcasting . . . can create in the home innumerable program adjuncts, from pictures of star performers or public figures, to sporting events or news shots and maps, to say nothing of illustrating products and packages and furnishing coupons or contest blanks.[26]

Perhaps the spatial ability required for designing electrical circuits made these two electronics entrepreneurs more aware of the map's role in news reporting than most of the journalists their inventions were intended to serve.

Maps might have attained greater prominence in the facsimile newspaper had this medium ever become commercially practicable. The wire services, no doubt, would have supplied maps, photos, and other news graphics to local facsimile papers, and national facsimile networks, similar to the major radio networks of the day, might have formed to furnish other local facsimile broadcasters with a complete package of national and foreign news and nationwide advertising. Keen competition would have encouraged facsimile publishers to use maps to locate unfamiliar sites of disasters, show battle lines in Korea, and dramatize feature stories on the threats of the cold war. Centrally produced cartography might then have reached readers' homes without having to pass through a local gatekeeper who could then print some wire-service illustrations and ignore others. Perhaps ahead of its time, the facsimile newspaper and its subtle promise for a more vigorous journalistic cartography wilted rapidly toward the end of the 1940s, when news publishers recognized that facsimile might at its best be merely a costly supplement to the full-size, full-length newspaper.[27]

TV News and Video Graphics

In its early years, ironically, television held less promise for the map than did the facsimile newspaper.[28] Portraying motion better than text, television was better able to entertain than to inform. As had been true with radio news, television news broadcasts were hardly significant until the national networks enrolled a sufficient number of affiliates to encourage and support regular news broadcasts. NBC originated the first network broadcast, carried by stations in New York City and Schenectady, New York, on February 1, 1940, but World War II delayed regular network programming until 1945, when NBC linked stations in New York, Philadelphia, and Schenectady. Two other radio networks, ABC and CBS, expanded to television in the late 1940s, and the market for network programming grew during the 1950s as the proportion of American households with TV increased from 7 percent in 1950 to 37 percent in 1952, and to 82 percent by 1957.[29] News programs became important because they not only added variety to the broadcast schedule but also delivered some prime-time viewers who, having switched on the TV for the evening news, were reluctant to turn off the receiver or change channels. Yet in a medium dominated by plays, quiz shows, situation comedies, and variety shows, the news broadcast clearly was a minor element, accounting for 7 percent of network evening programming in 1950, and only 5 percent later in the decade.[30]

Early TV newscasts consisted largely of "talking heads"—a reporter reading news bulletins or interviewing prominent politicians—but maps were frequently employed as visual aids, particularly for war news. Occasionally film from battlefronts and political campaign trips was added, but because of the time required to transport and process motion-picture film, the result was a broadcast newsreel just slightly more current than those shown in movie theaters. Because the map offered directors an opportunity for visual variety, a second camera sometimes was used to point out an obscure location or to illustrate the advance or retreat of military forces. The value of the map as a visual aid was demonstrated as early as December 1941, when CBS's experimental New York station WCBW reported the Japanese attack on Pearl Harbor. As noted in *Television News Reporting*, a text published by CBS in 1958,

> Although WCBW was only 160 days old in that December 7, [newscaster Richard] Hubbell and [writer Gilbert] Skedgell had been producing two fifteen-minute news programs a day, five days a week, and they had learned to make the most of what they had. It wasn't much. But maps were their forte, and they had learned something of how to make news visual.
>
> During that Pearl Harbor telecast, Hubbell showed on maps the location of islands like Wake and Midway, and pointed out the possible lines of attack against the Philippines and Singapore. The viewer saw the positions, at least as they were known on that day, of United States Pacific Fleet units.
>
> The program, through diagrams, arrows, and other symbols, defined news in terms of the visual. Expert analyses, again with maps as visual aids, were offered by Major George Fielding Eliot and Fletcher Pratt, while Linton Wells reported the fast-breaking political developments.[31]

The CBS textbook offered several other reasons to use maps: as a substitute for film, as impressive newsroom decor, and as a signal to the viewer of a change in locale.[32] But in a chapter on still pictures, the CBS writers clearly demonstrated an appreciation of the map as information, rather than mere decoration:

> Very few of us are walking geographies. More often than not a story gains meaning and interest if we are shown *where* it is taking place. Maps can be used to set the geographic scene for film, by first coming up on the map, then dissolving to film. A map showing Cyprus in relation to Greece and Turkey makes the tug of war over the island clearer. Talk of the Gaza Strip means little until a map shows it up for what it is—a no-man's land between Israel and Egypt.[33]

Two other texts from the 1950s and 1960s offered further testimony to the value of the map in TV journalism. *Radio and Television News*, published in 1954, included a chapter in which a Milwaukee news director included the map in a list of 15 items for which he compared radio, TV, and the newspaper:

> [Television] can present maps, charts, graphs, etc., to aid in the understanding of news, but these aids on television lack the permanency of the printed form. . . [whereas newspapers] can present . . . maps . . . for readers to study and digest at their leisure.[34]

More laudatory was Irving Fang, who worked at ABC News before joining the School of Journalism at the University of Minnesota. In the first edition of *Television News*, published in 1968, Fang noted the value of artists able to draw maps:

> A higher degree of drawing skill is required here. However, a newsman using a little care can draw a serviceable map or graph. When a map is needed, nothing else will do so well, and television newsrooms need maps often. A good world atlas and some state, county and city maps are a worthwhile investment for any television art department.[35]

Fang presented three examples, of maps illustrating stories on an auto accident, a market holdup, and an air crash, and demonstrated how "pull tabs" could be used to add motion to a map with white symbols on a black background.[36] The map in this case consisted of three layers: an upper layer with basemap features in white on transparent film, a removable black middle layer, and a bottom layer with selected symbols and type on a black card. As shown in Figure 5.2, the slow, steady removal of the pull tab in the middle revealed the information on the bottom layer as if an invisible hand were tracing, say, the path of a new Interstate highway or an ill-fated airliner. Television artists recognized early their medium's potential for dynamic maps.

Including a map in a newscast was quite simple: a studio camera could focus on a map drawn or pasted on a card, or a 35mm slide showing a map could be inserted in a slide projector connected to camera designed especially for still photographs. Rear-screen projection could show the map on the wall behind the newscaster or in a frame to one side, so that both map and reporter might appear on the screen simultaneously. Around 1970 television stations began to add still and motion pictures to a background by an electronic technique called *chroma keying*, a by-product of color television.[37] A news reader might sit or stand in front of a green wall. The director could then use the chroma keyer to replace the green portions of the scene with the corresponding parts of an image from a second video camera, say, connected to a slide projector. Of course, the newscaster must be careful not to wear clothing of the same shade of green as the background, and the artwork must be designed so that important features on the map do not appear hidden behind the reporter.

Television is a very confining medium for maps. The screen's aspect ratio—four units wide to three units high—might be ideal for presenting comparatively wide regions, such as Pennsylvania or the United States, but is ill-suited to the efficient portrayal of areas with a pronounced north-south trend, for example, Idaho or Chile. Moreover, geometric distortion at the edge of the screen, possibly aggravated by poor

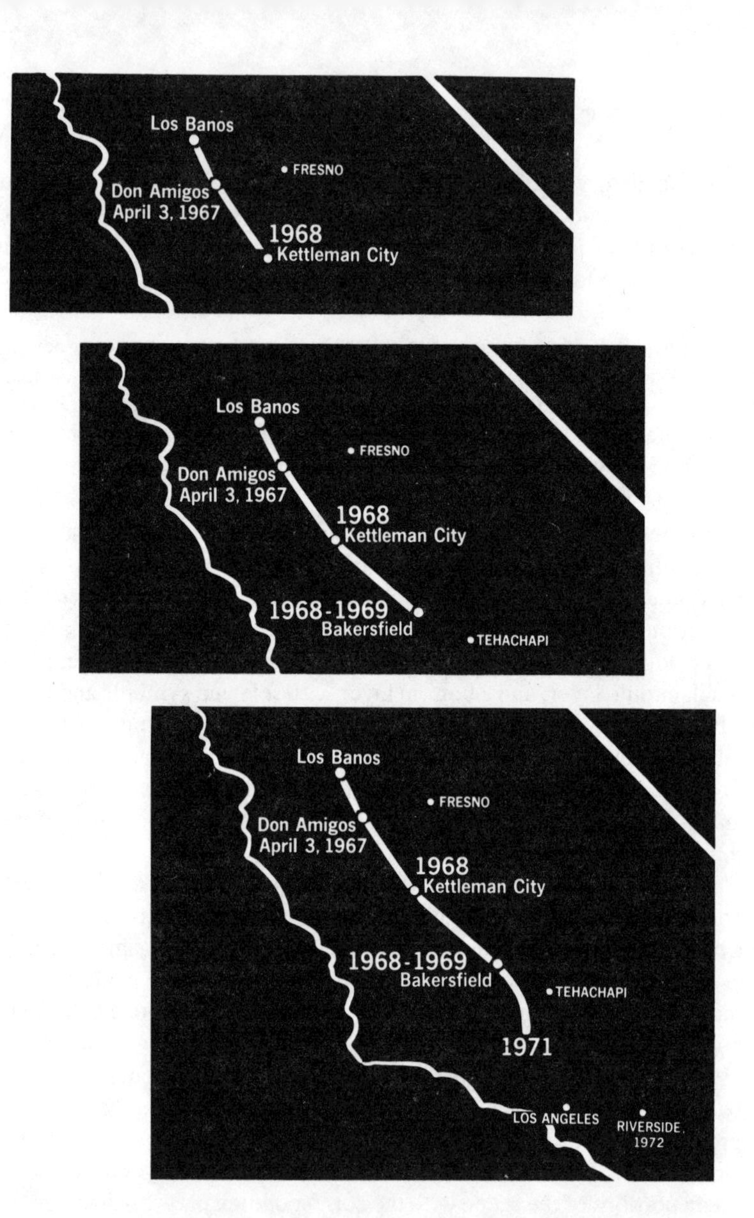

Figure 5.2. Pull-tab map showing highway development in California consists of three layers: (a) an outer layer of black cardboard containing the coastline, city names and symbols, and state border in white; (b) a middle, movable layer of black cardboard, and (c) an inner layer of white cardboard. The route alignment and numerals of the year-dates in this example have been cut out of the outer black layer. Pulling the middle, black pull-tab layer downward reveals the route and dates as white-on-black symbols so that viewers see the progressive southward extension of the highway to Kettleman City in 1968 (top), to Bakersfield in 1968-69 (center), and then to mid-way between Bakersfield and Los Angeles in 1971 (bottom). (Source: I. E. Fang, *Television News* (New York: Hastings House, 1968), 200-201.)

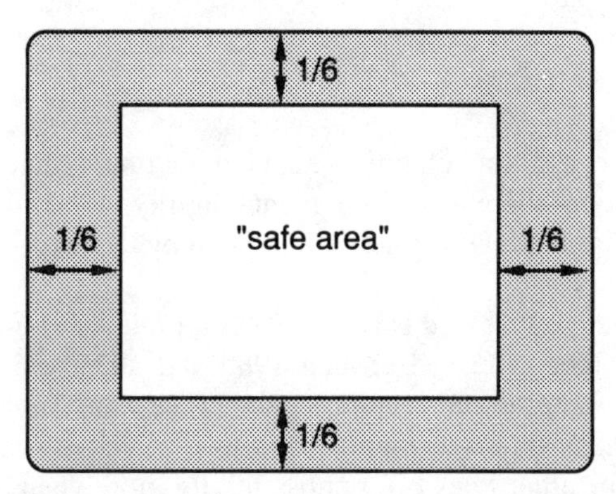

Figure 5.3. Television cameras scan an area four units wide for every three units tall, an aspect ratio designed to accommodate black-and-white motion pictures, ca. 1940. These proportions are slightly different from the 3:2 aspect ratio of 35 mm color slides. The "safe area," likely to appear entirely and without peripheral distortion on any screen or monitor, omits 1/6 the height at both top and bottom and 1/6 the width at both left and right.

adjustment of the viewer's receiver or a deteriorating picture tube, renders much of the televised picture useless or chancy for maps and their exterior labels. Patricia Caldwell, a cartographer who studied television news maps, agreed with other video designers that peripheral distortion reduces the "effective area" to the middle two-thirds of the screen's width and height—less than half the area scanned by the TV camera, as shown in Figure 5.3.[38] Type and symbols are constrained further by the low resolution of the televised image, which in North America consists of 525 horizontal scan lines lines, regardless of screen size.[39] Caldwell noted that, for clarity, television maps require larger type than other maps, and that place names are likely to cover a larger share of the image.[40] Printed maps commonly use the visual contrast between italic and Roman or between serif and sans-serif typefaces to differentiate features by their labels, but few viewers could detect such subtle differences on a televised map. Moreover, patterned area symbols tend to be coarser, and line symbols simpler, than on newspaper maps.

Even simple television maps require time for the viewer to recognize the image as a map, identify the region, and read labels. As Irving Fang noted in his text *Television News*,

> The first rule for drawing a television news map is: *keep it simple*. The second rule is: *hold it on camera longer than a photo would be held*, giving viewers plenty of time to read place names, understand the relative distances and absorb all the information even a simply drawn map conveys. A map should contain not one place name, not one highway line, not one

> mark more than is absolutely necessary to locate the spot where whatever happened happened. . . .[41]

Yet in practice, as Caldwell observed, viewing time is determined not by the complexity of the map and the mental agility of the audience, but by the time required to read the news illustrated by the map—13.5 seconds, on the average.[42]

In the late 1960s and early 1970s color television further complicated the designer's task. Between 1965 and 1975, the proportion of American households with a color receiver rose from 9 to 66 percent, but many black-and-white sets were still in use.[43] Symbols and type in red, green, blue, or other hues offer substantially more visual contrast than their graytone counterparts—but only for viewers with color sets. For black-and-white broadcasts, Caldwell noted two severe constraints: reliable contrast could be guaranteed only among black, white, and three to five intermediate shades of gray, and pure white should be employed only for small areas, and never for more than 20 percent of the screen.[44] Moreover, a televised white is less reflective than a moderately well-illuminated sheet of paper, and largely because of glare from ambient lighting, a televised black is perceptibly less dark than a symbol printed in solid black. Designers attempting to accommodate both color and monochrome receivers thus must avoid color combinations that appear similar on black-and-white receivers. Medium blue, lime green, medium red, and magenta, for instance, are a similar medium gray on monochrome receivers, and light blue and light red yield virtually indistinguishable light grays. Yet because of the greater aesthetic freedom of designing principally for viewers of color TV, who now represent a solid majority of the TV audience, designers of television graphics have drifted away from earlier, and often half-hearted, attempts to serve the shrinking but ever-present minority of viewers with black-and-white receivers or impaired color vision.[45] Textbooks on television production provide only mild warnings that contrasts based on hue alone will fail many viewers.[46]

Understanding the design challenge of video maps requires an appreciation of the electronic principles of color television display. The color video signal consists of a single brightness value and separate *chrominance* values for each of the three *additive primary* colors—red, green, and blue.[47] Chrominance measures the saturation of a hue—its position between a pure hue and a gray. These colors are called additive because adding together red, green, and blue light in various strengths can generate most other hues, and they are called primary because no combination of any two of them can yield the third. Combining red light with green light, for instance, yields yellow, adding red to blue produces magenta, and adding all three primaries creates white light. Using only the brightness component of the video signal, a black-and-white receiver varies the intensity of its single cathode ray beam, which scans the screen, from top to bottom, 30 times each second.[48] A color receiver converts the overall brightness value and the three chrominance values into separate brightness values for red, green, and blue. As shown in Figure 5.4, separate electron beams scan the screen for each primary color, but each cathode ray is partly blocked by a shadow mask that allows it to strike only selected phosphor dots on the screen, inside the picture tube.[49] The dots are arranged on the screen in triads, and each triad is represented by one of the half million holes in the mask. At a single instant during the scan, separate red, green, and blue beams are directed by a lens through the appropriate hole to their respective dots in the triad. When struck by their designated electron beam, these dots emit light of the corresponding hue—red, green, or blue—with a brightness controlled by the strength of the incident stream of electrons. These dots are tiny; each scan line holds about 1,000 dots for each color, yielding roughly 1.5 million dots overall. Hence, instead of perceiving a pattern of red, green, and blue dots, the viewer sees the composite hue represented by the mixture of these additive primary colors, in varying strengths.

As most owners of older color televisions will attest, not to mention many owners of newer models, colors on the screen frequently fail at representing hues in front of the camera, and adjusting the set often merely replaces a sickly green fleshtone with a prickly red. Moreover, picture-tube manufacturers recognize no single set of standards for color phosphors.[50] In their efforts to contend with the distorting effects upon color of signal interference and improperly tuned or poorly designed receivers, designers of television maps find little guidance in established principles of cartographic design. Before television, color was added to maps by applying individual "flat-tint" inks, one for each color, or by overprinting screened combinations of the *subtractive primary* colors—yellow, magenta, and cyan. Cartographic textbooks offering advice for offset-lithographic reproduction have largely ignored such considerations as the tendency of televised pastels to fade if pale and to intensify if bright.[51] Caldwell noted a number of potential conflicts between color television and traditional, print-oriented cartographic practice, including an insufficient contrast on video displays between aesthetic pastels preferred for color maps, and a tendency among television artists to achieve contrast by ignoring traditional associations of blue with water and green with vegetation.[52] She candidly recognized that traditional aesthetic values must often be sacrificed to promote a rapid differentiation of mapped features.

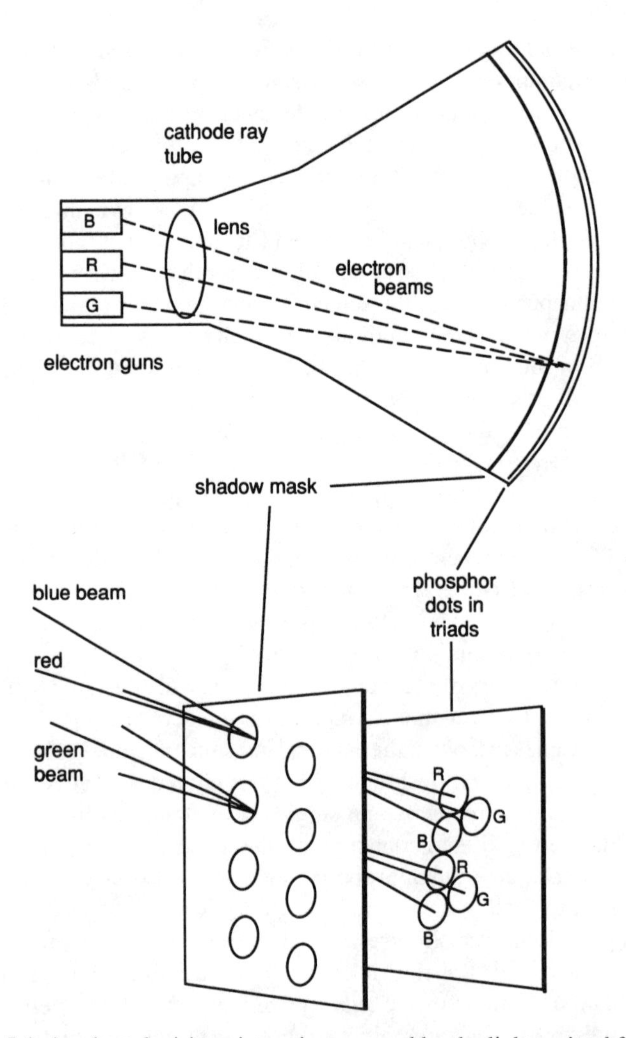

Figure 5.4. A color television picture is generated by the light emitted from about 1.5 million phosphor dots, divided equally among triads emitting red, green, and blue light. A separate cathode ray scans the screen for each color. A shadow mask with 0.5 million tiny holes prevents the cathode rays from striking the wrong phosphor dots. Lines show how the mask directs the three electron beams to appropriate dots in each triad.

That thoughtful graphic design can compensate for many of the pitfalls of televised color is apparent from the text *Professional Video Graphic Design*, published in 1986 by Ben Blank, managing director of ABC Broadcast Graphics and the acknowledged "Father of Television Graphics," and Mario Garcia, the map-conscious newspaper designer and text author mentioned in Chapter 4.[53] No doubt reflecting the high cost of color printing, their publisher's decision to print in black-and-white has not substantially impaired these authors' presentation of design principles. Maps comprise a fifth of their more than 200 example frames, and are divided about equally among title frames in which the map is a major element and stand-alone maps used to illustrate location, movement, or a regional trend. As Figure 5.5 shows, a title frame can combine the well-known outline of a country or state with type, a photo, pictorial illustration, or a simple statistical graphic to set the tone of a story in which place provides an important visual cue. As Figure 5.6 demonstrates, stand-alone television maps more important for their spatial information than their decorative or iconic value can illustrate the location of a country or

Figure 5.5. Title frame with map shows the location of Amsterdam, New York, where a New York Thruway bridge collapsed in April 1987. [Source: Courtesy of ABC TV Network Broadcast Graphics.]

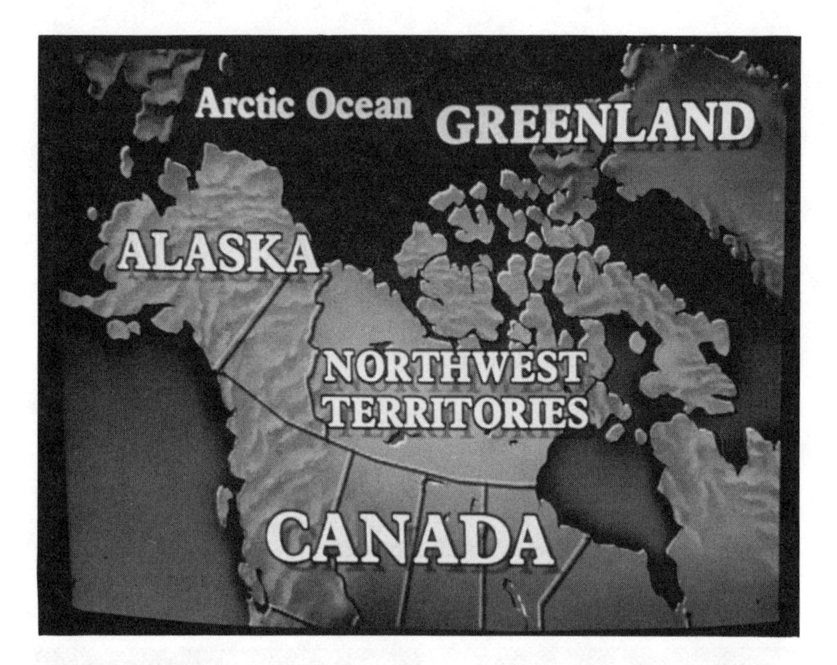

Figure 5.6. Stand-alone map shows Canada's Northwest Territories, between Alaska and Greenland. [Source: Courtesy of ABC TV Network Broadcast Graphics.]

region within a continent, a city within a country or state, a specific site within a city, or a route across a continent. Although color can improve contrast and enhance visual appeal, creative map symbols and three-dimensional effects are equally useful for designing video maps.

In addition to providing numerous examples of television maps throughout their book, Blank and Garcia also included "The Design and Use of Maps as Informational Graphics" as one of their nine chapters. This chapter's lead paragraphs reflect a clear appreciation of the unique contribution of the map to television news, as well as to general education and print journalism.

> Maps bring the viewer closer to the geographical specifics of a story, providing an opportunity to expand the viewer's perspective of an event in terms of its *location*, the *surrounding areas*, and *how it relates to the viewer's situation*.
>
> Maps have always held a special fascination for people. A person's first encounter with maps is in an early elementary-school textbook. Maps also appear in newspapers and magazines, and have made their contribution to highlighting television news.[54]

Blank and Garcia identified five essential properties of a useful television map: accuracy; legibility; geographic perspective; visual impact; and when necessary, topographic detail. Accuracy is the most important characteristic, although their criteria for an accurate television map consist of little more than spelling place names correctly, showing surrounding areas to provide a locational frame of reference, and portraying whatever aspects of geographic situation are pertinent to the story. Because television journalism places severe limits on viewing time and image resolution, the designer must omit largely peripheral details that might be acceptable in a book, an atlas, or even a newspaper. For legibility, labels must be larger than for a newspaper map. A geographic perspective such as an oblique three-dimensional view, in which a country or state rises above, and partly obscures, its neighbors, sometimes provides the required visual impact. Topographic illustration, which might be necessary for portraying a mountain pass or volcano, can provide a visual focus. When television news has spotlighted a place for many days, the designer might assume the viewer is familiar with the geographic situation, and can avoid monotony by invoking a "limbo map" similar to Figure 5.7. A regular viewing audience (if it is a valid assumption that

Figure 5.7. "Limbo map" of Australia floats on the screen in front of the Australian flag. Country outline and national flag provide mutually reinforcing geographic cues. Shaded relief highlights mountains along Australia's east coast. [Source: Courtesy of G & G Designs/Communications.]

such an audience exists) encourages the evolutionary generalization of an "accurate map" into a geographic icon. Brief viewing time also encourages overgeneralization.

A frequent criticism of television news maps is the short viewing time allowed. For instance, Britain's Geographical Association, which in November 1984 monitored television news programs as part of its Media Map Watch, noted that prerecorded and edited magazine-format news programs allowed viewers more time to examine maps than daily news broadcasts.[55] Longer viewing times for maps used in these special programs seemed related partly to the use of animated maps to illustrate routes of travel. Animated maps reflect a conscious decision to invest a more-than-average portion of the program's budget in an interesting display requiring a more-than-average share of the viewer's attention. Particularly appalling to these British geographers was the short airtime allocated to weather maps, which are among the most visually complex, information-rich television graphics. Viewing time seemed more adequate for grossly overgeneralized maps, such as a dot representing Warsaw at the center of a circle representing Poland. Misuse of maps in television journalism, as in print journalism, might reflect cartographic insensitivity not only among media designers but also among those responsible for format and packaging.

Even though geographers might deem 13 seconds inadequate for reading any decent map, however simple, highly ephemeral television news maps might well have an important cumulative effect, particularly when viewers encounter similar images several days in a row. After all, the human mind can absorb pictures and text unconsciously through repeated brief exposures, as demonstrated by experiments in the subliminal perception of advertisements. Indeed, the ever-so-subtle effectiveness of fleeting electronic images has raised ethical questions so serious that federal regulators were at one time asked to ban the advertising industry from using this form of "subliminal seduction."[56] Citing Marshall McLuhan's well-known theorizing about the far-reaching effects of passive television viewing, Patricia Caldwell suggested that during the 1970s television news maps were for millions of people the prime stimulus for developing a "mental map" of Vietnam and southeast Asia.[57] As Caldwell also noted, TV news maps and maplike symbols—at least in the Los Angeles area during the late-1970s—appeared not only to outnumber newspaper maps in frequency but also to reach a much greater audience.[58] Although their effect on geographic awareness and knowledge might be difficult to assess, television news maps appear to make an important contribution to the informal education of millions of people throughout the world.

For years the Associated Press has promoted map use by providing local television-station clients with slides of simple maps for rear-projection or chroma-key display.[59] G & G Designs/Communications, a southern California firm that designs studio sets for television stations, also provides maps such as Figure 5.7, in both photographic and electronic formats. Thus even a small-scale station might have professionally designed frames for locally produced late-evening and mid-day newscasts covering local, national, and world news. Network news departments, of course, each have an art staff to provide logos, titles, and news graphics. More recently, computer graphics systems have enabled both the networks and the larger local stations to expand their use of maps and other graphics. Ben Blank and Mario Garcia listed ten computer systems useful for television graphics, showed several example frames, and noted that in addition to having the overriding advantage of speed, these systems are particularly useful for displaying maps in a wide variety of perspectives.[60] Use of computer graphics for video animation is less common than the use of "electronic paintbox" systems to produce the single-frame maps, graphs, logos, and other illustrations displayed on an electronic frame, or viewport, to the side of the news anchor's head, as in Figure 5.8. When

Figure 5.8. Maps and other graphics appear in the framed electronic window added to the screen above the news anchor's left shoulder. [Source: Courtesy of ABC TV Network Broadcast Graphics.]

video pictures are unavailable for a foreign story or economic report meriting more than a three-sentence summary, these colorful graphics provide a visually interesting complement to the newscaster's "talking head".[61] Between 1981 and 1985 the number of firms offering television graphics systems more than doubled from 6 to 13, as new and existing firms recognized the industry's need for graphics when live pictures could not be shot.[62] As a result, in the years ahead television news maps are likely not only to be more numerous, more dynamic, and more complex but also to have appropriately longer viewing times.

Electronic Weather Graphics for TV News

Television broadcasters have been particularly enthusiastic in adopting computer graphics for weather reporting. Weather affects us all, and a local station might gain an important competitive edge by finding a more dramatic way to display weather data. Indeed, competition among local stations striving for a greater share of the viewing audience explains much of the increased use of computer graphics in TV weather reports. Not only are weathercasters an important part of the local TV news team, but 70 percent of the TV news audience is highly interested in the weather.[63] Because small improvements in audience share represent substantial increases in fees charged advertisers, even small stations seem willing to invest $20,000 to $50,000 in a computer graphics system, and to pay $3,000 a month to a weather data service.[64] These expenditures reflect viewer expectations higher than in television's early decades, when often a news director sought to boost ratings by hiring an attractive young woman whose principal qualification as a weather reporter was a spectacular bustline. Another ploy was the weather clown: dressed in fur coat or bathing trunks to suit the occasion, the station's utility announcer tried to enliven an otherwise dull weather report by squirting shaving cream onto the weather map if snow was forecast, or dousing the map with a hose when showers were anticipated. Although many stations still rely heavily on a "weatherperson" more distinguished for rapport with the audience than knowledge of meteorology, other stations emphasize technical sophistication by combining spectacular graphics with lucid explanations of past and present conditions and the forecast. Some weathercasters, in addition to being personable and articulate, have earned not only a college degree in meteorology but also a "Seal of Approval" for having completed the American Meteorological Society's training program for television meteorologists.[65] In regions where agriculture or outdoor recreation are highly important, many viewers rely upon weathercasters who use National Weather Service data to make their own forecasts for the local area.

These TV meteorologists can use electronic weather graphics most effectively, and their stations point proudly to their weather graphics systems as further evidence of a commitment to local journalism.

Today's electronic weather graphics are markedly more sophisticated than the hand-drawn weather charts of the 1950s and 1960s. Perhaps the most widely used weather map of that era was the preprinted outline map, about three feet by four feet to fill the screen during close-up shots, and showing state boundaries and sometimes terrain or major cities.[66] A weathercaster standing in front of the map would describe the forecast while using a bold marking pen to draw fronts, pressure cells, air flow, and temperatures. Artistic ability was an asset, particularly if the reporter could embellish the map with a few cartoon symbols—a snowman, an umbrella, a smiling sun, or a dismal cloud, as appropriate. More elaborate was the map of state boundaries on a transparent sheet of Plexiglas, separating camera and weathercaster.[67] So that the camera would record a right-reading image of the map, the reporter would draw (with an erasable marker) on the wrong-reading side of the map, with the east coast on the left and the west coast on the right. Ability to write the mirror images of words such as "warm," "cold", and "rain" was a useful skill. Other types of weather maps used before computer graphics included the simple erasable chalkboard with permanent state boundaries in white paint; the magnetic map board onto which the weathercaster could easily attach iron-backed arrows, rain clouds, suns, and fronts (Figure 5.9); the rear-projection map prepared specially for each day's telecast and used with an illuminated-arrow pointer; the map prepared before the show with marking pen on a large sheet of white paper, on which the reporter could refer to important features with a finger or pointing stick; the regional map with blocks containing electronically controlled numbers positioned to represent outlying towns for which volunteer observers reported temperatures; and the satellite photo received on an AP or UPI photofacsimile printer and mounted on an artboard in front of a camera.[68] Electronics sometimes made the weathercaster's job easier, as when a modified camera scanned a Plexiglas map backwards so that the reporter could write normally on the right-reading side. Engineers discovered how to enhance visual impact by placing a rotating polarity filter directly in front of the camera's lens to make specially treated symbols on the magnetic map board appear to pulsate.[69] Electronics added yet another type of map in the late 1950s when a few stations acquired a radar system that displayed a crude plot of current regional precipitation on a cathode ray tube in the studio.[70]

Computer graphics systems provide weather-conscious local stations with spectacular animated displays using images collected with their own Plan Position Indicator (PPI) radar systems. In 1978, for instance,

Minneapolis station KSTP-TV began recording radar images at 30- to 60-second intervals, and enhancing them with a radar colorizing unit, which added boundary lines and two-letter abbreviations to indicate cities and six colors to indicate variations in storm intensity.[71] The resulting time-lapse, animated video sequence, which presented viewers with a dramatic account of storms moving through the region, required a 50-m (150-ft) tower supporting a radar dish 2.6 m (8 ft) in diameter having a range as great as 450 km (280 mi). Yet such a substantial investment was fully reasonable for a station with its own staff of meteorologists serving a major metropolitan region in an important agricultural state. In the early 1980s, when computer graphics began to receive enormous benefits from

Figure 5.9. TV weather map with a blackboard for cartoon-like local forecast and magnetic symbols representing temperature, sky conditions, fronts, and pressure cells. [Source: Courtesy of G & G Designs/Communications, and Station KCST-TV, San Diego, Calif.]

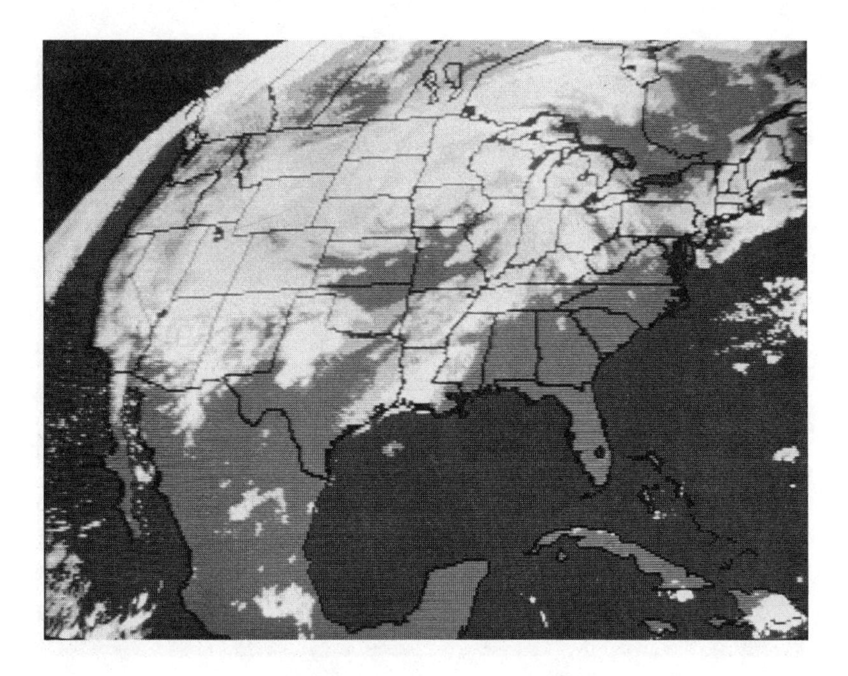

Figure 5.10. Satellite weather graphic showing cloud cover over North America. [Source: Courtesy of Alden Electronics, Inc., Westboro, Mass.]

powerful graphics chips and low-cost minicomputers, many other television broadcasters acquired not only their own radar systems coupled to colorizing units but also weather graphics systems that permitted the rapid display of satellite maps, forecast maps, temperature-band maps, jet stream maps, and local temperature maps.[72] Figure 5.10 illustrates an enhanced satellite cloud-cover image with map boundaries added to provide a geographic frame of reference, and Figure 5.11 shows a weather radar picture with computer-generated type and symbols. Television weather cartography is often both rich and dazzling—with as many as 18 different maps in a three-minute weather segment.[73]

Although some stations collect their own local radar imagery and many others have their own networks of temperatures observers, the National Weather Service collects most weather data used to make maps for TV weather reports. Since NASA launched the first weather satellite TIROS-1 in 1960, satellite data have played an important role in weather forecasting and storm tracking.[74] Geostationary satellites positioned 36,000 km (22,000 mi) above the same point on the equator collect imagery every half hour to provide spectacular earth-from-space time-lapse images of cloud cover. The public can receive weather maps and satellite photos at

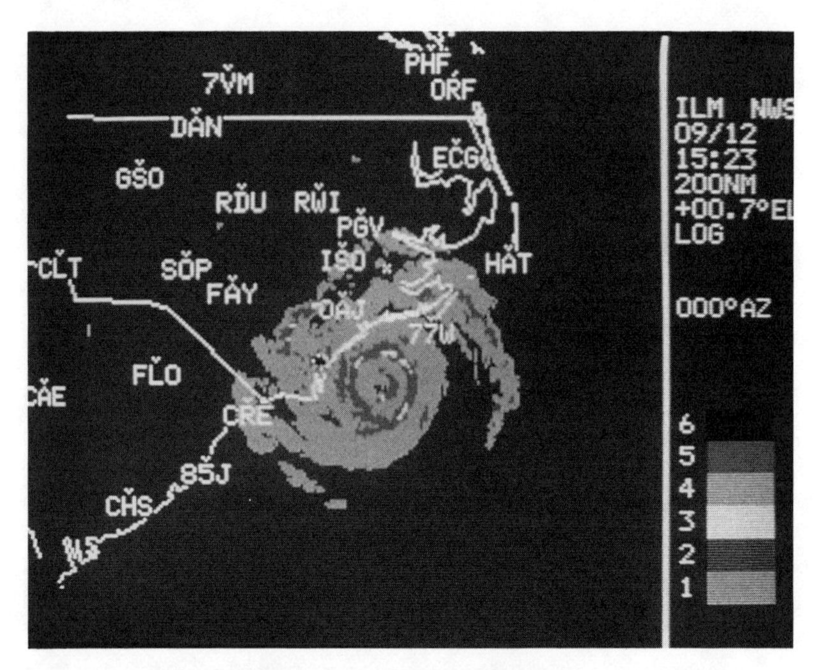

Figure 5.11. Color weather radar picture of Hurricane Diane, 12 September 1985, as seen on National Weather Service weather radar transmitter at Wilmington, North Carolina. [Source: Courtesy of Alden Electronics, Inc., Westboro, Mass.]

frequent intervals by monitoring NOAA shortwave facsimile broadcasts or by dialing a National Weather Service computer at any of 37 stations throughout the country and recording sound-encoded facsimile images.[75] A heavy user of computer graphics at its several forecasting and research centers, the National Weather Service has supported the development of several sophisticated image-processing systems tailored to its own needs.[76]

Although readily available to any citizen, National Weather Service data are not designed for direct use by weathercasters, and most television stations prefer to use weather forecasts and image maps developed from government data by private meteorological services. Commercial firms such as Weather Services Corporation, of Bedford, Massachusetts, and Accu-Weather, of State College, Pennsylvania, also serve radio broadcasters, newspapers, and nonmedia business clients requiring specialized forecasts.[77] Their television clients receive not only local forecasts but a variety of tailor-made television maps ready for use with minimal effort by station personnel. In many ways, these meteorological services play a gatekeeping role similar to that of the AP and UPI, which deliver generalized weather maps to hundreds of local newspapers. Like independent graphics networks that have formed recently to serve newspaper clients, several meteorological services use telephone lines and modems to deliver map images. WSI Corporation (founded as Weather Scan International, and later called Weather Services International) has operated a minicomputer-based dial-up network since 1977 and now uses high-speed, 2,400-bits-per-second modems to transmit satellite images and TV weather maps.[78] Colorgraphics and several other vendors have designed their weather graphics systems to receive these high-speed telephone transmissions, which the local station can easily modify and integrate into its news program.

The spread of cable television and specialized news-and-entertainment channels led to initiation in 1982 of "The Weather Channel," a national televised weather service now widely available to cable subscribers throughout the United States.[79] Meteorologists and graphics artists at the Weather Channel generate numerous weather maps for their 24-hour, seven-day broadcasts using a computer system developed by WSI Corporation. As Figure 5.12 illustrates, its recreational and other specialized forecast maps often address specific regions of the country. Founded by

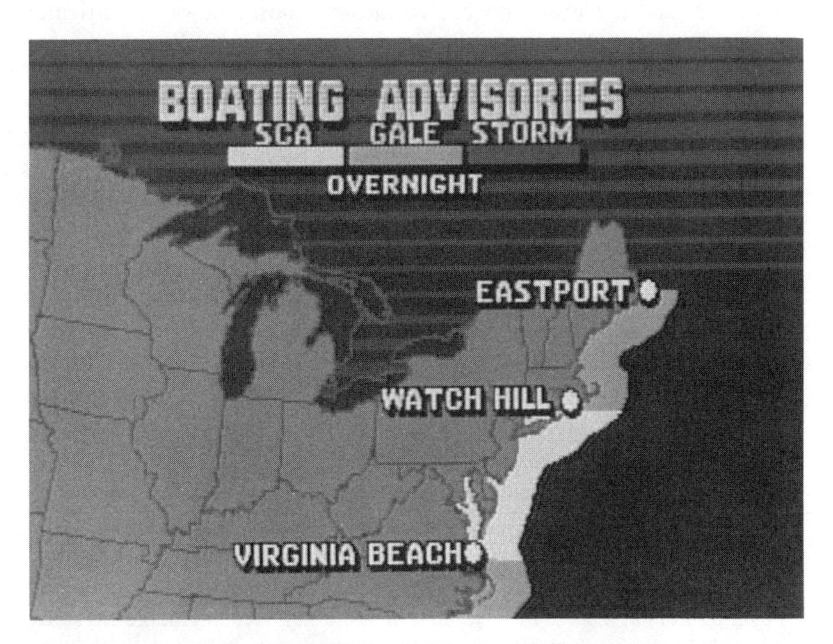

Figure 5.12. Regional recreation forecast weather graphic produced with a computer graphics system. [Source: Courtesy of the Weather Channel, a division of Landmark Communications, Inc.]

TV weathercaster John Coleman, who formerly reported the weather on "Good Morning America," on the ABC network, the Weather Channel is a further example of how advances in both computers and telecommunications have expanded the number, variety, and information content of news maps.

Videotex and Dynamic News Maps

Videotex is a type of interactive video information system developed in the 1970s for use with television receivers but expanded in the 1980s to include personal computers.[80] Like the facsimile newspaper of the 1930s and 1940s, videotex is a text-oriented electronic medium, with individual frames referred to as "pages." And like the facsimile newspaper, videotex has attracted the attention and, in some cases, the active involvement of newspaper publishers. More efficient than its stillborn older cousin, videotex has a much greater chance of becoming a mass medium: the cost of converting a television set to receive videotex would be well under $100 if a large number of subscribers could be recruited. Videotex is faster than the facsimile newspaper as well as interactive, rather than programmed, and the user can receive information or news of particular interest immediately and directly, instead of waiting while the receiver prints preceding pages. Moreover, some forms of videotex take advantage of the TV cable and telephone networks to offer subscribers a wide variety of services, such as banking, shopping, and advice on housekeeping and cooking.

Like the conventional newspaper, a videotex news system would use maps for both spatial description and visual impact. Yet, although some videotex systems can deliver colorful, timely maps and data graphics to complement their words and numbers, others are blatantly text-only systems. Because high-resolution images impose heavy demands upon display-communications systems designed principally for the delivery and depiction of electronically coded text, fairly simple maps, such as the weather map in Figure 5.13, are among the more complex graphics most videotex systems have been able to handle. Indeed, until videotex technology can efficiently integrate conventional television pictures with text and graphics frames, videotex publishers wanting to offer visual variety must rely principally upon maps, statistical diagrams, and simple flavor art.

Although many Americans might not know the word "videotex," they are likely at least to have seen, if not actually used, a videotex system. Figure 5.14 shows an electronic information system similar to those in use at thousands of shopping malls, hotels, and other public places. These systems present a *menu*, or index page, of choices on a video screen, and

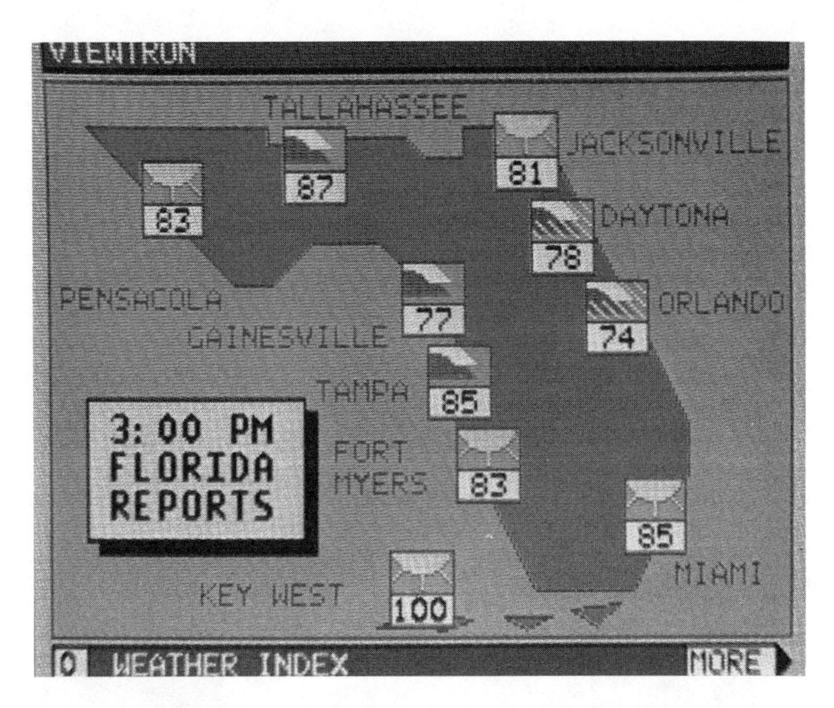

Figure 5.13. Simple videotex weather map used by Viewtron to show temperatures and sky conditions at major Florida cities. [Source: Courtesy of Knight-Ridder Newspapers, Inc.]

accept selections through a keypad, keyboard, or touch-sensitive screen. A *tree structure* similar to that in Figure 5.15 is the conceptual basis for the more primitive versions of videotex: the *root* of the tree is a master index page with several branches each headed by a more specialized menu. From these second-level index pages the user selects a specific information page or series of pages—or if not sufficiently interested in the content offered in the submenu, returns to the main menu. A videotex system covering a wide variety of topics might have third- or fourth-level index pages, or might present the master menu or individual submenus as a series of pages. When examining a series of information or index pages, the user progresses at his or her own speed to the next page, returns to the previous page, or jumps back to the submenu or master menu. When the user is finished with a single page or has completed a series, the system returns to the previous menu.

Early videotex operations were one-way systems that broadcast the entire set of 100 or more index and information pages in a sequence repeated every 30 seconds or so. A number indentifying its position in the

Figure 5.14. AT&T Ariel videotex system displays menu and information screens, including simple maps, in a medium-resolution color CRT positioned above a keypad, used to enter requests. The Ariel terminal was designed for a variety of public locations, to serve such uses as a tenant directory, hotel entertainment guide, and retail product and service guide. [Source: Courtesy of AT&T.].

tree accompanied each page so that, as Figure 5.15 shows, submenu index page 1.0 pointed to information pages 11.0, 12.0, and 13.0—as choices 1, 2, and 3—and the decimal portion of the number referred to the pages in a three-page sequence, such as 12.0, 12.1, and 12.2.[81] The decoder used to convert a standard television set to a videotex terminal included a *framegrabber* that examined the identifying number of each page and extracted from the broadcast cycle the page selected by the viewer. Videotex pages shared the signal of a conventional television broadcast, which could accommodate additional information in the *vertical blanking interval* used to separate successive video pictures. In North America, for instance, the video signal actually uses only 504 of the 525 lines in each television frame, and in western Europe, where the screen resolution is higher, the televised picture occupies only 600 of the 625 lines broadcast. Viewers sometimes "see" these VBI lines in the form of a thick black bar when the receiver is out of vertical synchronization. Although several VBI lines carry horizontal and synchronization pulses, others are available for data transmission.[82] Broadcast videotex typically can use two to four lines near the beginning of each field, or a total of four to eight lines for the pair of fields interlaced to form a single frame, so that a single videotex page is embedded in the VBI signal of several successive frames.[83] By adopting a standard page with, say, 24 rows of 40 characters each, videotex engineers have been able to pack far more than four to eight lines of picture into the VBI: the microprocessor used to adapt a television receiver for videotex reception not only grabs the appropriate data lines from the broadcast videotex cycle but also interprets these data and generates the corresponding characters in their required positions on the page. Packing allowed Ceefax, a one-way videotex service operating in Britain in the mid-1970s, to broadcast its 100 pages at a rate of four pages per second, so that a viewer need wait no longer than 25 seconds for a response.[84] An entire channel dedicated to videotex would increase the

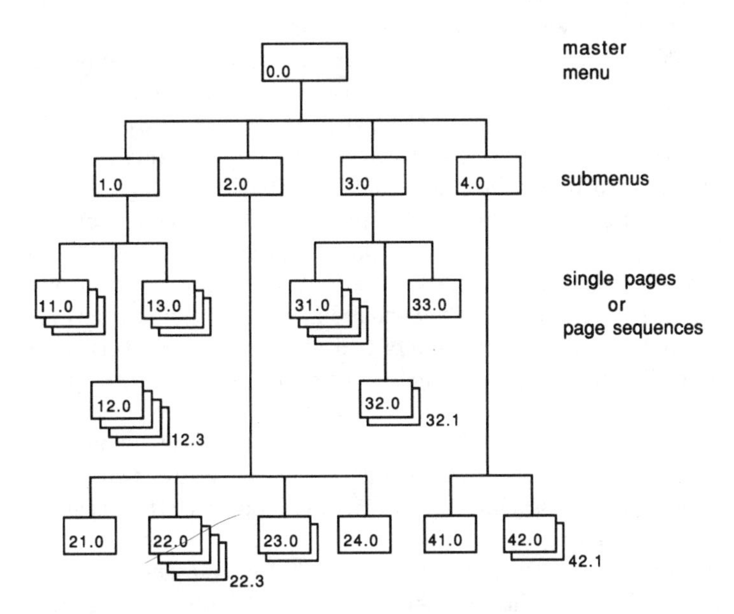

Figure 5.15. Tree structure of a simple videotex information system.

page-transmission rate and expand greatly the number of pages in the broadcast cycle. A demand-based cycle, with index and other high-use pages broadcast more frequently and low-use pages broadcast less frequently, would permit an even larger repertoire without increasing the average wait.

Truly interactive videotex requires two-way telecommunications, most readily available through the telephone network. Some cities have television cable systems that permit two-way communications, and electronic data transmission networks designed for business communications allow two-way message traffic. Two-way videotex can archive a vast number of pages in a computer database and can monitor demand for individual pages. Owners of home computers equipped with a modem—to convert the computer's digital signals to acoustic signals—can use the telephone network to connect to any of a vast variety of "on-line" database systems offering bibliographic, financial, legal, medical, news, travel, and general reference information. Videotex, which must be both easy to use and visually attractive for mass-market appeal, demands more of the telephone system than other on-line information services, which usually have less elegant, black-and-white displays and often respond to queries in the terse, economical codes of the stock exchange. Designed for the continual, more uniform flow of voice communications rather than the intensive, sporadic "bursts" of data communications, telephone service has a low data rate and can be expensive to use because charges are based on connect time, not information delivered.

More efficient for videotex service is two-way compatible television cable, developed originally for pay-television, fire and burglar alarm systems, and opinion polling.[85] Coaxial TV cable systems carrying signals in the band of frequencies between 5 and 500 MHz might allocate a portion of their bandwidth, say, from 5 to 40 MHz, to short "upstream" communications from individual subscribers to an emergency service dispatcher, a pay-TV channel, a survey-research firm, or a videotex service.[86] Broadband two-way cable television systems are at the heart of futurist's predictions of an "information age" in which citizens have a fuller control over the operation of their government and workers "telecommute" to the office from their "electronic cottages."[87]

Compact coding of video frames not only can compensate for the otherwise slow delivery of high-resolution pictures by telephone but also can increase the number of videotex customers served by a two-way cable system. On-line information systems use the simple but efficient approach of coding each unique letter, number, and punctuation mark as a unique string of *bits*, or *bi*nary digi*ts*, such as 1001000 to represent the upper-case *H*. The receiving terminal microprocessor includes a *character generator*,

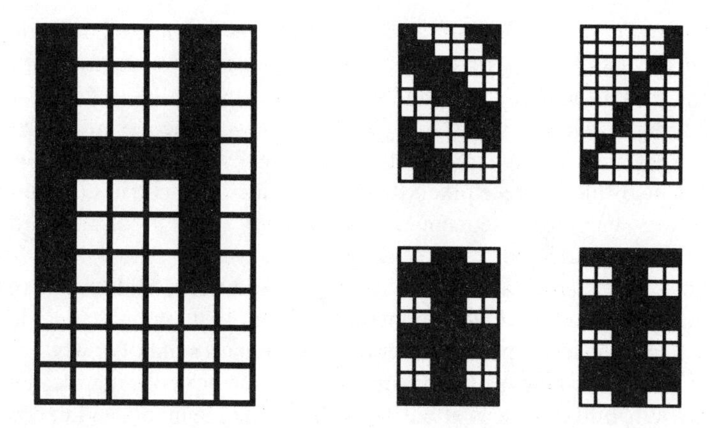

Figure 5.16. Representation of the letter *H* (left) and a few graphics patterns (right) in an alphamosaic pattern with a mosaic cell having 10 rows and 6 columns. Blank cells below the *H* allow for descenders on lowercase letters and space between lines, and blank cells to the right of the *H* provide space preceding the next character on the same line.

which displays the corresponding character on the screen by darkening appropriate cells within a subgrid of, say, 7 rows by 5 columns, as the left side of Figure 5.16 shows. To separate adjoining characters and lines as well as to allow sufficient space for the lower-case *j*, *p*, and other letters with descenders, each alphanumeric character in this example actually occupies a 60-cell subgrid of 10 rows by 6 columns. To transmit each character as a separate uncoded image would thus require 60 bits—instead of a mere 7—for black-and-white and several times as many for color. *Alphamosaic coding* extends this strategy by adding a few non-alphanumeric symbols, such as those in the right half of Figure 5.16. These primitive mosaic patterns can serve as building blocks for simple graphics.

More versatile, though, is *alphageometric coding*, which uses *picture description instructions* (PDIs) to achieve high screen resolution with limited data.[88] Some PDIs are "graphic primitives"—a point symbol described by its diameter and coordinates, a single- or multiple-segment line described by its width and a list of points, a polygon described by the width of its perimeter and a list of vertex coordinates, a rectangle described by the thickness of its perimeter and the coordinates of opposite corners, or an elliptical arc specified by its orientation and ellipticity and the coordinates of its center and endpoints. Other PDIs specify the color, pattern, or texture of an entire shape, its perimeter, or its interior. The sequence of PDIs is important, for an alphageometric decoder paints a picture in the screen by first filling the display memory with the color code of the background and then adding whatever texture—say, a grid—the artist might specify. Figure 5.17 illustrates how a videotex microprocessor might draw a simple picture using a sequence of PDIs, each of which fills in many more screen pixels than those coinciding with its coordinates. Working in a layered sequence, from the back of the picture toward the front, the decoder paints the background sky, and covers it with the ground, a sun, a tree, a flagpole, and a flag. Additional PDIs might signal a pause in this additive development of the drawing—to give the viewer time to note absorb a particular feature—or make a symbol or word blink.

A cartographically promising aspect of videotex is the map designer's increased ability to focus attention with blinking symbols and to control the sequence in which the viewer reads the map.[89] For example, in displaying a choropleth map showing the variation of county-unit arrest rates throughout a state, a videotex system might first write the map's full title over the background—to define the distribution portrayed and allow an early abort should the viewer deem another distribution more appropriate. The screen might next show the state outline and internal county boundaries, and follow these with the map key, showing the number of categories and the range of values within each class. To direct attention

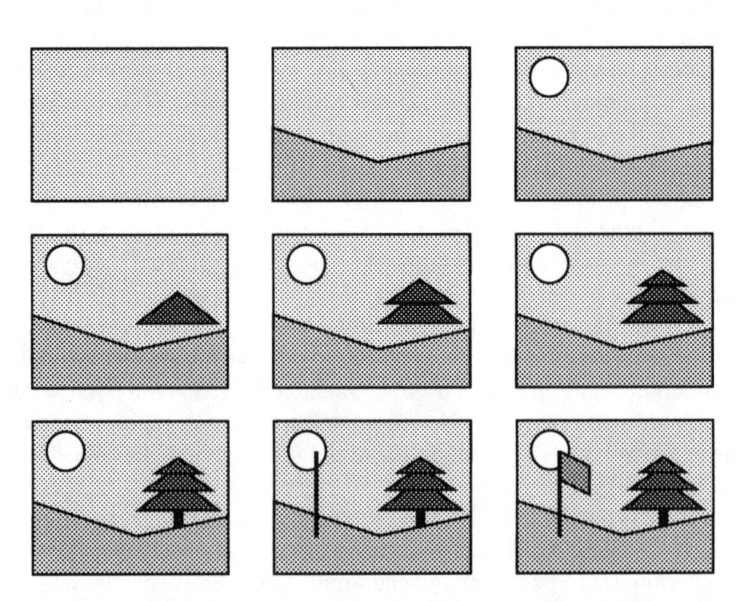

Figure 5.17. Sequential development of a simple picture represented by picture description instructions (PDIs) used in alphageometric videotex coding.

toward high-crime areas—or those with efficient police forces—the system would paint in the symbols for the highest class, and then jump to the lowest class to examine the other extreme. Portraying intermediate categories last, this cartographic videotex program might pause between categories and include flashing text to identify particular places or highlight relevant facts. More sophisticated decoders might well give future videotex cartographers a fuller use of animation.

No single PDI coding scheme and decoder design is an obvious, inherently superior, "natural" standard, yet for videotex to gain wide acceptance as a medium of mass communications requires the establishment of a single standard within a nation, if not throughout a continent. A common set of specifications ensures compatibility among videotex publishers and their clients, promotes competition and lowers unit prices by providing guidance for a variety of electronics manufacturers, and allays apprehensions about rapid obsolescence. Indeed, without a national standard, broadcast television would not have achieved its nearly total penetration of the American home entertainment market.

Although the United States adopted a wait-and-see attitude on videotex standards and lagged behind through the 1970s and 1980s, other technologically advanced nations not only developed official or de facto standards but often provided substantial government support for research and development, partly in hope of exporting potentially profitable videotex technology to other countries, including the United States.[90] Britain was the acknowledged leader: in 1976 the BBC and the Independent TeleVision network (ITV) introduced one-way broadcast television services called, respectively, Ceefax and Oracle, and in 1979 the British Post Office (at the time operating the country's telephone system) initiated a two-way telephone-request videotex service called Prestel, which could display information on a television receiver or a personal computer.[91] In 1982 France's state-run telecommunications monopoly, the PTT, introduced a two-way-telephone videotex service, called Minitel, after its low-cost computer terminal, which included a keyboard, a CRT monitor, and a telephone handset for voice communications.[92] Based on a videotex standard called Antiope, the Minitel service provided telephone directory information and served as a gateway to over 2,000 information and home-shopping services. Ceefax, Oracle, Prestel, Antiope, and other videotex standards used in Western Europe are alphamosaic systems—efficient for delivering text but poor for displaying graphics.

Asian electronics manufacturers could envy the sparse western alphabet—or rise to the graphics challenge of several thousand intricate text characters. In the 1970s several Japanese firms experimented with videotex systems designed to accommodate the more than 2,000 pictographs in the graphically complex Japanese alphabet.[93] In 1978, for instance, the Nippon Hoso Kyokai television network introduced the experimental Character Information Broadcast Station (CIBS), which transmitted in the VBI but treated the screen as a medium-resolution matrix of 250 lines, each with 332 dots. In 1979 the Nippon Telephone and Telegraph Company introduced a trial service, the Character and Pattern Telephone Access Network System (CAPTAIN), which was similar to Prestel but combined sound with visual videotex images.

Particularly concerned with computer-generated graphics and the need for transmitting presentable graphics of at least medium resolution, Canada promoted its own standard, called Telidon, through the government Department of Communications.[94] Canada's goal was not a specific terminal or converter but a non-patentable *communications protocol*—a standard for the efficient communication of high-quality graphics for videotex or other facsimile applications. Drawing largely upon Telidon but incorporating elements from Prestel, Antiope, and an experimental standard developed by AT&T, Canadian communications experts developed the North American Presentation Level Protocol Syntax (NAPLPS), which allows for the flexible representation of videotex pages in three levels—a crude but economical alphamosaic code, a graphically superior alphageometric code based on PDIs, and a highly demanding high-resolution photographic code with separate specifications for each screen pixel.[95] In 1981 Grassroots, a two-way-telephone videotex system serving agricultural businesses in Manitoba, became Canada's first commercial Telidon service. Operated by Infomart, a videotex publishing enterprise formed by the large Canadian media conglomerates Southam Inc. and Torstar Inc., Grassroots expanded to several other provinces and added banking services to its menu.[96] Among Grassroots services are regional weather maps such as Figure 5.18.

Despite a comparatively slow and less certain start in the United States, videotex's many promising trials and few demonstrated successes in Europe have made American newspaper publishers both wary of and interested in this new word-oriented mass-communications technology. Indeed, the prospect of commercially successful videotex contains both the threat of obsolescence or new competition and the promise of reduced costs or new business. Among the principal threats are a possible loss of control of the news publisher's own distribution network—by the late 1970s cable television served most American metropolitan areas, and the possible electronic delivery of print over these government-regulated local telecommunications monopolies raised a variety of First Amendment issues.[97] Moreover, by providing local customers with direct access to national newspapers, videotex might lower a local newspaper publisher's

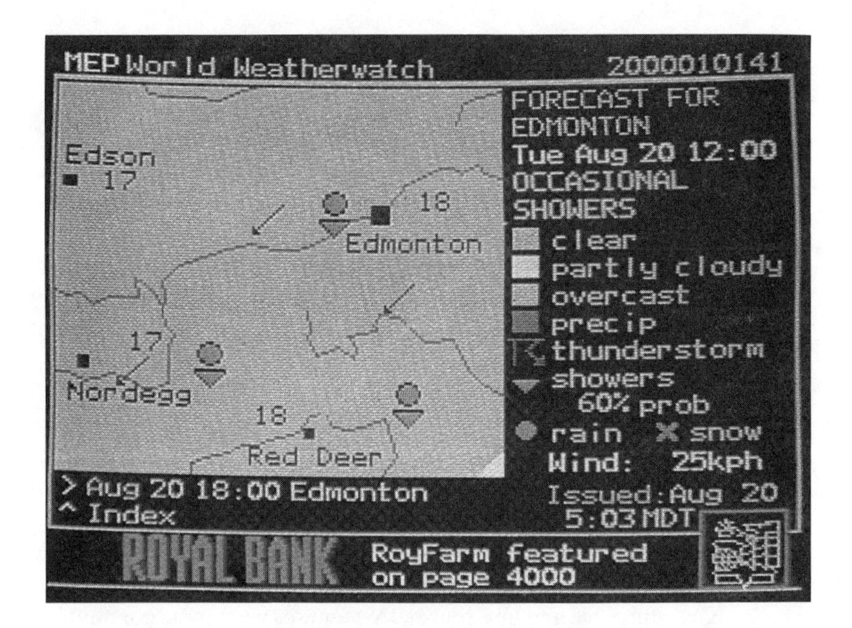

Figure 5.18. Videotex weather map regenerated from Telidon picture description instructions portrays wind direction, temperature, sky conditions, and precipitation for the Edmonton, Alberta, area. [Source: Courtesy of Grassroots Information Services Ltd.]

circulation in the same way that cable systems are reducing the local television broadcaster's market share by "importing" not only movie, sports, and other special channels but also the signals of distant independent stations. Kathleen Criner, director of telecommunications affairs at the American Newspaper Publishers Association (ANPA), observed succinctly that "yesterday's colleagues [might become] today's competitors as new technologies erode traditional geographic market boundaries."[98] For news publishers eager to innovate and move aggressively into new areas and new media, videotex holds the promise of not only returning a significantly greater profit from the electronic editorial database that newspapers prepare daily to feed their electronic typesetting systems but also reducing the costs and annoyances of a distribution system plagued by high-cost newsprint, unreliable child labor, and disruptive strikes. One scenario called for tailor-made newspapers printed in the subscriber's home on a low-cost ink-jet printer—an updated, computerized version of the facsimile newspaper of the 1930s and 1940s.[99]

Particularly significant as both a reflection of concern about videotex within the newspaper industry and a stimulus to greater interest and experimentation among news publishers was communications scholar Anthony Smith's provocatively titled book *Goodbye Gutenberg*, published in 1980.[100] Smith considered a videotex newspaper of some form the natural and inevitable result not only of technological progress but also of a trend from "massified" information systems such as radio, television, and the traditional newspaper to more individualized, customized information services. His masterful synthesis recognized the additional potential of videotex as an electronic library housing fiction as well as nonfiction, and compared the challenge of radio, television, and videotex to the traditional newspaper with the challenge of the automobile and aviation to the railroad. In Smith's analogy, as the train has changed and survived, so will the newspaper.

In the early 1980s many news publishers experimented with electronic delivery through one- or two-way videotex, cable television, or low-power short-range broadcast television. A 1983 survey by the ANPA and the Newspaper Advertising Bureau suggested that almost half of U.S. and Canadian daily newspapers might have been experimenting with these new media.[101] Involvement was greatest among large-circulation newspapers, and cable and low-power TV ventures were more common—and presumably more immediately profitable—than videotex enterprises. Cable delivery was a low-cost option attractive to a number of small newspapers, which merely rented a channel on the local cable system and broadcast news headlines, brief stories and features, and advertising through a video camera trained on a continuous roll of copy or a set of self-changing flip charts.[102] Larger markets could support more sophisticated technology, such as the computer page-generation and page-storage system on which the *Milwaukee Journal* and *Sentinel* produced in color a 20-minute cycle of 60 information pages, available 24 hours a day in four major daily editions.[103] Although the 20-second "dwell rate" might have irritated fast readers, this noninteractive cable newspaper required no special adapter or decoder, and carried news and advertising to 27,000 suburban cable subscribers. In contrast, interactive videotex—which required a decoder but lacked a large, ready audience—attracted more promises than practitioners despite a number of widely publicized affiliation agreements between local newspapers and such national videotex syndicates as that organized by the Times-Mirror Company to develop computer systems and distribute national advertising.[104] Local general-purpose videotex systems proved far less successful than more specialized nationwide information services, such as the Dow Jones News Retrieval Service, which had attracted over 55,000 paying customers by 1983.[105]

Most significant among the videotex trials of the early 1980s was Knight-Ridder Newspapers' Viewtron service, initiated in 1980 in south

Florida, home of the group's Miami *Herald*.[106] The Viewtron trial was a serious attempt to introduce an American audience to European commercial videotex: the startup cost alone was $36 million, and its staff once numbered 210. Delivered by telephone to TV sets equipped with an adapter, Viewtron attracted no more than 3,000 local subscribers—well below its first-year objective of 5,000—despite a series of reductions in the monthly fee from $40 to $25 to $15. Expanding in its final months to a nationwide service available to personal computers with modems, Viewtron increased its subscribers to 20,000. But in early 1986, after investing a reported $50 million, Knight-Ridder announced the end of Viewtron as president James Batten observed, "Videotex is not likely to be a threat to newspapers in the foreseeable future."[107]

Whatever threat might exist, the newspaper industry is both vigilant and defensive. Acknowledging a wait-and-see attitude among news publishers, ANPA's Kathleen Criner, in a mid-decade report on newspapers and telecommunications, observed that "the stakes are high, and the game has just begun."[108] In addition to watching the course of state-subsidized videotex in England and France, American communications companies and newspaper groups have invested in advertiser-supported public access videotex (PAV) systems that provide information and electronic shopping services in airports, hotels, and shopping malls.[109] Although the unsuccessful attempts of Knight-Ridder and Times-Mirror videotex systems to overcome the high cost of terminals has dampened interest in graphics-oriented videotex, at least temporarily, information vendors are still enthusiastic about less sophisticated "text only" services.[110] Continued uncertainty, if not an immediate threat, inspires traditional journalists to argue strenuously—and rightly—that the strength of the newspaper is in its content, and that journalism must be independent of its medium of delivery.[111] Concerned about both new competition and equal access to consumers, the ANPA lobbies against proposed regulations that would permit AT&T and its former regional subsidiaries, the now-independent Bell operating companies, to provide electronic information services, and remains wary of the telephone companies' hopes for broadband local fiber-optics networks.[112]

Will videotex suffer the fate of the facsimile newspaper? After all, a more entertaining electronic competitor eclipsed both media in the home electronics market—television devastating demand for facsimile receivers and the video cassette recorder (VCR), 24-hour cable TV news, and the personal computer, in a sense, collectively preempting consumer demand for videotex. Yet videotex can provide a wide array of services, and might easily be integrated with personal computers, which are likely to become within a generation at least as common as, say, the typewriter was in 1980.

Coexistence of some sort seems inevitable, with the existing media ceding more territory to interactive videotex, in the same way that the newspaper learned to share the market with radio, and later with television. And as television developed its own graphic and cartographic potential, so too videotex might develop its unique brand of sequenced, lively news maps.

6 Concluding Remarks

Several factors acting in concert account for the fuller integration of maps with the presentation of news. Among them, advances in telecommunications and graphics printing have been crucial in making news maps more timely and less costly. Equally important is a fundamental change in attitude among editors and publishers, who have become more fully aware of the intrinsic value and appropriate use of news maps. Underlying these proximal causes is the inherent utility of the map for showing location and communicating an explanation with spatial dimensions. Yet, as the parts of the preceding discussion have shown, the versatile map's attractions for news publishers include its effectiveness as both art and icon.

Technological innovation in the reproduction and transmission of news pictures did more than help make news maps timely and commonplace. Technological change also altered the look of news maps, with each new technology leaving a characteristic imprint on the cartographic image and its symbols. Wood-engraved news maps, for instance, reflected the engraver's straight cuts and the wood block's incapacity for thin lines and finely textured area symbols. In contrast, photoengraving, which liberally accepted all but the finest marks an artist—or anyone else—could make with black ink on white paper, fostered crude lettering and vague, overly generalized representations of geographic features. To the particular benefit of small and medium-size newspapers, stereotyping and the railway network permitted syndicates to distribute widely printing plates or cardboard mats bearing text and photoengraved illustrations, including an occasional map. Electronic facsimile equipment and photowire networks made possible the almost instantaneous transmission of maps represented by a fine grid of black or white specks. Photowire maps characteristically had bold lettering and coarse, parallel-line area symbols because photowire systems, designed for "moving" photographic images reproduced as halftones, could not cope with fine-dot area symbols and intricate type.

Microcomputers and laser printers, widely adopted by both large and small newspapers in the late 1980s, constitute a major break from photowire transmission systems, which have tended to degrade the cartographic image. Through satellite transmission and electronic bulletin boards, syndicates and wire services can deliver news maps and news photos over separate channels suited to the unique needs of line and halftone images. But the effect of this new technology on the appearance of news maps is more than freedom from bold type and coarse area symbols. Like any graphic medium, MacDraw and the LaserWriter constrain as well as encourage the artist. Rigidly horizontal or vertical type (as in Figures 3.32 and 4.10) and a limited selection of "fill patterns" for area symbols merely reflect the immature development of line-drawing software, whereas enthusiasm for the Macintosh's easily generated drop shadows (as in Figures 3.19 and 4.10) and gray lines (as in Figure 1.17) suggests the permanent adoption of these devices as either standard symbols or graphic clichés.[1] More significant, though, is the speed with which clip-art files of boundaries and map symbols permit the neophyte to produce more or less convincing news maps. Like photographic line-engraving in the late nineteenth century, microcomputer graphics is almost equally at the service of the skilled hand of the knowledgeable journalistic cartographer and the uncritical eye of the aesthetically insensitive, geographically ignorant graphics hacker.

Transition in the geographic pattern of production and distribution of news maps is at least equally significant, given its implications for both design and information content. Figure 6.1 shows the three principal stages in the development of centrally produced newspaper maps. The pre-electronic stage, before the photowire, depended upon the railroad for intercity transport, and news graphics were less timely than verbal accounts, which were sent by telegraph. For the most part, news maps distributed by syndicates were simple locator maps prepared as illustrations for feature stories. With electronic facsimile transmission, and in particular the inauguration of the Associated Press Wirephoto network in 1935, news maps could address breaking news. Satellite transmission of digital images further increased the capacity of the photowires by reducing transmission time from almost ten minutes to one minute.[2]

Until the graphics revolution of the 1970s, photowire cartography was secondary to the goals and attitudes of the photojournalists who ran and used the news picture services at the Associated Press and United Press International. Industry demand for more and better news graphics led to the establishment of separate wire-service graphics departments as well as to the formation of several specialized graphics syndicates. As an attitudinal turning point marked the mid-1970s, the mid-1980s witnessed a massive techological revolution when newspapers, newspaper groups, and wire services adopted microcomputers, laser printers, and satellite transmission. Competition among the wire services and graphics syndicates fostered the development of specialized departments producing information graphics, with investments in specialized computer hardware, software, and graphics databases. Satellite delivery strengthened the institutional separation of photos and maps because the efficient transmission of high-resolution digital images required separate systems for news photos and information graphics, treated as raster data and vector data, respectively.[3] Divorced from the head shot and action photograph, the map now is remarried—more happily—to the statistical chart and explanatory diagram. This new relationship inherently is more stable, for the partners are highly compatible in their graphic requirements as well as more equal in their importance to editors and

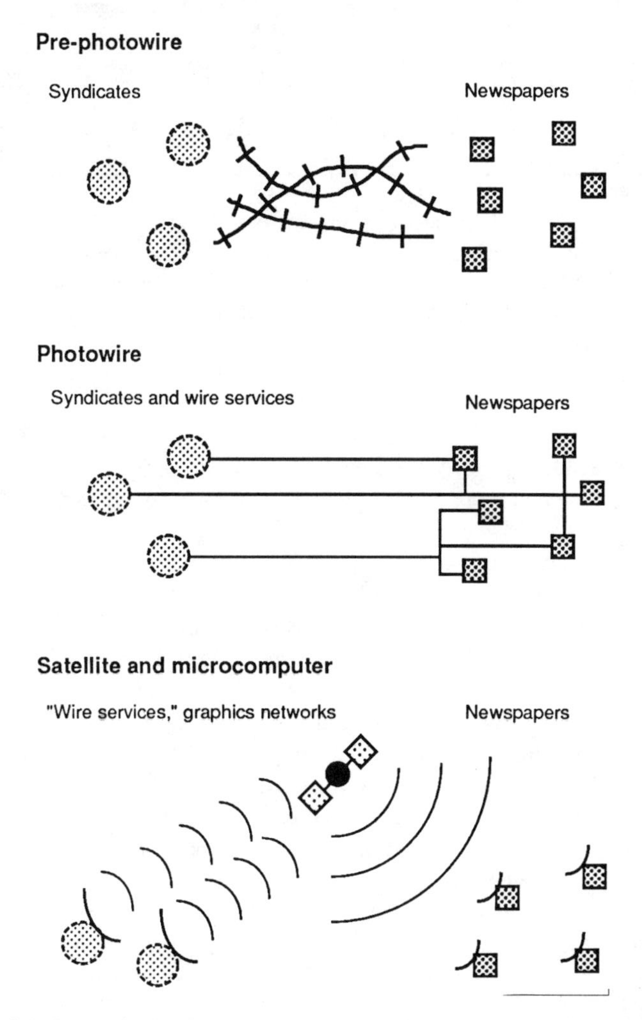

Figure 6.1. Stages in the development of centrally produced journalistic cartography.

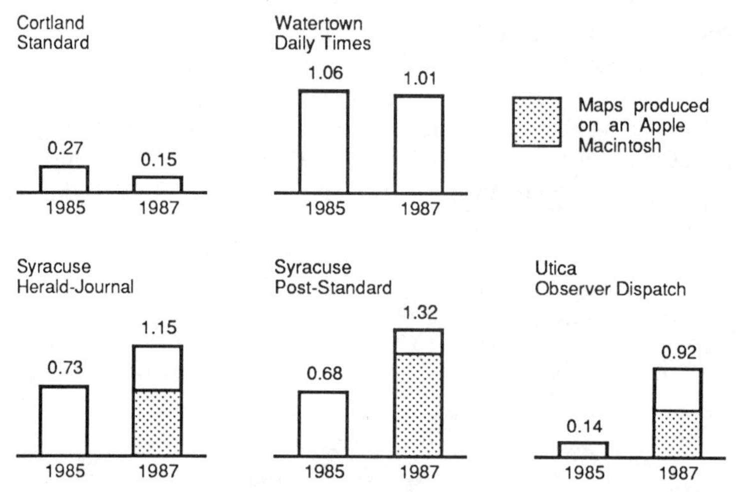

Figure 6.2. Average daily rates of mapped articles published Monday through Saturday during January and July by five central New York daily newspapers, 1985 and 1987. An Apple Macintosh and LaserWriter were used during both months in 1987 by the two Syracuse dailies and during July 1987 by the Utica *Observer Dispatch.*

publishers. Moreover, with its own microcomputer and laser printer, the local newspaper not only can adapt centrally produced maps to its own format but also can create new maps of its own, to support local news coverage.

That microcomputers and graphics networks encourage the use of news maps is apparent in Figures 6.2 and 6.3, whcih are based upon a follow-up sampling of map use by five of the twelve central New York newspapers examined in Chapter 3. Figure 6.2, which presents average daily rates for Monday through Saturday, for January and July, compares 1985 and 1987; three of the five newspapers were using an Apple Macintosh system to produce news graphics during at least part of 1987. Among the two papers not using the Macintosh, there was little change: the *Cortland Standard* still used only an occasional map, and the *Watertown Daily Times* still averaged approximately one mapped article per issue. In contrast, gains were notable among the three papers using the Macintosh: the two Syracuse dailies nearly doubled their frequency of news maps, and the Utica *Observer Dispatch,* which had not yet begun to use the Macintosh during January 1987, rose from minimal cartographic coverage in 1985 to register nearly a map each day for the 1987 sample.

Figure 6.3, which examines change separately in the number of mapped articles addressing local, domestic, and foreign news, demonstrates significant advances among the Macintosh users. At the Syracuse and Utica newspapers staff artists using the Macintosh increased the number of mapped articles with a local focus. Using Macintosh maps from the AP—and for the Syracuse newspapers, from the Knight-Ridder Graphics Network as well—these three papers also added significantly to the two nonlocal categories.[4] Most notable, though, is the increased cartographic coverage of local news, especially at the Utica paper, which had added a staff artist since July 1985. In a sense these three newspapers moved closer in their use of news maps to the *Watertown Daily Times*, which had a staff artist and a strong cartographic commitment in both years. With a new Macintosh and LaserWriter, the Syracuse newspapers used their larger art staff and larger news holes to surpass the Watertown paper's use of news maps.[5]

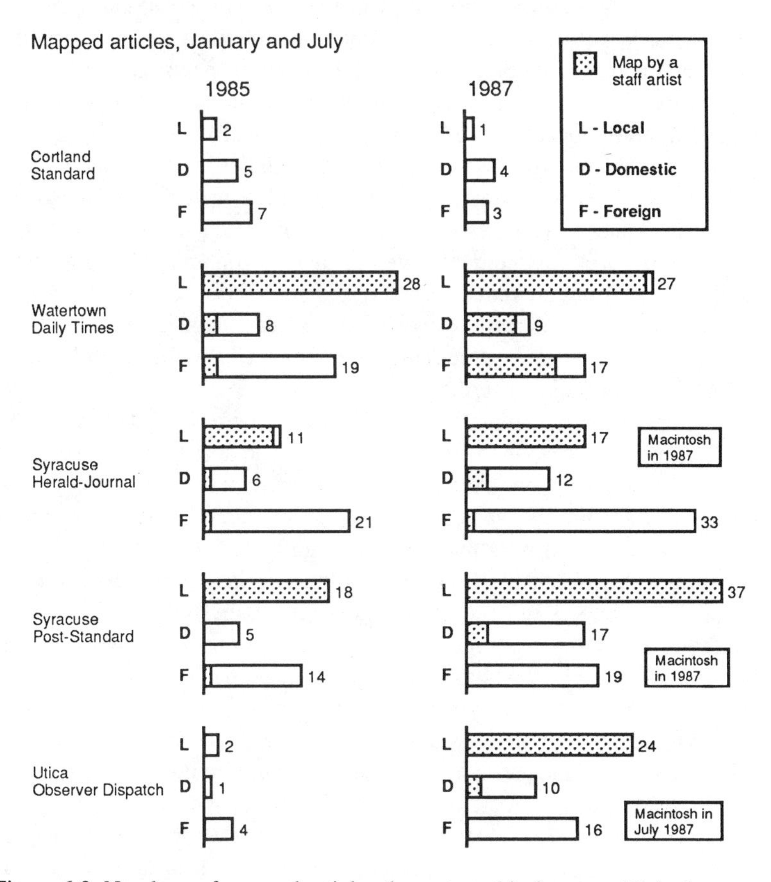

Figure 6.3. Numbers of mapped articles, by geographic focus, published Monday through Saturday during January and July by five central New York daily newspapers, 1985 and 1987.

Of no less importance is the development of television and television networks, the electronic media's extension from broadcast television to cable, and the enthusiastic adoption of computer graphics systems for titles, animation, and maps. Despite obvious differences between newspaper maps and television maps in image size, graphic resolution, and the use of color, the electronic media have developed an appreciation for journalistic cartography faintly similar, at least at the network level, to the print media's recent enthusiasm for news maps. Somewhat equivalent institutional relationships have evolved as well, with the television networks akin to the wire services, and with graphics services such as G & G Designs/Communications and specialized cable channels such as Cable News Network and the Weather Channel playing the roles of news graphics syndicates. The principal difference is that the local broadcast affiliate and the local cable system exercise far less editorial control over the selection of national and foreign topics and their presentation than does the local daily newspaper. Although a few television stations in lucrative markets occasionally employ satellite uplinks to allow local newscasters to cover nationally prominent political and sporting events, for the most part local affiliates merely decide when, within a narrow range of times, to air the network's evening news program. Their gatekeeper role consists largely of deciding which, if any, network stories to excerpt for their late-evening news programs covering local, national, and foreign events. Cable systems are even less involved with program content—their principal role as cartographic gatekeeper has been the decision as to whether to purchase the Weather Channel and similar map-rich services for their subscribers.

Coexistence of the print and electronic media reflects the division of the news market into a number of overlapping niche-spaces.[6] During the past seventy years, starting with the advent of commercial radio, technology and legislation have reorganized these niche-spaces several times. The typical American household now may choose among a single local newspaper (possibly with different morning and evening editions), several national and regional newspapers, several weekly news magazines, several local radio and television stations, a single cable franchise, and several national television networks and specialized cable channels. The newspaper survives because it is richer than television in information and more transportable in both time and space; unlike the television viewer, the newspaper reader can conveniently take the paper into the bathroom or onto the back porch, and read its stories in any order, at whatever pace he or she chooses. Yet most newspaper readers are also television viewers, for television offers not only dramatic entertainment and live sports coverage but also timely, concentrated news summaries, which also have become a form of passive entertainment. Proliferation of video options and the concurrent decline in the number of local newspapers in a sense reflect the capacity of a single well-managed newspaper to present many columnists, more thoughtful commentary, many features, many more news stories, and more thorough reports than even a two-hour sequence of expanded local and network television news programs.

Maps have their own niche in media news products. Except for weather reports, which are well suited to a dazzling and information-rich array of fleeting local, national and near-hemispherical images, that niche is far more prominent in the print media. In part, this reflects the print consumer's ability to set his or her own pace, and take as much time as necessary to examine a map in detail—or to ignore it completely. A plausible reason why local television broadcasters seem reluctant to exploit computer graphics systems and offer more abundant local news maps is that a viewing time sufficient to allow the average interested viewer to absorb key reference features might well provoke uninterested viewers to switch channels. Moreover, maps in newspapers serve a dual purpose: as noted in Chapter 4, they not only provide spatially organized information and explain geographically complex stories, but also decorate the page and help diminish the monotony of drab columns of bodytype. An additional, iconic, role involves the use of maps of local phenomena to reinforce the local newspaper's image as the best, most thorough single source of information about local occurrences and problems.[7]

Recognition, appreciation, and expansion of cartography's niche in the newspaper and the television newscast has led to the creation of map-related niche-spaces in news organizations. At the national level, as examples, the graphics departments at the AP and UPI, ABC TV Network Broadcast Graphics, the *Chicago Tribune*, and the Knight-Ridder Graphics Network have emerged as central producers and wide distributors of maps and other information graphics. At the local level, large, medium-size, and even some small newspapers have established separate graphics departments or hired artists concerned principally with the design and preparation of information graphics. Once a newspaper makes this level of commitment, as demonstrated by the Utica *Observer Dispatch* between 1985 and 1987 and by the *Watertown Daily Times* between 1980 and 1985, cartographic coverage of local news becomes regular and prominent.

How will technology and legislation again restructure the news business, and how will these reformed niche-spaces affect journalistic cartography? Just as inexpensive newsprint and public education led to the development of mass communications in the latter part of the nineteenth century, electronic telecommunications and supportive government policies could well bring about a demassified public communications serving a mass market in the early twenty-first century.[8] Indeed, demonstrations of local general-purpose videotex systems clearly indicate the technology's ability to serve large numbers of users with informative text and attractive graphics. In the late 1970s and early 1980s several American newspapers—some possibly feeling threatened by the hyperbole of videotex promoters and communications theorists, some aggravated by the unreliability of "little merchant" child labor for delivery service, and still others attracted by possible profits and the competitive advantage of establishing the first foothold in a multi-newspaper city—eagerly participated in what proved to be technologically successful but financially disastrous videotex trials. Although these demonstrations showed that considerably more news and related information might be offered than even a large metropolitan newspaper could conveniently accommodate, demand was insufficient to make the service self-sustaining. Perhaps the most cartographically significant result was the development of the Knight-Ridder Graphics Network, using expertise salvaged from the Viewtron experiment.

These early videotex trials were similar in outcome to the facsimile newspaper experiments of the 1930s and 1940s, except that in the 1980s television already provided a more versatile, less expensive distribution system. In a society of newspaper readers conditioned to inexpensive, advertiser-supported newspapers and of television viewers reluctant to spend more than they currently pay for cable service, videotex seems not commercially feasible without the support of advertisers—who, in turn, are reluctant to support a medium with a small audience less expensively reached through more established channels. Moreover, although ideal for classified advertising, videotex query languages would tend to exclude advertising similar to the lucrative display ads abundantly distributed throughout the American newspaper. It is doubtful that large advertisers such as department stores and automobile manufacturers would be willing to alter demonstrably effective marketing strategies based upon eye-catching full-page display ads and TV commercials aired before a more or less captive audience. Although advertisers might possibly subsidize a videotex news system in exchange for commercial messages either displayed in a part of the video screen or printed on whatever paper copies the subscriber chose to make, these relatively unobtrusive pitches seem ill-suited to the aggressive merchandising of furniture, home improvements, and deodorants.

Nonetheless, a videotex service rich in local as well as national and foreign news and information clearly could find a profitable niche-space if only a means other than subscriber fees were available to support a news-gathering and editorial staff at least as large as that of the daily newspaper with which it might compete—or from which it might evolve. Demand might even develop without display advertising or other subsidies—the telephone, home electrification, wireless broadcasting, and television had similarly doubtful futures early in their histories. Favorable legislation, shrewd marketing, excessive costs of newsprint, increased restrictions on child labor, increased aversion to ink rub-off, and a wide availability of broadband cable connections and inexpensive household devices for displaying and printing electronic information might well combine early in the next century to promote an electronic newspaper vastly superior to that available at the newsstand or behind the shrubbery. Adoption of an electronic newspaper would most certainly mean an expanded niche-space for news maps, which could address economically a far wider range of topics than at present, as well as provide the attraction of dynamic displays.

Were such a videotex system to be supported by powerful home microcomputers with large amounts of memory, the role of art and design in journalistic cartography might well be diminished by the simultaneous availability of an electronic atlas. Most news maps, after all, are simple locator maps—their value to the newspaper reader lies principally in their proximity on the page or screen to the verbal account they support. A conscientious, persevering reader usually could obtain the same information from a reasonably current atlas or local street map in the same way that many other nongraphic facts contained in a news report could be extracted from an encyclopedia, an unabridged dictionary, or a statistical handbook. Thus a "locate" function, activated with a **Where?** button and possibly a pointing device to indicate a specific place mentioned in the text on the screen, might afford the user an almost limitless supply of readily retrievable geographic information and obviate having to store in the videotex system's news database little more than the geographic coordinates of specific points, lines, and areas relevant to the particular news story. Journalistic cartography would thus be split between the reporter's task of gathering and coding of geographic positions and, for a lesser proportion of stories, the editorial artist's or gifted reporter's task of designing explanatory maps addressing not only *Where?* but also *How?* and *Why?*

How the news business will evolve in the near future is suggested by the eagerness with which news publishers in the 1980s invested heavily in expensive new printing presses, amortized over decades, not years.[9] Electronic technology will support the writing of news stories and the capture of news pictures, the distribution of centrally produced text and images, the design of advertising and the layout of pages, and the maintenance and analysis of circulation lists. Direct-to-plate pagination and printing systems might eliminate platemaking, the last photographic phase of newspaper manufacturing, but for the financially foreseeable future the product of this technology will be images applied in ink to newsprint, but distributed by truck and carrier. Until videotex or something like it becomes economically sound, the demassified newspaper seems likely to progress no further than the regional editions and zoned inserts with which metropolitan newspapers compete with small-city dailies and suburban weeklies.[10] In these endeavors, news maps are proving most useful in describing the intricacies of proposed development projects, zoning changes, traffic and other accidents, parade routes, robberies and murders, administrative and election districts, and other facts useful to informed citizens. Journalistic cartography and the local newspaper are symbiotic.

Comparatively minor additional investments in technology can further improve the informativeness of news publications. McLuhan's well-known doctrine of print as a nonpictorial, linear medium, with facts and explanations coded in characters of rows of type, is a residue of movable type, which is now rarely used.[11] Computer typesetting and computer graphics still display typographic images with a linear structure, but the thin lines of raster-mode laser printers and film writers now accommodate halftones and line drawings as well as letters and numbers. The technological integration of text printing and graphics printing has made possible workstations useful for preparing both articles and supporting illustrations, and might soon make obsolete the term "word processor."[12] Might not news publishers economically provide all reporters and editors with such workstations? Might not mapping software for microcomputers serve as a design filter, to obviate the more predictable cartographic and aesthetic blunders at the expense of a somewhat bland standardization? Might not journalists—and those who employ them and those who train them—recognize that good reporting requires nonverbal skills in describing phenomena with geographic dimensions?

Clearly, American news publishers already have made significant strides, although not with the purest of motives. The recent advance of journalistic cartography is an externality, or side-effect, of the perceived need to improve the packaging of newspapers and news magazines, in meeting the challenge of television and in adjusting to the one-newspaper metropolis. For the reader, though, the result is a more thorough, more efficient news report particularly apparent in the high proportion of comparatively simple locator maps employed in the American press.[13] Although critics might denounce most news maps as poor representatives of the genre's potential, it is important to note that most news, after all, is largely an organized collection of facts, with very little penetrating analysis of relationships and implications.[14] *Where* an event occurred is of fundamental interest to many readers, and a locator map showing political units, highways or streets, and relevant landmarks addresses the need for locational facts. Most often, a locator map also supplements the headline by including in a text block a concise summary of the story, such as, "Boulder Hits Bus." Moreover, editors now recognize that a "stand-alone graphic" consisting of a map, an explanatory diagram, and a few text blocks sometimes provides a more efficient explanation than the standard written account, with headline and supporting illustrations.

A broader concern for literacy might build upon these gains, especially if leaders at all levels in American education would look beyond the narrow view of literacy as pertaining only to words formatted into text. Two important corollaries of a broadened literacy are geographic literacy and graphic literacy, also called graphicacy. Geographic literacy refers not only to an understanding of fundamental world patterns such as temperature, rainfall, and urbanized populations but also to the basic place-name vocabulary shared by literate people in a society. In his best-selling essay on cultural literacy, humanist Eric Hirsch recognizes the importance of knowing facts about places: over 5 percent of the approximately 4,500 items in his list of "What Literate Americans Know" are the names of countries, the states of the U.S., large cities, state and national capitals, oceans, continents, mountain ranges, and other prominent geographic features.[15] Knowledge of significant places and their locations provides a basic geographic framework for cultural literacy—a framework needed for both intelligent discourse and the effective use of news maps. This rudimentary geographic knowledge is closely tied to graphicacy, the ability to interpret and manipulate maps, diagrams, and statistical charts. A culturally literate person needs to know how to read and interpret graphics and how, when appropriate, to design and execute them as well.

Electronic publishing and computer graphics have fostered an increased use of information graphics not only in the news media but also in business and scientific research. As with much of newspaper and television graphics, most business graphics and many scientific charts are straightforward and unsophisticated. Often used to restate the obvious, these graphics become dramatic devices used to both inform and heighten interest—a dual role that carefully crafted prose has played for centuries. Use of maps and charts in business, science, and journalism will most certainly increase, for the utility of graphics far outweighs whatever ephemeral trendiness time may erode. Moreover, use of graphics seems destined to increase in other forms of written discourse: in textbooks, technical manuals, political commentary, popular fiction, novels, and even poetry. Through their wide availability and their highly visible use of information graphics, the news media are an influential force in this diffusion, perhaps second only to the computer industry.

Many writers, particularly avowed word-people, hold the view that maps and data graphics are simplistic, and that no special training is required to understand or even produce them.[16] Perhaps they are correct, in the sense that a parent or teacher usually can infer correctly most of what a stylistically inelegant, grammatically incorrect, poorly spelled student essay has to say. Yet good graphics respect rules and principles akin to the grammatical and stylistic guidelines that promote lucid, pleasing prose.[17] Unfortunately, what little is said in our schools about maps and data graphics is included with geography, mathematics, or science, rather than with writing, and students tend to see graphics as a means of analysis, not of communication. The stubborn effectiveness of locator maps produced by graphic artists or journalists lacking formal training in map design reflects a robustness of simple cartographic illustrations not unlike that of third-grade prose. How much more elegant and richer the media might be if journalistic illustrators and their audiences more fully understood maps and data graphics.

In persistence, if not sophistication, the news media have long been among the leaders in the use of information graphics. Indeed, today's fuller use of journalistic cartography can be seen as a wider adoption by newspapers and weekly news magazines of a graphic device carefully exploited in the last century by the editors of such important journals as the *Times* of London and the *New York Times*. It is not by accident that these prestigious newspapers have covered the life-threatening struggles of war and geopolitics with not only their longest stories and biggest headlines but also their fullest use of news maps.

Journalism is ideally equipped to demonstrate still further the integration of text and graphics. With the requisite hardware and software, a receptive audience, economic incentives, and a rising professional appreciation of graphic excellence, reporters and editors might strive toward a fuller amalgamation of words and images. Further progress lies in verbal linkages to maps and diagrams, and in page layouts designed to attract the eye from text to graphic.[18] In their ability to focus the viewer's attention on specific symbols and reinforce the visual presentation with simultaneous description or interpretation, the video media enjoy a distinct advantage as long as the graphic and its accompanying discussion remain intelligible to a mass audience.

In integrating text and graphics, the print media must be more subtle than television and videotex. Perhaps the most effective, most basic strategy is through an integration of graphic and verbal reasoning in the minds of reporters and editors. If expected to be a versatile communicator of facts and explanations rather than just a writer of words, the journalist might better convey the where and the why of a story. At present, far too few stories that might benefit from a map or chart receive the attention of an artist, and even the artist who reads the story lacks the reporter's full insights and perceptions. Technology can compensate for much of the journalist's probable lack of drawing skill, and an artistically talented editor might easily make needed refinements—as verbally talented editors currently screen and polish reporters' prose. Attitudes, of course, must change if journalists are to accept a merger of the once-specialized roles of news writer and news illustrator.[19] Newsrooms and schools of journalism will be the theater for the next battles in the news graphics revolution.

Notes

Chapter 1: Introduction

1. For a collection of essays addressing the use of maps as decoration and art, see David Woodward, ed., *Art and Cartography: Six Historical Essays* (Chicago: University of Chicago Press, 1987).

2. For a general introduction to cartographic iconography, see M. J. Blakemore and J. B. Harley, *Concepts in the History of Cartography: A Review and Perspective*, Cartographica monograph no. 26 (Toronto: University of Toronto Press, 1980), 76-86; and Denis Wood and John Fels, "Designs on Signs: Myth and Meaning in Maps," *Cartographica* 23, no. 3 (Autumn 1986), 54-103.

3. Among the nation's 170 million persons 18 years old and over, 50.6 percent (86 million) are estimated to watch early evening television news, 41.6 percent (71 million) watch late evening news, and 60.5 percent (103 million) read a daily newspaper. U. S. Bureau of the Census, *Statistical Abstract of the United States: 1987*, 107th ed. (Washington, D.C., 1986), 531.

4. The Federal Highway Administration estimates the number of licensed drivers in 1985 as over 158 million. Probably only a small proportion of these use a road map regularly. Ibid., 585.

5. Library of Congress, Geography and Map Division, *Bibliography of Cartography*, 5 vols. (Boston: G. K. Hall and Co., 1973); and Walter W. Ristow, "Journalistic Cartography," *Surveying and Mapping* 17, no. 4 (October 1957), 369-90.

6. Library of Congress, Geography and Map Division, *Bibliography of Cartography*, first supplement, 2 vols. (Boston: G. K. Hall and Co., 1980).

7. See, for example, the one-paragraph comment in David Rhind's periodic column in *Geographical Magazine*. Rhind acknowledged the numerical dominance of media maps and confessed an appreciation of television weather maps. Yet his assessment of the printing quality and simplicity of newspaper maps was highly negative: "Newspaper maps are almost always printed in one colour, with a printing quality which would drive most cartographers to suicide. . . .many newspapers do not even include a weather map but specialize in front-page maps showing items such as the location of exploding North Sea oil platforms, in relation to surrounding countries. . . . Are newspaper maps, in particular, merely a space-filling device?" David Rhind, "Cartography: Metropolitan Canada in Maps," *Geographical Magazine* 49, no. 10 (July 1977), 664-65.

8. Map-making technology and its effects are an important element in the conceptual framework advanced by David Woodward for the scholarly study of

the history of cartography. Woodward notes the importance not only of the product (the map) but also the personnel, techniques, and tools of production, and the processes of information gathering, information processing, document distribution, and document use. See David Woodward, "The Study of the History of Cartography: A Suggested Framework," *American Cartographer* 1, no. 2 (October 1974), 101-15.

9. Topographic maps have, of course, benefited from a variety of advances in surveying and measurement technology, including photogrammetry, stereocompilation, electronic distance measurement, satellite positioning, and inertial surveying. Land-cover mapping has been advanced by camouflage detection film, satellite and aerial platforms, electronic remote sensing systems, and computer-based image processing. See Mark Monmonier, *Technological Transition in Cartography* (Madison, Wis.: University of Wisconsin Press, 1985), 30-34, 71-74, 85-107.

10. Experimental studies of the effectiveness of news maps are fraught with the need to minimize experimenter bias (as when a geography professor might survey students about their attitudes toward or use of maps), control for previous geographic knowledge, and disentangle the effects of various other sources of information also relevant to the formation of opinions and insights. Yet a promising short-term approach is the examination of subjects' recall of news maps. See, for example, Patricia Gilmartin, "The Recall of Journalistic Maps and Other Graphics," (Paper presented at the annual meeting of the Canadian Cartographic Association, July 1986, Burnaby, British Columbia). For an earlier, methodologically significant study of the memorability of news pictures, see Alan Booth, "The Recall of News Items," *Public Opinion Quarterly* 34, no. 4 (Winter 1970-71), 604-10.

11. Precision journalism as an adaptation of social science statistical methods to journalistic analysis and presentation is advocated in Philip Meyer, *Precision Journalism* (Bloomington, Ind.: Indiana University Press, 1973).

12. Several studies are reported in Lucia Solarzano, "Why Johnny Can't Read Maps, Either," *U.S. News and World Report* 98, no. 11 (25 March 1985), 50; and "TV Tests Reflect Pervasive Geographic Ignorance," *Update,* National Geographic Society, Educational Media Division, no. 7 (Spring 1987), 2. Also see Richard J. Kopec, "Geography: No 'Where' in North Carolina " (Department of Geography, University of North Carolina, 1984, Photocopy).

13. These results of the Gallup polls are reported in Chilton R. Bush, *News Research for Better Newspapers* (New York: American Newspaper Publishers Foundation, 1969), 4:17.

14. For a concise appraisal of the problem of American geographic education as well as some workable solutions, see Association of American Geographers, Committee on Geography and International Studies, *Geography and International Knowledge* (Washington, D.C., 1982); and Joint Committee on Geographic Education of the National Council for Geographic Education and the Association of American Geographers, *Guidelines for Geographic Education: Elementary and Secondary Schools* (Washington, D.C., 1984).

15. See Simon's essay "Designing Organizations for an Information Rich World," in Herbert A. Simon, *Models of Bounded Rationality* (Cambridge, Mass.: M.I.T. Press, 1982), 171-84 (ref. pp. 173-75).

16. Its own photographically rich but cartographically prominent monthly magazine, *National Geographic*, generates such large revenues that the society, in addition to its News Service Division, undertakes a number of public education campaigns to preserve its tax status as a not-for-profit educational society. See Howard S. Abramson, *National Geographic: Behind Americas Lens on the World* (New York: Crown, 1987), 219-32.

17. British educator William Balchin used the term "graphicacy" to describe the graphic equivalent of literacy. It embraces skills in using maps, statistical diagrams, and explanatory illustrations. See, for example, W. G. V. Balchin, "Graphicacy," *American Cartographer* 3, no. 1 (April 1976), 33-38.

Chapter 2: Engraving Technology and the Rise of Journalistic Cartography

1. Andre Blum, *The Origins of Printing and Engraving* (New York: Charles Scribner's Sons, 1940), 68.

2. Ibid., 92-110; and Colin Clair, *A History of European Printing* (London: Academic Press, 1976), 3-6.

3. S. H. Steinberg, *Five Hundred Years of Printing*, 2nd ed., rev. (Harmondsworth, Middlesex: Penguin Books, 1966), 17-26.

4. Folke Dahl, *A Bibliography of English Corantos and Periodical Newsbooks, 1620-1642* (London: The Bibliographical Society, 1952), 18.

5. Ibid., 18. For a concise history of newspapers, with particular attention to the British and American press, see Anthony Smith, *The Newspaper: An International History* (London: Thames and Hudson, 1979).

6. Allen Hutt, *The Changing Newspaper: Typographic Trends in Britain and America, 1622-1972* (London: Gordon Fraser, 1973), 9, 12. The Star Chamber was an English court with broad, absolute powers over criminal and civil matters; established in the sixteenth century, it lasted until 1641, when Parliament began to take greater control of the British government.

7. Dahl, *Bibliography of English Corantos*, 19.

8. Hutt, *Changing Newspaper*, 12-17.

9. Frederick W. Hamilton, *A Brief History of Printing in America* (Chicago: Committee on Education, United Typothetae of America, 1918), 23-25.

10. Isaiah Thomas, *The History of Printing in America* (1810; reprint, New York: Weathervane Books, 1975), 17.

11. The Times (London), *The History of The Times*, vol. 1, *"The Thunderer" in the Making, 1785-1841* (London: Office of The Times, 1935), 115-17.

12. Alfred McClung Lee, *The Daily Newspaper in America: The Evolution of a Social Instrument* (New York: Macmillan, 1937), 117. In this first Hoe rotary press, the type frames were not curved; four or six type frames revolved with the main cylinder, and each was fed by a separate pressman. See Frank E. Comparato, *Chronicles of Genius and Folly: R. Hoe and Company and the Printing Press As a Service to Democracy* (Culver City, Cal.: Labyrinthos, 1979), 269-70.

13. George A. Kubler, *A New History of Stereotyping* (New York, 1941), 174-78.

14. Joel Munsell, *Chronology of the Origin and Progress of Paper-Making*, 5th ed. (Albany, N. Y.: Munsell, 1876), 60-61.

15. The large size of some British "broadsheet" newspapers, which is reflected in most present-day American papers, can be traced to this per-sheet newspaper duty. See W. Turner Berry, "Printing and Related Trades," in *A History of Technology*, vol. 5, *The Late Nineteenth Century* (Oxford: Clarendon Press, 1958), 683-715 (ref. p. 699-700).

16. James Moran, *Printing Presses: History and Development from the Fifteenth Century to Modern Times* (Berkeley and Los Angeles: University of California Press, 1973), 190-91.

17. Walter's grandfather was coke merchant John Walter, who in 1785 founded the *Daily Universal Register*, renamed the *Times* in 1788. The *Times* had an extensive mechanical department, whose engineers designed much of the printing equipment. See The Times (London), *The History of The Times*, vol. 2, *The Tradition is Established, 1841-1885* (London: Office of The Times, 1939), 347-49.

18. See Lyman Horace Weeks, *A History of Paper-Manufacturing in the United States, 1690-1916* (New York: Lockwood Trade Journal Company, 1916), 226-28.

19. S. N. D. North, *History and Present Condition of the Newspaper and Periodical Press of the United States* U. S. Department Dept. of the Interior, Census Office, 10th Census, 1880 (Washington: Government Printing Office, 1884), 82.

20. Edmund G. Gress, *The American Handbook of Printing*, 3rd ed. (New York: Oswald Publishing Company, 1913), 100-104.

21. Data on historical trends in the newspaper industry are from William A. Dill, *Growth of Newspapers in the United States* (Lawrence, Kans.: Department of Journalism, University of Kansas, 1928), esp. p. 28.

22. For a short account of the New York *Morning Post* and the price of competing papers, see Hamilton, *Brief History of Printing*, 28-29. For estimates of wages during the period 1830-32, see Stanley Lebergott, "Wage Trends, 1800-1900," in *Trends in the American Economy in the Nineteenth Century*, Conference on Research in Income and Wealth (Princeton, N.J.: Princeton University Press, 1960), 449-99 (ref. p. 462).

23. For a history of Benjamin Day and the New York *Daily Sun*, see Frank M. O'Brien, *The Story of the Sun* (New York: D. Appleton and Company, 1928; reprint, New York: Greenwood Press, 1968).

24. For an account of the man who, according to historian Will Irwin, "invented news as we know it," see Oliver Carlson, *The Man Who Made News* (New York: Duell, Sloan and Pearce, 1942).

25. Ralph Ray Fahrney, *Horace Greeley and the Tribune in the Civil War* (Cedar Rapids, Ia.: Torch Press, 1936), 149. For a concise bibliography of Greeley, see Henry Luther Stoddard, *Horace Greeley: Printer, Editor, Crusader* (New York: G. P. Putnam's Sons, 1946).

26. Gay Talese, *The Kingdom and the Power* (New York: World Publishing Company, 1969), 149. Also see Francis Brown, *Raymond of the Times* (New York: W. W. Norton and Company, 1951).

27. Dill, *Growth of Newspapers*, 22, 30.

28. Lebergott, "Wage Trends, 1800-1900," 493.

29. Dill, *Growth of Newspapers*, 21.

30. U. S. Bureau of the Census, *Historical Statistics of the United States, Colonial Times to 1970* (Washington, D.C.: Government Printing Office, 1975), 839.

31. Dill, *Growth of Newspapers*, 76.

32. See Thomas, *History of Printing in America*, 335.

33. In his monograph on the evolution of newspaper design, Allen Hutt reported the use of wood engravings in news publications as early as 1643, when the *Intelligencer* used an illustration of the king and queen on its title page and Samuel Pecke's *Perfect Diurnall* employed a front-page engraving of the House of Commons. See Hutt, *Changing Newspaper*, 13. For a general history of wood engraving, see Arthur M. Hind, *An Introduction to the History of Woodcut*, 2 vols. (Boston: Houghton Mifflin, 1935; reprint New York: Dover Publications, 1963). For an essay on the applications of wood engraving in cartography, see David Woodward, "The Woodcut Map," in David Woodward, ed., *Five Centuries of Map Printing* (Chicago: University of Chicago Press, 1975), 25-50.

34. Albert Matthews, "The Snake Devices, 1754-1776, and the Constitutional Courant," *Publications of the Colonial Society of Massachusetts* 11 (1906-7), 409-53.

35. See, for example, Stephen Hess and Milton Kaplan, *The Ungentlemanly Art: A History of American Political Cartoons* (New York: Macmillan, 1968), 52-53; William Murrell, *A History of American Graphic Humor, vol. 1, 1747-1865* (New York: Whitney Museum of American Art, 1933), 11-12; and Charles Press, *The Political Cartoon* (East Brunswick, N.J.: Associated University Presses, 1981), 29, 209.

36. See, for example, Hess and Kaplan, *The Ungentlemanly Art*, 61; and Frank Weitenkampf, *Political Caricature in the United States* (New York: The New York Public Library, 1953), 17.

37. John Ward Dean, "The Gerrymander," *The New England Historical and Geneological Register* 46, no. 184 (October 1892), 374-83; and Elmer C. Griffith, *The Rise and Development of the Gerrymander* (Chicago: Scott, Foresman and Company, 1907), 16-19.

38. Woodward, "The Woodcut Map," 45-48.

39. T. F. W. Hurley, "Wood-cuts and Photo-engravings," *The Month* 68, no. 308 (February 1890), 191-97.

40. Mason Jackson, *The Pictorial Press: Its Origin and Progress* (London: Hurst and Blackett, 1885), 316-17.

41. North, *History and Present Condition of the Newspaper and Periodical Press*, 125-26; and John Tebbel, *The American Magazine: A Compact History* (New York: Hawthorne Books, 1969), 107-12, 115-17.

42. R. H. Smith, "All the 'Firsts' of the *Illustrated London News*," *Penrose Annual* 67 (1974), 101-12.

43. Ibid., 104. For further discussion of photographic transfer for wood engraving, see St. Vincent Beechey, "Photography Applied to Engraving on Wood," *The Art Journal,* n.s., 6 (1854), 244; and Thomas W. Smillie, "Photographing on Wood for Engraving," *Smithsonian Miscellaneous Collections* 47 (1905), 497-99.

44. Smith, "All the 'Firsts'," 104. For a description of wood-block electroplating, see Hind, *Introduction to the History of Woodcut*, 15-16.

45. Albert A. Sutton, *Design and Makeup of the Newspaper* (New York: Prentice-Hall, 1948), 235-37.

46. See Pat Hodgson, *The War Illustrators* (New York: Macmillan, 1977), 34; and Smith, "All the 'Firsts'," 103-14.

47. Richard W. Stephenson, "Maps for the General Public: Commercial Cartography of the American Civil War," (Paper delivered at the Eleventh International Conference on the History of Cartography, Ottawa, 8-12 July 1985), 12-15.

48. For a discussion of a variety of methods adopted during the nineteenth century for the production of area shading, see Karen Severud Pearson, "Mechanization and the Area Symbol: Techniques in the 19th Century Geographical Journals," *Cartographica* 20, no. 4 (1983), 1-34.

49. A British scientific writer was impressed with the speed of photoengraving and news illustration in general: "An accident happens at noon; a small photograph is taken of the scene. This is enlarged, a drawing made over the enlargement, the photograph bleached out under the drawing, a negative made from the drawing, a plate coated, exposed, developed, bitten, trimmed, stereotyped, and blocked, put on the press about midnight and a hundred thousand impressions or so distributed by six o'clock the next morning. We are so surrounded by wonderful things that they cease to be wonderful." See S. R. Koehler, "The Photo-Mechanical Processes," *Technological Quarterly and Proceedings of the Society of Arts* 5, no. 3 (October 1892), 164-204 (ref. p. 179). The manager of the art department of the Dublin, Ireland, *Freeman's Journal* described the process: "Our time for making these blocks, including the drawing, is two and a half hours, and you will see that they are fairly deep; otherwise they would print very dirty on a rapid printing machine. The drawings are made on ordinary transfer paper and transferred to zinc." See "Newspaper Illustrations," *Penrose Annual* 3 (1897), 17-32 (ref. p. 29). A photoengraving time of 30 minutes for half-tone photographs was noted a decade later, in John L. Given, *Making a Newspaper* (New York: Henry Holt and Company, 1907), 248.

50. See, for examples, John Bakeless, *Magazine Making* (New York: Viking Press, 1931), 69-70; Given, *Making a Newspaper*, 283-84; and Carl Schraubstadter, Jr., "Newspaper Illustration. No. 2," *Inland Printer* 5, no. 3 (December 1887), 173-74.

51. Louis Edward Levy, "Development and Recent Advances of the Techno-Graphic Arts," *Journal of the Franklin Institute* 180, no. 4 (October 1915), 387-408; and Leo Hagedoorn, "The First Century of Photo-Engraving," *Penrose Annual* 19 (1913-14), 189-97.

52. See, for example, William Gamble, "Photographic Processes of To-Day," *Penrose Annual* 1 (1895), 5-11.

53. For a concise summary of improvements during the twentieth century, see W. B. Hislop, "Photo-engraving: A Survey of Six Decades," *Penrose Annual* 56 (1962), 112-15.

54. For a general history of the Yellow Press, see Sidney Kobre, *The Yellow Press and Gilded Age Journalism* (Tallahassee, Fla.: Florida State University, 1964). For a comprehensive biography of Joseph Pulitzer, see W. A. Swanberg, *Pulitzer* (New York: Charles Scribner's Sons, 1967).

55. Robert E. Park, "The Natural History of the Newspaper." In Robert E. Park, Ernest W. Burgess, and Roderick D. McKenzie, *The City* (Chicago: University of Chicago Press, 1925), 80-98 (ref. p. 95).

56. Sheridan Ford, "Illustrated Daily Journalism," *Inland Printer* 13 (July 1894), 346-47

57. Stephen Henry Horgan, "The World's First Illustrated Newspaper," *Penrose Annual* 35 (1933), 23-24; and Hutt, *Changing Newspaper*, 83-84.

58. See, for example, Stephen Henry Horgan, *Horgan's Half-tone and Photomechanical Processes* (Chicago: Inland Printer Co., 1913); and Stephen Henry Horgan, "Journalism's Greatest Alliance," in Louis Flader, ed., *Achievement in Photo-engraving and Letterpress Printing* (Chicago: American Photo-Engravers Association, 1927), 61-62.

59. Frank Colebrook, "Newspaper Half-Tones," *Penrose Annual* 11 (1905), 93-96; W. B. Hislop, "The Work of Frederick E. Ives: An Appreciation," *Penrose Annual* 40 (1933), 105-7; and "Making a Half-tone Engraving," *Scientific American*, n.s., 83, no. 10 (8 September 1900), 145, 153-54.

60. R. Smith Schuneman, "Art or Photography: A Question for Newspaper Editors of the 1890s," *Journalism Quarterly* 42, no.1 (Winter 1965), 43-52. A similar delay in the adoption of the halftone photograph affected American magazines in the 1880s; see Robert S. Kahn, "Magazine Photography Begins: An Editorial Negative," *Journalism Quarterly* 42, no. 1 (Winter 1965), 53-59.

61. David Woodward, "The Decline of Commercial Wood-Engraving in Nineteenth-century America," *Journal of the Printing Historical Society* no. 10 (1974-75), 56-83.

62. "Modern Newspaper Illustration," *The Newspaper Maker* 12, no. 314 (28 March 1901), 5.

63. Arthur H. Robinson, "Mapmaking and Map Printing: The Evolution of a Working Relationship," in David Woodward, ed., *Five Centuries of Map Printing* (Chicago: University of Chicago Press, 1975), 1-23 (ref. pp. 20-22).

64. For a discussion of the decline in number of newspapers, following the rapid expansion of television broadcasting, see Ben H. Bagdikian, *The Media Monopoly* (Boston: Beacon Press, 1983), 197-209; Mark Monmonier, "The Geography of Change in the Newspaper Industry of the Northeast United States, 1940-1980," *Proceedings of the Pennsylvania Academy of Science* 60, no. 1 (1986), 55-59; and Anthony Smith, *Goodbye Gutenberg: The Newspaper Revolution of the 1980s* (New York: Oxford University Press, 1980), 27-41.

65. John P. Robinson, "Daily News Habits of the American Public," *ANPA News Research Report* (American Newspaper Publishers Association) no. 15 (22 September 1978), 5.

66. Victor Strauss, *The Printing Industry* (Washington, D.C.: Printing Industries of America; New York: R. R. Bowker Company, 1967), 93-148.

67. For further discussion of recent changes in newspaper layout and design, see Thomas J. Holbein, "Trends in Newspaper Editing and Design," *Design, Journal of the Society of Newspaper Design* no. 16 (Summer 1984), 16-17; and Gerald C. Stone, John C. Schweitzer, and David H. Weaver, "Adoption of Modern Newspaper Design," *Journalism Quarterly* 55, no. 4 (Winter 1978), 761-66.

68. Ernest C. Hynds, *American Newspapers in the 1980s* (New York: Hastings House, 1980), 265-66. Some newspapers converted their letter presses for flexographic printing, in which inked rubber-like flexible relief plates transfer the image to the surface of the paper with a very light, "kiss" impression. See Mark Fitzgerald, "Flexo Is Passing the Test," *Editor and Publisher* 177, no. 45 (10 November 1984), 28-29.

69. For a discussion of the conceptual basis of the electronic scanner engraver, see O. Hassing and J. Oskar Nielson, "The Hassing Electro Optical Engraving Machine," *Penrose Annual* 40 (1983), 108-11. For discussion of the adoption and uses of models offered in the 1950s by Fairchild, see Ray Erwin, "Fairchild Improves Its Scan-A-Graver," *Editor and Publisher* 88, no. 24 (4 June 1955), 10, 69; Walter Leckrone, "Those New Plastic Cuts," *Bulletin of the American Society of Newspaper Editors* no. 342 (1 April 1952), 14-15; and Harold W. Wilson, "Electronics and Plastics in Photo-Journalism," *Journalism Quarterly* 29, no. 3 (Summer 1952), 316-19.

70. See, for examples of color-scanning systems, Brian Chapman, "Electronic Scanning," *Penrose Annual* 68 (1975), 130-36; Clive Goodacre, "Response 300: On-line Retouching," *Penrose Annual* 72 (1980), 97-100; and Peter Pugsley, "A New High Speed Digital Colour Scanner," *Penrose Annual* 69 (1976), 79-86.

71. For a concise overview of the use of color in newspaper design, see Edmund C. Arnold, *Designing the Total Newspaper* (New York: Harper and Row, 1981), 124-25.

72. For discussion of *USA Today*'s impact on newspapers, see Ruth Clark, "Relating to Readers in the '80s." In *ASNE-84* (Washington, D.C.: American Society of Newspaper Editors, 1985), 107-21; and Philip C. Geraci, "Comparison of Graphic Design and Illustration Use in Three Washington, D.C., Newspapers," *Newspaper Research Journal* 5, no. 2 (Winter 1983), 29-39.

73. For a short history of Sunday newspapers, see William Peterfield Trent and others, *The Cambridge History of American Literature* (New York: G. P. Putnam's Sons, 1921), 3:330-31; and Lee, *The Daily Newspaper in America*, 376-407.

74. Stanley E. Kalish and Clifton C. Edom, *Picture Editing* (New York: Rinehart and Company, 1951), 6; and Talese, *The Kingdom and the Power*, 199.

75. For a description of rotogravure printing in the early twentieth century, see "The Rotogravure Quick-Printing Process and Its Possibilities," *Scientific American Supplement* 75, no. 1946 (19 April 1913), 248-49.

76. For a concise evaluation of the quality of these five newspapers, see John C. Merrill, *The Elite Press: Great Newspapers of the World* (New York: Pitman, 1969), 11-17, 44-54, 111-15, 121-24, 170-77, 263-72. Merrill describes

an elite newspaper in an open society as "a courageous, independent, news-views-oriented journal." For a more complete examination of historical trends in map use by these papers, see Mark Monmonier, "The Rise of Map Use by Elite Newspapers in England, Canada, and the United States," *Imago Mundi* 38 (1986), 46-60.

77. The *Times* of London was founded in 1785, and the *Sunday Times* in 1822. They have always had separate editorial staffs, but in 1967 they came under the common ownership of Lord Thompson. In the early 1980s both papers were owned by Rupert Murdoch. For a history of the *Sunday Times*, see Harold Hobson, Phillip Knightley, and Leonard Russell, *The Pearl of Days: An Intimate Memoir of the Sunday Times, 1822-1972* (London: Hamilton, 1972).

78. For a general history of the *New York Times*, see Meyer Berger, *The Story of the New York Times, 1851-1951* (New York: Simon and Schuster, 1951), or Talese, *The Kingdom and the Power*. For a discussion of map-making at the *New York Times*, see Neil MacNeil, "The Presentation of the News." In *The Newspaper: Its Making and Its Meaning*, ed. staff of the New York Times (New York: Charles Scribner's Sons, 1945), 125-45; and Andrew Sabbatini, (comments on setting up a map department), *APME Red Book* 31 (1978), 168-69.

79. For a comprehensive history of the Wall Street Journal, see Lloyd Wendt, *The Wall Street Journal* (Chicago: Rand McNally, 1982). For a discussion of the newspaper's design and traditional limited use of graphics, see Kathy McClelland, "Sprucing Up an Old Gray Lady," *Design, Journal of the Society of Newspaper Design* no. 17 (Fall 1984), 23-26.

80. For a brief history of the Toronto *Globe and Mail*, see W. H. Kesterton, *A History of Journalism in Canada* (Toronto: McClelland and Stewart, 1967), esp. pp. 85-86; and Merrill, *The Elite Press*, 121-24.

81. For a history of the *Christian Science Monitor* and discussion of its editorial philosophy, readership, and distribution, see Erwin D. Canham, *Commitment to Freedom: The Story of the Christian Science Monitor* (Boston: Houghton Mifflin, 1958). Canham stated, "No other daily newspaper reaches so many of its readers by mail—fully 90 percent of its total circulation." (ref. p. 396) Christian Science Reading Rooms worldwide serve as outlets for sales to nonsubscribers.

82. Joan Forbes, Notes from an interview held at Boston, 19 November 1984.

83. Louis Silverstein, while art director at the *New York Times*, implemented the concept of packing the news in special weekday sections. For a discussion of strategies for section organization and layout, see Mario R. Garcia, *Contemporary Newspaper Design: A Structural Approach* (Englewood Cliffs, N.J.: Prentice-Hall, 1981), 109-10, 192-97.

84. Andrew Sabbatini, Notes from an interview held at New York, 1 October 1984.

85. For Evans's account of Murdoch's changes to the *Times*, see Harold Evans, *Good Times, Bad Times* (New York: Atheneum, 1981). For a brief review of design changes promoted by Evans and design director Edwin Taylor, see Laurie Parker, "Meeting the Needs of Changing Times," *Design, Journal of the Society of Newspaper Design* no. 8 (December 1981), 5-6. Perhaps the most striking change in the design of the *Times* had occurred one and a half decades earlier, when the paper first devoted its front page to news. Previously the first few

pages had been used exclusively for advertising. See Michael Hides, "Fleet Street's New Face," *IPI Report* 15, no. 2 (June 1966), 6.

86. *Willings Press Guide, 1985*, 111th ed. (West Sussex: Thomas Skinner Directories, 1985), 1097.

87. Weekly news magazines are treated as national newspapers in James Playsted Wood, *Magazines in the United States* (New York: Ronald Press, 1971), 202-36.

88. At mid-decade, for instance, *Time* had a circulation of about 4.6 million, more than twice the *Wall Street Journal*'s almost 2.0 million. *Time* was the most widely circulated news weekly, and the *Journal* was the most widely circulated daily newspaper. In 1986 *USA Today*'s circulation approached that of the *Journal*. Circulation data for American news magazines are from *IMS/Ayer Directory of Publications* (Fort Washington, Pa.: IMS Press, 1985); and *Ulrich's International Periodicals Directory*, 24th ed. (New York: R. R. Bowker, 1985).

89. "Washington Post Sets Modest Circulation Goal for Weekly," *Editor and Publisher* 116, no. 18 (30 April 1983), 45.

90. Wood, *Magazines in the United States*, 3-11. For a history of Cave and his publication, see C. Lennart Carlson, *The First Magazine: A History of The Gentleman's Magazine* (Providence, R.I.: Brown University, 1938).

91. Theodore Peterson, *Magazines in the Twentieth Century* (Urbana, Ill.: University of Illinois Press, 1964), 5-15.

92. American news magazines that preceded *Time* included the *Connecticut Magazine* and the *Worcester Magazine*, both founded in 1786; the *Balance and Columbian Repository*, published from 1802 through 1808; *Niles' Weekly Register*, published from 1811 to 1848; and the *Pathfinder*, a marginally successful news magazine published from 1894 to 1957. *Leslie's* and *Harper's* were largely picture magazines, akin to *Life*. The *Literary Digest*, published from 1890 to 1937, was more a digest than a news magazine. See Peterson, *Magazines in the Twentieth Century*, 324-26.

93. See Wood, *Magazines in the United States*, 205-11. For an official history, see Robert T. Elson, *Time Inc.: The Intimate History of a Publishing Enterprise, 1923-1941*, ed. Duncan Norton-Taylor (New York: Atheneum, 1968), (esp. pp.3-93).

94. Peterson, *Magazines in the Twentieth Century*, 326-29.

95. Ibid., 330-32.

96. Ibid., 333-34.

97. *Business Week*'s wide scope and significance have been recognized by at least one prominent media historian. See Frank Luther Mott, *The News In America* (Cambridge, Mass.: Harvard University Press, 1962), 16-17.

98. William H. Taft, *American Magazines for the 1980s* (New York: Hastings House, 1982), 26-28, 92-93.

99. For a general history, see *The Economist, 1843-1943: A Centenary Volume* (London: Oxford University Press, 1943). Like articles in the magazine, chapters in this book written by editorial personnel provide no indication of their authors. The *Economist* does not identify its editors and writers in a masthead.

100. Margaret Jay, "The *Economist* Formula: Elitism, Anonymity, 'Successful Coziness'," *Washington Journalism Review* 7, no. 4 (April 1985), 52-54.

101. Peter C. Newman, "Each Venture Is a New Beginning, a Raid on the Inarticulate," *Maclean's* 91, no. 19 (18 September 1978), 2. Also see "Canada's Own," *Newsweek* 64, no. 8 (24 August 1964), 51, 53; and Peter C. Newman, "Maclean's at 75: Still Keeping Shared Dreams Alive," *Maclean's* 93, no. 39 (29 September 1980), 3. For a history of *Maclean's* early years and its founder, John Bayne Maclean, see Floyd S. Chalmers, *A Gentleman of the Press* (Garden City, N.Y.: Doubleday and Company, 1969).

102. *Canadian Advertising Rates and Data* 58, no. 10 (October 1985), 133-38.

103. Design critic Edward Tufte discussed embellishments that detract from "graphical integrity." See Edward Tufte, *The Visual Display of Quantitative Information* (Cheshire, Conn.: Graphics Press, 1983), esp. pp. 53-71 and 79-80. For a survey and critique of information graphics used in *Newsweek* and *Time* during 1981 and 1982, see Magella-J. Gauthier and Claude Chamberland, "Communication graphique et magazines: les cas de Newsweek et de Time," *Revue de Carto-Quebec* 5, no. 1 (1984) 11, 21.

Chapter 3: Wire Services, News Syndicates, and Graphics Gatekeepers

1. For a history of early postal communications and use of the mails for administration, commerce, and military intelligence, see Alvin F. Harlow, *Old Post Bags* (New York: D. Appleton and Company, 1928). For histories of postal service in Britain and the United States, respectively, see F. George Kay, *Royal Mail: The Story of the Posts in England from the Time of Edward IVth to the Present Day* (London: Rockliff, 1951); and Carl H. Scheele, *A Short History of the Mail Service* (Washington, D.C.: Smithsonian Institution Press, 1970).

2. See George T. Matthews, ed., *News and Rumor in Renaissance Europe* (New York: G. P. Putnam's Sons, 1959), 13-21. Matthews provides a selection of 268 excerpts from Fugger news-letters published between 1568 and 1604.

3. C. F. Dendy Marshall, *The British Post Office from Its Beginnings to the End of 1925* (London: Oxford University Press, 1926), 17-27.

4. The *Boston News-Letter*, begun in 1704, when Campbell was postmaster, was the first newspaper regularly published in the American colonies. Franklin's career as a printer and newspaper publisher began well before his appointment in 1753 as deputy postmaster general. The *Pennsylvania Gazette* was begun in 1728 by Samuel Keimer, who relinquished the paper to Franklin in 1729. Franklin held an active interest in the newspaper until 1765. With the Revolution, he became the first Postmaster General of the United States. See Thomas, *History of Printing in America*, 215, 368, 434-35.

5. Allan R. Pred, "Urban Systems Development and the Long-Distance Flow of Information through Preelectronic U.S. Newspapers," *Economic Geography* 47, no. 4 (October 1971), 498-524.

6. For a concise survey of semaphore telegraphy, see Alvin F. Harlow, *Old Wires and New Waves: The History of the Telegraph, Telephone, and Wireless*

(New York: D. Appleton-Century Company, 1936), 13-34; and Alfred Still, *Communications Through the Ages: From Sign Language to Television* (New York: McGraw-Hill, 1946), 32-37.

7. Robert W. Desmond, *The Press and World Affairs* (New York: D. Appleton-Century Company, 1937), 55-57; and United Nations Educational, Scientific and Cultural Organization (UNESCO), *News Agencies: Their Structure and Operation* (New York: Greenwood Press, 1969), 11.

8. The Times (London), *"The Thunderer" in the Making*, 106-8. Although the *Times* just earlier had had a bitter dispute with the British government over the quality of the international postal service, the Government approved and appreciated the newspaper's smuggling intelligence about Napolean's exploits on the Continent.

9. G. R. M. Garratt, "Telegraphy," in *The Industrial Revolution*, vol. 4 of *A History of Technology*, ed. Charles Singer and others, (Oxford: Clarendon Press, 1958), 644-62.

10. Robert Luther Thompson, *Wiring a Continent: The History of the Telegraph Industry in the United States, 1832-1866* (Princeton, N.J.: Princeton University Press, 1947), 440. Morse's demonstration used two wire pairs between Baltimore and Washington—one pair for each direction. See Herbert W. Meyer, *A History of Electricity and Magnetism* (Cambridge, Mass.: M. I. T. Press, 1971), 107. The telegraph demonstrated in 1837 by Cooke and Wheatstone would have required five wire pairs for each direction. Later telegraphs, using Morse code, could send and receive through a single wire pair.

11. For a general history of trans-Atlantic cables, both telegraph and telephone, see Arthur C. Clarke, *Voices Across the Sea*, rev. ed. (New York: Harper and Row, 1974).

12. For a discussion of wire news practices and tensions between newspapers and telegraph companies in the first years of commercial telegraphy, see Thompson, *Wiring a Continent*, 220-23.

13. Harlow, *Old Wires and New Waves*, 199-200.

14. See Victor Rosewater, *History of Cooperative News-Gathering in the United States* (New York: D. Appleton and Company, 1930), 57-98; and Richard A. Schwarzlose, "Harbor News Association: The Formal Origin of the AP," *Journalism Quarterley* 45, no. 2 (Summer 1968), 253-60. An examination of the similarity of news dispatches in several newspapers suggests the existence of a cooperative press association in New York City several years earlier. See Richard A. Schwarzlose, "Early Telegraphic News Dispatches: Forerunner of the AP," *Journalism Quarterly* 51, no. 4 (Winter 1974), 596-601. For an account of the early operations of the New York Associated Press, as well as a detailed description of the use of telegraphic news in a typical daily in mid-1860s America, see "How We Get Our News," *Harper's New Monthly Magazine* 34, no. 102 (March 1867), 511-22.

15. For a detailed official history of the AP, see Oliver Gramling, *AP: The Story of the News* (New York: Farrar and Rinehart, 1940), 19-31, 60-68, 148-57.

16. North, *History and Present Condition of the Newspaper and Periodical Press*, 107.

17. Desmond, *The Press and World Affairs*, 56-60. Also see Doris Willens, "Reuters Celebrates Its 100th Anniversary, London, July 11," *Editor and Publisher* 84, no. 24 (9 July 1951), 9, 75-76.

18. John Hohenberg, *The Professional Journalist: A Guide to the Principles and Practices of the News Media*, 2nd ed. (New York: Holt, Rinehart and Winston, 1969), 111-14; and Grant Milnor Hyde, *Journalistic Writing*, 2nd ed. (New York: D. Appleton-Century Company, 1929), 163-72.

19. See Ben H. Bagdikian, *The Information Machines: Their Impact on Men and the Media* (New York: Harper and Row, 1971), 271-74; and Roger Burlingame, *Engines of Democracy: Inventions and Society in Mature America* (New York: Charles Scribner's Sons, 1940), 158-60.

20. For a study of the effects of the teletypesetter on the uncritical use of wire-service copy, see Scott M. Cutlip, "Content and Flow of AP News—From Trunk to TTS to Reader," *Journalism Quarterly* 31, no. 4 (Fall 1954), 434-46.

21. Susan R. Brooker-Gross, "The Changing Concept of Place in the News," in *Geography, the Media, and Popular Culture*, ed. Jacqeline Burgess and John R. Gold (London: Croom Helm, 1985), 63-85.

22. Some historians of journalism consider Tillotson the first news syndicate. See, for example, Lee, *The Daily Newspaper in America*, 580; and S. S. McClure, "Newspaper 'Syndicates'," *The Critic* 8, no. 186 (23 July 1887), 42-43.

23. McClure announced plans to increase the weekly service to 100,000 words by the end of 1887. See McClure, "Newspaper 'Syndicates'," 43. For a history of *McClure's Magazine*, see Harold S. Wilson, *Magazine and the Muckrakers* (Princeton, N.J.: Princeton University Press, 1970), 70.

24. For an account of Kellogg's activities, see Elmo Scott Watson, *A History of Newspaper Syndicates in the United States, 1865-1935* (Chicago, 1936), 5-25.

25. Preprints are still an important part of the American newspaper, especially on Sunday, in the form of advertising inserts and syndicated weekly magazines, such as *Parade* and *USA Weekend*. For a discussion of Sunday supplement magazines, see Hynds, *American Newspapers in the 1980s*, 209-12.

26. Lee, *Daily Newspaper in America*, 580.

27. Kubler, *New History of Stereotyping*, 318-20.

28. Elmo Scott Watson, *History of Auxiliary Newspaper Service in the United States* (Champaign, Ill.: Illini Publishing Company, 1923), 29-34.

29. Watson, *History of Newspaper Syndicates*, 34-35.

30. Watson, *History of Auxiliary Newspaper Service*, 33.

31. For a concise description of flat stereotyping from mats, see Kenneth E. Olson, *Typography and Mechanics of the Newspaper* (New York: D. Appleton and Company, 1930), 383-85.

32. Kubler, *New History of Stereotyping*, 324-31.

33. Hynds, *American Newspapers in the 1980s*, 72.

34. Information on the Scripps chain and its news services and feature syndicates is from Lee, *Daily Newspaper in America*, 212-16, 520-21, 589-90.

35. Watson, *History of Newspaper Syndicates*, 57.

36. Ibid., 57-58, 64-66.

37. For a discussion of Hearst's newspaper chain and news agencies, see Lee, *Daily Newspaper in America*, 215-20, 538-40.

38. Sidney H. MacKean, "How a Modern News Service Operates," *The Publisher's Guide* 21, no. 6 (June 1914), 24-28.

39. "Newspictures Ignore Profit and Loss," *Fortune* 11, no. 3 (September 1930), 60-67, 107-8.

40. Kent Cooper, *Kent Cooper and the Associated Press: An Autobiography* (New York: Random House, 1959), 128-40.

41. "Newspictures Ignore Profit and Loss," 107.

42. Lee, *Daily Newspaper in America*, 530-31.

43. Gramling, *AP: Story of the News*, 328-30.

44. Carl Hodge, "APN's Pictorial Maps a Showcase for News," *Editor and Publisher* 81, no. 16 (10 April 1948), 50.

45. For a brief history of photojournalism, see Tim Nachum Gidal, *Modern Photojournalism: Origin and Evolution*, vol. 1 of *Photography: Men and Movements* (New York: Macmillan, 1973), 6-13; and R. Smith Schuneman, *Photographic Communication* (New York: Hastings House, 1972), 35-41.

46. For a concise history of facsimile telecommunications, see Clarence R. Jones, *Facsimile* (New York: Murray Hill Books, 1949), 2-23.

47. For a description of a vector-mode "writing telegraph" apparatus, see "The Telautograph," *Nature* 64, no. 1648 (30 May 1901), 107-9.

48. "The Gray Telautograph," *Engineering News* 29, no. 12 (23 March 1893), 271-73.

49. For a description of Bain's raster-mode "automatic electrochemical recording telegraph," see Daniel M. Costigan, *FAX: The Principles and Practice of Facsimile Communication* (Philadelphia: Chilton Book Company, 1971), 2-5.

50. For a description of Korn's facsimile system, see T. Thorne Baker, *The Telegraphic Transmission of Photographs* (London: Constable and Company, 1910), 22-60; and Robert Grimshaw, "Korn's Photographic Fac-simile Telegraph," *Scientific American* 96, no. 7 (16 February 1907), 148.

51. Most of the examples were handwritten greetings from politicians. The entire first three pages of this issue were devoted to Gray's Telautograph. Yet the newspaper appears to have made little or no use of it in the weeks following.

52. "Pictures by Wire," *The Newspaper Maker* 9, no. 214 (27 April 1899), 5. This article also reports the purchase of rights to Hummel's invention by a syndicate of five newspapers—the Boston *Herald*, the Chicago *Herald-Times*, the New York *Herald*, the Philadelphia *Inquirer*, and the St. Louis *Republic*.

53. "An Electric Picture Telegraph," *Engineering News* 47, no. 18 (1 May 1902), 354-55.

54. Grimshaw, "Korn's Photographic Fac-simile Telegraph."

55. "Korn's New Telephotographic System," *Scientific American Supplement* 66, no. 1696 (4 July 1908), 14-16.

56. T. Thorne Baker, "The Telegraphy of Photographs, Wireless and by Wire," *Nature* 84, no. 2129 (18 August 1910), 220-26. Also see E. R. Stewart, "Telegraphing Pictures to the 'Daily Mirror'," *Penrose Annual* 13 (1907-8), 147-48.

57. "The Transmission of Photographs Over Telephone Wires," *Scientific American* 106, no. 22 (1 June 1912), 493-94.

58. Austin C. Lescarboura, "Sending Photographs Over Wires," *Scientific American* 123, no. 19 (6 November 1920), 474, 483-84.

59. "Telephoning Our Press Photographs," *Scientific American* 131, no. 2 (April 1924), 87, 139.

60. For descriptions of AT&T's commercial facsimile service operated during the late 1920s and early 1930s, see A. Dinsdale, "Commercial Picture Transmission," *Wireless World* 20, no. 17 (27 April 1924), 510-16; and F. W. Reynolds, "A New Telautograph System," *Electrical Engineering* 55, no. 9 (September 1936), 996-1007, esp. pp. 996-97.

61. For Kent Cooper's account of the AP's decision to develop a photowire network, see Cooper, *Kent Cooper and the Associated Press*, 212-21. Also see "(AP)," *Fortune* 15, no. 2 (February 1937), 88-93, 148-62, esp. 88-89, 91-92, 158; William E. Berchtold, "More Fodder for Photomaniacs," *North American Review* 239, no. 1 (January 1935), 19-30; Lee, *Daily Newspaper in America*, 532-35; and Anthony North, "No, But I Saw the Pictures," *New Outlook* 163, no. 6 (June 1934), 17-21.

62. John W. Perry, "A.P. Members Approve Telephoto Service After Heated Debate in Convention," *Editor and Publisher* 66, no. 50 (28 April 1934), 5, 110.

63. For a description of the transmitting and receiving equipment, see "A.P. Explains New Telephoto Service," *Editor and Publisher* 66, no. 50 (28 April 1934), 6, 111; and Reynolds, "A New Telephotograph System."

64. For a brief description of operations and procedures of the early AP Wirephoto network, see John W. Perry, "Wirephoto Found Flexible and Speedy; Operates with Uncanny Precision," *Editor and Publisher* 67, no. 35 (12 January 1935), III, XI.

65. Cartographer Richard Edes Harrison provided an instructive graphic explanation of the portable transmitter and the stationary receiver; see "(AP)," 92-93. The portable transmitter scanned a 10 by 12.5 cm (4 by 5 in.) photo at 79 lines per cm (200 lines per in.), to yield a 10 by 25 cm (8 by 10 in.) negative on the receiving unit.

66. The log for a single day—a Saturday, unfortunately—is presented in Perry, "Wirephoto Found Flexible and Speedy," XI.

67. "A.P. Wirephotos Flash Across Nation," *Editor and Publisher* 67, no. 34 (5 January 1935), 7.

68. Bice Clemow, "Picture Services Rushing into Field of Telephotograph Transmission," *Editor and Publisher* 69, no. 9 (29 February 1936), 3-4; Lee, *Daily Newspaper in America*, 535-36; and "News Pictures by Wire," *Electronics* 10, no. 11 (November 1937), 12-17, 82-83.

69. For a concise appraisal of news picture services in the early 1950s, see Kalish and Edom, *Picture Editing*, 65-69.

70. At its inauguration in 1935, AP Wirephoto served 39 dailies directly, and 10 to 12 other papers through "expedited delivery"; see "A.P. Wirephotos Flash Across Nation." Estimates for the early 1980s are from Jonathan Fenby, *The*

International News Services (New York: Schocken Books, 1986), 104. The AP declined to provide more precise counts.

71. Fenby, *International News Services.* UPI declined to provide more precise counts. In 1984 UPI sold its foreign news picture operations to Reuters, and its number of domestic clients undoubtedly declined throughout the 1980s.

72. Although the AP does not divulge its charges to member papers for various services, several news executives, who prefer to remain anonymous, corroborated this estimate of the cost of Laserphoto service. The AP generally has assessed members for a share of operating expenses prorated on the basis of population served. For a concise explanation of the AP's assessment formula, see "History of AP Assessment Formula," *Editor and Publisher* 117, no. 12 (24 March 1984), 12.

73. The Associated Press established a research laboratory in 1933, under the direction of Harold Carlson. For a concise summary of the laboratory's contribution to AP Wirephoto, see "Technical Progress," *APME Red Book* 11 (1958), 40-44.

74. For a concise description of the scanner-engraver and its operation, see C. A. Harrison, "The Fairchild Photo-Electric Engraver," *Penrose Annual* 45 (1951), 104-5.

75. See, for example, W. Curtis Ross, "More Pictures at Less Cost for this Small Town Paper," *Bulletin of the American Society of Newspaper Editors* no. 370 (1 November 1954), 10.

76. The report of the AP Managing Editors' Technical Progress Committee for 1954 described the efficiency of the Photofax receiver for newspaper members as well as for television stations receiving Wirephoto service. The report noted that "The biggest advantage for smaller newspapers and TV stations is that facsimile reception requires no attendant or darkroom facilities." See "Facsimile," *APME Red Book* 7 (1954), 100-103.

77. For discussion of electrolytic recording paper and the helix-and-blade process used with Photofax, see Costigan, *FAX: Principles and Practice*, 78-80.

78. Ray Erwin, "Facsimile Photo Machines Used by 300 Newspapers," *Editor and Publisher* 89, no. 30 (21 July 1956), 11, 50.

79. See ibid., and "Photos Edited Like News Copy by Use of Fax," *Editor and Publisher* 87, no. 12 (13 March 1954), 56.

80. "AP Photofax Device Hits at Moisture," *Editor and Publisher* 89, no. 40 (29 September 1956), 63.

81. Some newspapers used a Photofax recorder to monitor the photowire and a photographic Wirephoto recorder to produce negatives for high-quality reproduction. Muirhead and Company, Ltd., of London, manufactured the new AutoPhoto recorder. See "Dry 8 x 10 Glossies Delivered by Wire," *Editor and Publisher* 97, no. 47 (21 November 1964), 48; and "Wirephoto Glossies from New Receiver," *Editor and Publisher* 98, no. 14 (3 April 1965), 67. During the 1940s and 1950s the AP had purchased its facsimile transmitters from Times Facsimile Corporation, with manufacturing facilities on the seventh floor of the Times Building, in Mid-town Manhattan. Although the New York Times Company sold Wide World Photo to the AP in 1940, it retained the facsimile manufacturing business, which produced transmitters and receivers for newsphotos, weather

maps, and military and diplomatic uses. See Ray Erwin, "N.Y. Times Developing New 'Fax' Machines," *Editor and Publisher* 83, no. 25 (17 June 1950), 18.

82. For a description of the portable transmitter, see "UPI Photo Transmission All Automatic," *Editor and Publisher* 104, no. 6 (6 February 1971), 13. For a discussion of the Unifax II receiver, see "Unifax II Ranks As One of Best New Products," *Editor and Publisher* 108, no. 40 (4 October 1975), 13.

83. UPI used a lesser resolution of 43 lines per cm (109 lines per in.) for international transmission by radio or cable. See Daniel M. Costigan, *Electronic Delivery of Documents and Graphics* (New York: Van Nostrand Reinhold Company, 1978), 195.

84. For discussion of the use of laser light in the AP Laserphoto recorder, see ibid., 77-78, 87-89.

85. For a description of the Electronic Darkroom, see David E. Herbert, "Digitizing and Storing Graphics in the AP Electronic Darkroom," *Editor and Publisher* 115, no. 10 (6 March 1982), 26-29.

86. Jenk Jones, Jr., and others, "AP Laserphoto: How It Works," In *Report of the APME Photo and Graphics Committee*, ca. 1978, esp. pp. 7-8.

87. Photowire transmission of color separations of news photographs had been used at least as early as 1938, when Scotland's *Glasgow Record* experimented with sending color photos from London. See F. W. Plews, "The Electrical Transmission of Colour Photographs for Newspaper Printing," *Penrose Annual* 40 (1938), 166-67. Color photowire transmission was practicable but rare in 1956, when AP Wirephoto sent color pictures from a political convention for the first time. See "Convention 'First' Color via Wirephoto," *Editor and Publisher* 89, no. 34 (18 August 1956), 12.

88. Solar noise can interfere with satellite signals around the Spring and Fall equinoxes, when for periods of several minutes the satellite might lie in a straight line between the sun and either the receiving dish or the transmitting antenna. But calculations based on the precise locations of the satellite, the ground transmitter, and the receiving stations can predict these mini-eclipses years ahead with accuracy to the second. See "2 Wire Services Comment on Solar Noise Interference to Satellites," *Editor and Publisher* 115, no. 15 (10 April 1982), 28.

89. The AP, for instance, had projected a 69-percent increase between 1983 and 1984 in the cost of land-line telecommunications. See "AP to Expand Satellite System," *Editor and Publisher* 116, no. 52 (24 December 1983), 34.

90. Bill Gloede, "UPI: Goodbye AT&T, Hello Federal?" *Editor and Publisher* 116, no. 30 (23 July 1983), 11, 36.

91. See "A Marriage Made in Heaven?" *Editor and Publisher* 117, no. 9 (3 March 1984), 42; Lloyd Carver, "AP Buys Two Transponders from Western Union," *Editor and Publisher* 117, no. 28 (14 July 1984), 11; "London-based Firm Acquires Telecom Interest," *Editor and Publisher* 118, no. 8 (23 February 1985), 34; and "UPI Wants to Void Guild Contract," *Editor and Publisher* 118, no. 30 (27 July 1985), 14.

92. "UPI Distributes Accu-Weather," *Editor and Publisher* 118, no. 3 (19 January 1985), 9.

93. "AP Enters the Commercial Market," *Editor and Publisher* 118, no. 16 (20 April 1985), 31.

94. *APME Red Book* 2 (1949), 109.
95. *APME Red Book* 27 (1974), 212.
96. *APME Red Book* 28 (1975), 183.
97. *APME Red Book* 3 (1950), 90.
98. *APME Red Book* 7 (1954), 103-4.
99. *APME Red Book* 17 (1964) , 177-78.
100. *APME Red Book* 18 (1965), 165.
101. *APME Red Book* 19 (1966), 162.
102. *APME Red Book* 21 (1968), 45.
103. *APME Red Book* 24 (1971), 175-79.
104. *APME Red Book* 31 (1978), 168-71.
105. *APME Red Book* 32 (1979), 296.
106. Andrew Barnes, "AP's Graphics," *APME Red Book* 33 (1980), 242-44.
107. *APME Red Book* 34 (1981), 73-74.
108. Rod Carwell, "Open Cleaner Style on AP Graphics Gets Rave Notices," in *Report of the APME Photo and Graphics Committee*, Toronto, 1981, 15-16.
109. *APME Red Book* 36 (1983), 212-17.
110. Lockwood taught commercial art and served as art director of the Allentown *Morning Call*. He redesigned several newspapers and was the first president of the Society of Newspaper Design. He founded News Graphics in 1983, with the AP as his principal client. With three partners, he operates a second firm, Computer News Graphics, which has developed a cartographic and news-art database as well as software for the interactive design of information graphics that can be used directly or transmitted over a photowire.
111. See "AP Announces Three Major Service Enhancements," *Editor and Publisher* 118, no. 19 (11 May 1985), 20, 48.
112. "UPI Expands Graphics Dept.," *Editor and Publisher* 117, no. 33 (18 August 1984), 10, 35.
113. For background on UPI's financial difficulties in the 1980s, see Roger Tatarian, "UPI's Woes: A Historical Perspective," *Editor and Publisher* 118, no. 36 (7 September 1985), 52, 42, 31.
114. For a comparison of AP and UPI facsimile transmission quality, see Gerald B. Healey, "Photo Transmission Hit by Production Managers," *Editor and Publisher* 110, no. 37 (10 September 1977), 26, 50.
115. "UPI Adds New 'Special Sections Packages'," *Editor and Publisher* 117, no. 42 (20 October 1984), 26.
116. "UPI Chairman Seeks Newspapers' Understanding," *Editor and Publisher* 118, no. 19 (11 May 1985), 34.
117. Margaret Genovese, "UPI: Asking for a Year," *Presstime* 8, no. 12 (December 1986), 6-9.
118. The *Cortland Standard*, the Rome *Daily Sentinel* and the *Watertown Daily Times* were independent. Gannett Newspapers owned the *Ithaca Journal*, the Utica *Daily Press*, and the Utica *Observer Dispatch*. Newhouse Newspapers owned the Syracuse *Post-Standard* and the Syracuse *Herald-Journal*. Howard Publications owned the Auburn *Citizen*, Ingersoll Publications Co. owned the *Oneida Daily Dispatch*, Park Communications owned the Norwich *Evening Sun*,

and Thomson Newspapers owned the Oswego *Palladium Times*. As shown here, some newspapers include the name of the city of publication in their official name, whereas others do not.

119. The *Editor and Publisher Yearbook* groups newspapers into eight circulation categories: less than 5,000; 5,000 to 10,000; 10,001 to 25,000; 25,001 to 50,000; 50,001 to 100,000; 100,001 to 250,000; 250,001 to 500,000; and over 500,000. *Circulation*, an annual compilation of circulation data for advertisers, includes maps with symbols representing newspapers grouped into three categories, with breaks at 10,000 and 100,000. *Presstime*, a monthly magazine of the American Newspaper Publishers Association, often uses only three circulation groups: up to 25,000; 25,001 to 100,000; and over 100,000. A recent article on small daily newspapers discussed those with circulations under 40,000; see Paul Kruglinski, "Color in Small Dailies," *Presstime* 7, no. 11 (November 1985), 10-11.

120. The author visited each newspaper in 1984 or 1985, to interview news editors and art personnel, if any.

121. Most of the newspapers studied were unaffected by mergers, although a few changed their names slightly once or twice. For example, the Norwich *Morning Sun*, founded in 1891, had become the *Norwich Sun* by 1910. It changed its name again, in the 1960s, to the Norwich *Evening Sun*.

122. The mediocre positive relationship described graphically in Figure 3.22 is described numerically by a product-moment correlation coefficient of 0.65 and a Spearman rank-order correlation coefficient of 0.67. According to the product-moment correlation, only 42 percent of the variation in map use can be accounted for, in a statistical sense, by circulation.

123. See, for example, Sverre Petterssen, *Introduction to Meteorology*, 3rd ed. (New York: McGraw-Hill, 1969), 10. Earlier maps of weather elements exist, though, and at least one historian of cartography gives this famous first a much earlier date. Arthur Robinson cited as the "first meteorological chart" the map of tradewinds and monsoons that Edmund Halley published in 1686. See Arthur H. Robinson, *Early Thematic Mapping in the History of Cartography* (Chicago: University of Chicago Press, 1982), 46-49, 69-79; and Norman J. W. Thrower, "Edmund Halley As a Thematic Geo-Cartographer," *Annals of the Association of American Geographers* 59, no. 4 (December 1969), 652-76.

124. The Smithsonian Institution, under the direction of Joseph Henry, collected weather data by telegraph in 1849, and distributed daily weather maps in 1854. See Roy Popkin, *The Environmental Science Services Administration* (New York: Frederick A. Praeger, 1967), 52-55. The earliest published telegraphic weather data appear to be wind-direction and sky-condition data for August 30, 1848, presented in the London *Daily News* for August 31, 1948. The newspaper arranged and paid for collection of these data. See William Marriott, "The Earliest Telegraphic Daily Meteorological Reports and Weather Maps," *Quarterly Journal of the Royal Meteorological Society* 29, no. 126 (April 1903), 123-31.

125. Marriott, "Earliest Telegraphic Reports"; and Mark W. Harrington, "History of the Weather Map," *U.S. Weather Bureau Bulletin* no. 11, pt. 2 (1894), 327-35.

126. See R. H. Scott, "Weather Charts in Newspapers," *Journal of the Society of Arts* 23, no. 1183 (23 July 1875), 776-82; and "The 'Times' Weather Chart," *Nature* 11, no. 285 (15 April 1875), 473-74.

127. Edgar B. Calvert, "Development of the Daily Weather Map," *U.S. Weather Bureau Bulletin* no. 24 (1899), 144-50; and Robert De Courcy Ward, "The Newspaper Weather Maps of the United States," *American Meteorological Journal* 11, no. 3 (July 1894), 96-107.

128. Ward, "Newspaper Weather Maps."

129. Henry L. Heiskell, "The Commercial Weather Map of the United States Weather Bureau," in *Yearbook of Agriculture, 1912*, pp. 537-39.

130. U. S. Weather Bureau, *Report of the Chief of the Weather Bureau, 1911-1912* (Washington, D.C. : Government Printing Office, 1913), 24.

131. The article, by a Weather Bureau official, appeared eight days after the first *New York Times* weather map. See C. F. Talman, "The Elements of Forecasting," *New York Times*, 12 August 1934, section 8, p. 2.

132. Patrick Hughes, *A Century of Weather Service: A History of the Birth and Growth of the National Weather Service, 1870-1970* (New York: Gordon and Breach, 1970), 61.

133. For discussion of how *USA Today* prepared its weather package, see Jim Norman, "How USA Today Does It," *Design, Journal of the Society of Newspaper Design* no. 12 (Summer 1983), 4-6; and Edward F. Taylor, "Telling the Weather Story," *Weatherwise* 36, no. 2 (April 1983), 52-59.

134. Douglas A. Anderson and Claudia J. Anderson, "Weather Coverage in Dailies," *Journalism Quarterly* 63, no. 2 (Summer 1986), 382-85; and Mark Monmonier and Val Pipps, "Weather Maps and Newspaper Design: Response to *USA Today*?" *Newspaper Research Journal* 8, no. 4 (Summer 1987), 31-42.

135. George Garneau, "Weather Graphics Via PCs," *Editor and Publisher* 119, no. 40 (4 October 1986), 62-64.

136. Barry Bradley, "You Can Be a Weather Writer—and a Newsroom Star," *Bulletin of the American Society of Newspaper Editors* no. 664 (November 1983), 18-19.

137. For an overview of weather graphics services, see Gene Goltz, "Weather Pages," *Presstime* 9, no. 3 (March 1987), 16-19.

138. Menus of AP Access graphics distributed over the Laserphoto network are available 24 hours a day to AP members through an electronic bulletin board at Accu-Weather's VAX computer in State College, Pennsylvania. See George Garneau, "Improving Wire Service Graphics," *Editor and Publisher* 120, no. 12 (21 March 1987), 44.

Chapter 4: Map Use in Print Journalism

1. Each flap is a photographic mask, on which the orange areas indicate where a specified graytone is to appear on the finished illustration. A contact exposure taken from the flap yields a negative mask, which is opaque except in areas covered by orange on the flap. These orange areas thus have been replaced

by "open windows" through which a screened image can be added to the final "composite positive," on which all line elements and screened graytone appear. For a discussion of the use of flaps and tint screens in the production of negative cartographic artwork, see David J. Cuff and Mark T. Matson, *Thematic Maps: Their Design and Production* (New York: Methuen, 1982), 117-21.

2. A writer discussing labor and management problems of art departments described a situation in which "printers held an intolerable jurisdiction over the assembly of maps." Although he mentioned no paper by name, the conditions described were identical to those at the *Globe*. See Phillip Ritzenberg, "Backshop Blues Got You Down?" *Design, The Journal of the Society of Newspaper Design* no. 12 (Summer 1983), 30-31.

3. Artist Deborah Perugi, who drew many of the maps in the *Boston Globe*, has won several awards in the annual map design contest sponsored by the American Congress on Surveying and Mapping.

4. Nancy Tobin, *Understanding Changes in the Growth and Shape of Newspaper Art Departments* (Reston, Va.: Society of Newspaper Design, 1985). Tobin's study is based upon a 64-question survey distributed to over 300 registrants at the 1984 annual workshop of the Society of Newspaper Design.

5. Dick Furno's turnkey system was marketed as the Azimuth Computer Mapping System. See Craig L. Webb, "Syndicated and Computer Graphics," *Presstime* 7, no. 8 (August 1985), 26-28.

6. Leimer examined 207 maps found in 231 newspaper issues representing a week's publication by 36 newspapers. See Judith A. Leimer, "The Influences of the Technical Constraints and Personnel Limitations on the Quality of Maps in American Newspapers" (M.S. thesis, University of Wisconsin-Madison, 1982), 18-19.

7. Also see Mark Fitzgerald, "Freelance Graphics," *Editor and Publisher* 118, no. 45 (9 November 1985), 18; and Edward Canale, "Finding an Artist," *Design, The Journal of the Society of Newspaper Design* no. 17 (Fall 1984), 30-31.

8. For a brief review of the duties of art directors, and short biographical sketches of the careers and current responsibilities of five art directors, see William C. Sexton, "The Explosion in Graphics," *Bulletin of the American Society of Newspaper Editors* no. 547 (January 1971), 16-22. For additional discussion of the operation and management of art departments, see Robert J. Betcher, "How to Produce Art and Graphics," *Design, The Journal of the Society of Newspaper Design* no. 3 (September 1980), 19-20; and Bill Ostendorf, "Graphics on a Budget," *Design, The Journal of the Society of Newspaper Design* no. 16 (Summer 1984), 26.

9. At the time of the author's visit in 1985, the top drawers in each of three 5-drawer lateral filing cabinets contained only graphs and charts. The *Post* also had flat map files for oversize artwork and reference materials. According to several staff members, additional materials were stored "in a closet in the ladies' room." At that time, the *Post*'s art department, like many elsewhere, was pressed for space.

10. Michael Kidron and Ronald Segal, *The State of the World Atlas* (New York: Simon and Schuster, 1981), and GAIA Ltd. staff and Norman Myers, *GAIA: An Atlas of Planet Management* (New York: Doubleday, 1984).

11. The *National Atlas of the United States* was published in 1970, with a dedication page signed by then-president Richard Nixon, and a foreword by Walter J. Hickel, the Secretary of the Interior "dismissed" by Nixon after Hickel conceded that youthful protesters of the Vietnam War should be heard. The *National Atlas* was out-of-print by 1974, and was never reissued, despite several efforts by the Geological Survey to initiate a second edition. See Peter Henry Girard, "A Comparison of National Atlas Programs in the United States and Canada" (M.A. thesis, Syracuse University, 1986), 41-45, 48-50.

12. For discussion of the role of the graphics editors, see Howard I. Finberg, "In the Beginning, Graphics Editors Were As Popular As OPEC," *Design, The Journal of the Society of Newspaper Design* no. 17 (Fall 1984), 4-6; Lenora Williamson, "Management and Staff Must Care," *Editor and Publisher* 119, no. 23 (7 June 1986), 18-19, 41; and Ray Wong, "The Clarion Ledger," *Design, The Journal of the Society of Newspaper Design* no. 7 (September 1981), 13-14.

13. Dick D'Agostino and Michael Dresser, *The Baltimore Sun Typographic Design Stylebook* (Baltimore: The Sun, 1984), 34.

14. Ibid., 35.

15. Some individual pages had been revised, and a rewrite was planned at the time of my visit. I was allowed to examine, but asked not to quote from, the *Tribune* stylebook.

16. Fine screens often become blotchy or overly dark because of ink spread, in which the tiny dots of the screen grow larger during printing. For a discussion of how ink spread can devastate maps with screened graytone area symbols, see Mark Monmonier, "The Hopeless Pursuit of Purification in Cartographic Communication: A Comparison of Graphic-Arts and Perceptual Distortions of Graytone Symbols," *Cartographica* 17, no. 1 (1980), 24-39.

17. For discussion of the problems experienced during the 1960s and 1970s by newspapers adopting computers for editorial text processing, see Ernest C. Hynds, *American Newspapers in the 1970s* (New York: Hastings House, 1975), 242-51; and Joseph M. Ungaro, "The Electronic Newsroom," *Bulletin of the American Society of Newspaper Editors* no. 568 (April 1973), 9-11.

18. See, for example, Stanley Klein, "Coping with the Doldrums," *S. Klein Computer Graphics Review* 2, no. 1 (Fall 1986), 11-12; and Stephen T. McClellan, *The Coming Computer Industry Shakeout* (New York: John Wiley and Sons, 1984).

19. See Ron Couture, "Designing Graphics with a Computer," *Design, The Journal of the Society of Newspaper Design* no. 12 (Summer 1983), 28-29. For a review of SAS/GRAPH, see Alan M. Baker, Rowan Faludi, and David R. Green, "An Evaluation of the SAS/GRAPH Software for Computer Cartography," *Professional Geographer* 37, no. 2 (March 1985), 204-14.

20. For discussion of the objectives and operation of electronic pagination systems, see Paul Kruglinski, "Pagination," *Presstime* 7, no. 3 (March 1985), 30-31; William C. Porter, "Eventually—Pagination Saves Time," *Editor and Publisher* 119, no. 25 (21 June 1986), 32, 123; and Phillip Ritzenberg, "The Coming Effects of Technology," *Design, The Journal of the Society of Newspaper Design* no. 3 (September 1980), 8-9.

21. For discussion of the difficulties of including photos on pages composed with early-1980s pagination systems, see Bob Bradley, "Paging the Eighties,"

Penrose Annual 72 (1980), 77-84; Mark Fitzgerald, "Photo Scanners Move into Pagination," *Editor and Publisher* 119, no. 27 (5 July 1986), 26-29; and David B. Gray, "Pagination Puzzle Still Missing Pieces," *Design, The Journal of the Society of Newspaper Design* no. 18 (Winter 1985), 7-8.

22. For accounts of on-site trials of pagination systems, including that at Gannett's Utica newspapers, see George Garneau, "Another Computer-to-Plate Test in the Works," *Editor and Publisher* 118, no. 12 (23 March 1985), 45; George Garneau, "Computer-to-Plate Project to End," *Editor and Publisher* 118, no. 12 (23 March 1985), 32-33; and Tom Walker, "Pagination Breakthrough May Finally Be at Hand," *Presstime* 4, no. 12 (December 1982), 38-40.

23. For discussion of the development of the personal computer in the late 1970s and early 1980s, see Hoo-Min D. Toong and Amar Gupta, "The Computer Age Gets Personal," *Technology Review* 86, no. 1 (January 1983), 26-27. For an examination of the steady increase in efficiency of digital computers, see Humberto Gerola and Ralph E. Gomory, "Computers in Science and Technology," *Science* 225, no. 4657 (6 July 1984), 11-18.

24. For discussion of IBM's role during the early 1980s in the development of microcomputer technology, see "Personal Computers: And the Winner is IBM," *Business Week* no. 2810 (3 October 1983), 76-95; and M. David Stone, "Has IBM Finally Done It?" *Science Digest* 93, no. 2 (February 1985), 78-79. For a concise evaluation of the PC/2, see Tom Moran and Ed Foster, "IBM Models Offer Diverse Capabilities," *Info World* 9, no. 14 (6 April 1987), 3, 80.

25. Furno's system is discussed in Stuart Silverstone, "Newspapers Turn to Mac's Graphics," *Computer Graphics World* 10, no. 5 (May 1987), 81-86.

26. For discussion of the Computer News Graphics project, see "Coming Attraction: Designer Graphics by Wire," *Design, The Journal of the Society of Newspaper Design* no. 15 (Spring 1984), 7; and Webb, "Syndicated and Computer Graphics," 26-28. The AP contributed to the development of the database, for which it received a copy.

27. For a brief history of Apple Computer, Inc., and its role in the development of microcomputer technology, see "Apple Computer Reaches for Its Macintosh," *Economist* 290, no. 7326 (28 January 1984), 68-69; Stephen Kindel, "Applesauce," *Forbes* 133, no. 4 (13 February 1984), 39-41; Phillip Robinson, "The Macintosh Plus," *Byte* 11, no. 6 (June 1986), 85-90; and Ed Yasaki, "Big Mac Attack," *Datamation* 30, no. 2 (February 1984), 61-64. For an overview of the Macintosh SE and the Macintosh II, see Brenton R. Schlender, "Apple to Unveil Two Macintosh Models to Step Up Role in Business Computers," *Wall Street Journal*, 2 March 1987, 4.

28. For a lucid introduction to MacDraw and to MacPaint, which produces comparatively coarse pictures of little use in newspaper graphics, see Vahé Guzelimian, *Becoming a MacArtist* (Greensboro, N.C.: Compute! Publications, 1985).

29. Microsoft Chart is a product of Microsoft Corporation. MacAtlas is a product of Micro:Maps, a firm founded by geographer Robert Dahl. MacAtlas is one of several clip-art files mentioned in Erfert Nielson, "Art to Go," *Macworld* 3, no. 12 (December 1986), 130-37. Another geographer, Michael Peterson, developed MacChoro, a program for making choropleth maps on the Macintosh.

Eugene Turner reviewed MacChoro in *American Cartographer* 14, no. 1 (January 1987), 69-71.

30. For discussion of the LaserWriter's development and operation, see Dwight B. Davis, "Business Turns to In-House Publishing," *High Technology* 6, no. 4 (April 1986), 18-26; "Enter the Business Mac—Apple in the Office," *Modern Office Technology* 30, no. 3 (March 1985), 34-41; Danny Goodman, "The Laser's Edge," *Macworld* 2, no. 2 (February 1985), 70-79; and Cary Lu, "Laser Printers Zap the Price Barrier," *High Technology* 4, no. 9 (September 1984), 52-57.

31. George Garneau, "Using Personal Computers to Put Out a Newspaper," *Editor and Publisher* 119, no. 11 (15 March 1986), 32; Stuart Silverstone, "Newsroom Graphics," *Macworld* 4, no. 2 (February 1987), 130-35; and Stuart Silverstone and Craig L. Webb, "Many Newspapers, Small and Large, Turn to 'The Mac'," *Presstime* 8, no. 4 (April 1986), 23-26.

32. Roger Fidler, "Knight-Ridder Network On-line," *Deadline Mac* 1, no. 4 (July 1986), 2. Also see "Knight-Ridder Launches Computer Graphics Network," Editor and Publisher 118, no. 36 (7 September 1985), 30; and Louis Mintz, "Graphics Networks," *Presstime* 8, no. 9 (September 1986), 12-14.

33. Jeffrey L. Albert, "Gannett's Macintosh Menu," *Deadline Mac* 1, no. 4 (July 1986), 2; and Paul Lacy, "USA TODAY Artists Try Their Hand at New Computer Graphics System," *Gannetteer* (March 1985), 16-17.

34. Marcy Eckroth Mullins and Buzzy Albert, "USA TODAY: How to Draw Maps," *Deadline Mac* 1, no. 2 (May 1986), 1-2.

35. Richard A. Curtis (interviewed), "The Evolution of USA Today," *Design, The Journal of the Society of Newspaper Design* no. 9 (Fall 1982), 5-12.

36. McClelland, "Sprucing Up an Old Gray Lady," 23-26. McClelland attributes the increased use of art in the *Wall Street Journal* to a deliberate decision of top management, specifically associate publisher Peter Kann and managing editor Norman Pearlstine.

37. Arnold Thackray and Robert K. Merton, "On Discipline Building: the Paradoxes of George Sarton," *Isis* 63, no. 219 (December 1972), 473-95.

38. "Newspaper Designers, Editors Give Birth to a New Organization," *Newspaper Design Notebook* 1, no. 2 (March/April 1979), 10-11; and "Society of Newspaper Designers on the Drawing Board," *Newspaper Design Notebook* 1, no. 1 (January/February 1979), 3.

39. *Society of Newspaper Design 1986 Membership Directory* (Reston, Va.: 1986), 45-46.

40. Bill Ostendorf, "Who Are We?" *Design, The Journal of the Society of Newspaper Design* no. 18 (Winter 1985), 12-13. Because of concentrations in college towns, such as Columbia, Missouri, many of those without job titles were believed to be students.

41. At a regional meeting of the Society of Newspaper Design, in Syracuse in April 1986, and at a seminar for artists and illustrators, in St. Petersburg, Florida, at the Poynter Institute, in December 1985, the author met a number of newspaper artists who were not SND members.

42. See Howard I. Finberg, "Mapmaker, Mapmaker, Make Me a Map," *Design, The Journal of the Society of Newspaper Design* no. 15 (Spring 1984), 9-11; and Kate Newton Anthony, "Base Maps, Files and Simple Rules Guide USA

Today's Graphics," *Design, The Journal of the Society of Newspaper Design* no. 18 (Winter 1985), 9.

43. *The Best of Newspaper Design*, 6th ed. (Washington, D.C.: Society of Newspaper Design, 1985), 286-300.

44. See Gary Bradford Chappell, "Newsweekly Magazine Maps: Coverage of the Falkland Islands Crisis in *Newsweek, Time*, and *U.S. News and World Report*" (M.S. thesis, University of Wisconsin-Madison, 1983), 163-72.

45. Nigel Holmes, *Designer's Guide to Creating Maps and Charts* (New York: Watson-Guptill Publications, 1984). Also see Nigel Holmes and Rose DeNeve, *Designing Pictorial Symbols* (New York: Watson-Guptill Publications, 1985). For discussion of Holmes's style and influence, see Richard Brown, "Nigel Holmes: Time's Graphic Statistician," *Penrose Annual* 73 (1981), 9-24.

46. Richard Natkiel, *Decisive Battles of the Twentieth Century: Land-Sea-Air* (London: Sidgwick and Jackson, 1976).

47. Daniel Beaudat and Serge Bonin, *Cartes et Figures de la Terre* (Paris: Centre Georges Pompidou, 1980), 455-58.

48. Leimer, "Influences of Technical Constraints and Personnel Limitations," 109-15.

49. Chappell, "Newsweekly Magazine Maps," 141-47.

50. Patricia Gilmartin, "The Design of Journalistic Maps: Purposes, Parameters, and Prospects," *Cartographica* 22, no. 4 (Winter 1985), 1-18. For a journalist's perspective on cartographic coverage of the KAL 007 disaster, see Mario R. Garcia and Lynn Price, "The Front Page: Flexing to Fit the Big News," *Washington Journalism Review* 6, no. 2 (March 1984), 40-45.

51. See, for example, W. G. V. Balchin, "Media Map Watch: A Report," *Geography* 70, no. 309, pt. 4 (October 1985), 339-43.

52. A list of expected participants furnished by Christopher Board, one of the organizers, included 29 persons, only eight of whom worked in the news media. The remainder were academicians and school teachers, non-news media cartographers, and map enthusiasts. Three of the eight journalists were on the program, and according to Board, at least one of the other five expected participants failed to attend. Also see W. G. V. Balchin, "Media Map Watch," *Geographical Magazine* 57, no. 8 (August 1985), 408-9.

53. For comments on Nigel Holmes and the newspaper design and graphics workshops at the Rhode Island School of Design, see David B. Gray, "Information Please," *Design, The Journal of the Society of Newspaper Design* no. 15 (Spring 1984), 3-8.

54. Holmes, *Designer's Guide to Creating Maps and Charts*, 9.

55. Edward R. Tufte, *The Visual Display of Quantitative Information* (Cheshire, Conn.: Graphics Press, 1983), 80.

56. Tufte's sample is based on issues selected at random from those published between 1974 and 1980. Tufte counted as "relational" any graphic involving more than one variable, but excluded maps and time-series charts. Ibid., 83-84.

57. For a critical appraisal of local coverage by Gannett papers, see Bagdikian, *The Media Monopoly* , 69-91, esp. p. 79-80.

58. For discussion of *USA Today*'s marketing concepts and graphics, see Richard A. Curtis, "Doing Graphics for Your Readers," *Proceedings of the American*

Society of Newspaper Editors, 1983 convention, 291-302; "GANSAT: Another Different Voice Takes Shape," *Gannetteer* (June 1981), 4-7; George Garneau, "Color Quality Control: How It's Done at USA Today," *Editor and Publisher* 118, no. 3 (19 January 1985), 11, 22-33; Geraci, "Comparison of Graphic Design and Illustration Use in Three Washington, D.C., Newspapers"; and "Looking at USA Today" panel discussion, *APME Red Book* 36 (1983), 46-56.

59. Edward D. Miller, "Newspaper Design: Visual and Visionary," *Bulletin the American Society of Newspaper Editors* no. 622 (March 1979), 3-5.

60. Edward D. Miller, Robert Lockwood, and Jeff Lindenmuth, "Allentown: Designing for Each Day's News," *Newspaper Design Notebook* 1, no. 2 (March/April 1979), 1, 13-15.

61. The effect of Miller and Lockwood upon newspaper design seems more lasting than their imprint upon the *Morning Call*. Miller became its publisher in 1979, and left in 1981, and Lockwood formed his own graphics service in 1980. The *Morning Call* and the *Evening Chronicle* were combined into a single paper in 1980, and the less graphically indoctrinated staff of the afternoon paper displaced many of those who had been hired by Miller, or had otherwise learned to appreciate information graphics. In 1984, Miller's father sold the paper for $108 million—a huge profit. The *Morning Call*, with a circulation of about 128 thousand, still appears to use more graphics than the average paper of its size, but its design and use of information graphics is notably less innovative than in the late 1970s.

62. Edward D. Miller, "Newspaper Design: Visual and Visionary," *Bulletin of the American Society of Newspaper Editors* no. 622 (March 1979), 3-5 (ref. p. 5).

63. See, for example, Sid J. Segalowitz, *Two Sides of the Brain: Brain Lateralization Explored* (Englewood Cliffs, N.J.: Prentice-Hall, 1983); and Sally P. Springer and Georg Deutsch, *Left Brain, Right Brain* (San Francisco: W. H. Freeman, 1981). For a short article by a journalist who recognized a separation of his peers into verbal and visual people, see Harry Stapler, "Attitudes on Design," *Design, The Journal of the Society of Newspaper Design* no. 3 (September 1980), 17-18. Rejection or neglect of graphics by news editors was recognized as a problem much earlier. See, for example, George S. Crandall, "Now, As to the Matter of Newspaper Pictures," *Bulletin of the American Society of Newspaper Editors* no. 86 (16 November 1934), 3-4.

64. Edwin L. Shuman, *Practical Journalism* (New York: D. Appleton and Company, 1903). For his chapter on newspaper art, according to his preface, Shuman, an instructor at Northwestern University, apparently relied heavily upon William Schmedtgen, art director at the Chicago *Record-Herald*.

65. George C. Bastian, *Editing the Day's News* (New York: Macmillan, 1924), 148. Bastian was an instructor at Northwestern University's Medill School of Journalism.

66. Ibid., 9.

67. John J. Floherty, *Your Daily Paper* (Philadelphia: J. B. Lippincott, 1938).

68. "Troy Member Sees Need for Morgue of Maps," *Bulletin of the American Society of Newspaper Editors* no. 133 (1 November 1936), 2.

69. Theodore M. Bernstein, "Maps: An Important Reader Tool," *Bulletin of the American Society of Newspaper Editors* no. 483 (1 February 1965), 1-3.

70. AP Photo and Graphics Committee, "Continuing Study Report," *APME Red Book* 31 (1978), 168-71. The report included suggestions from Andrew Sabbatini, of the *New York Times*; Gus Hartoonian, of the *Chicago Tribune*; and Frank Peters, of the *St. Petersburg Times*.

71. Howard S. Shapiro, "Giving a Graphic Example: The Increased Use of Charts and Maps," *Nieman Reports* 36, no. 1 (Spring 1982), 4-7.

72. Edmund C. Arnold, "Map Out Future Use of Expository Art," *Publisher's Auxiliary* 121, no. 8 (22 April 1985), 5.

73. Edmund C. Arnold, *Functional Newspaper Design* (New York: Harper and Row, 1956), 146-47.

74. John E. Allen, *Newspaper Makeup* (New York: Harper and Brothers, 1936).

75. Ibid., 157. The New York *Daily Graphic* printed a daily weather map between May 9, 1879, and September 14, 1882, but the *Times* (London) had included a daily weather map since 1875. See Ward, "Newspaper Weather Maps."

76. Sutton, *Design and Makeup of the Newspaper*, 459-60.

77. Edmund C. Arnold, *Modern Newspaper Design* (New York: Harper and Row, 1969), 191-95.

78. Arnold, *Designing the Total Newspaper*, 108-10.

79. See, for example, Edmund C. Arnold, *Ink on Paper 2: A Handbook for the Graphic Arts* (New York: Harper and Row, 1972). An earlier version had been published as *Ink on Paper*, in 1963.

80. Harold Evans, *Editing and Design*, vol. 4, *Pictures on a Page: Photo-Journalism, Graphics and Picture-Editing* (London: Heinemann, 1978), 290.

81. Garcia, *Contemporary Newspaper Design: A Structural Approach*, 103.

82. Mario R. Garcia, *Contemporary Newspaper Design: A Structural Approach*, 2nd ed. (Englewood Cliffs, N.J.: Prentice-Hall, 1987), 151.

83. Daryl R. Moen, *Newspaper Layout and Design* (Ames, Ia.: Iowa State University Press, 1984), 99-100.

84. Louis Alexander, *Beyond the Facts: A Guide to the Art of Feature Writing*, 2nd ed. (Houston: Gulf Publishing Company, 1982), 137-38.

85. See Harry Franklin Harrington and Elmo Scott Watson, *Modern Feature Writing* (New York: Harper and Brothers, 1935), 146, 340, 391-92.

86. Bruce H. Westley, *News Editing*, 3rd ed. (Boston: Houghton Mifflin, 1980), 314-15.

87. Harry W. Stonecipher, Edward C. Nicholls, and Douglas A. Anderson, *Electronic Age News Editing* (Chicago: Nelson–Hall, 1981), 163-65, 180.

Chapter 5: Maps in the Electronic Media

1. W. G. Fitz-Gerald, "A Telephone Newspaper," *Scientific American* 96, no. 25 (22 June 1907), 507.

2. Arthur F. Colton, "Telephone Newspaper—A New Marvel," *Technical World Magazine* 16, no. 2 (February 1912), 666-69.

3. Irving Settel, *A Pictorial History of Radio* (New York: Grosset and Dunlap, 1967), 36-41.

4. Curtis Mitchell, *Cavalcade of Broadcasting* (New York: Rutledge Books, 1970), 79.

5. For discussion of early newspaper involvement in the operation of commercial radio stations, see Erik Barnouw, *A Tower in Babel* (New York: Oxford University Press, 1966), 61-64, 92-99, 131-42; and James C. Young, "Is the Radio Newspaper Next?" *Radio Broadcast* 7, no. 5 (September 1925), 575-80.

6. "FACSIMILE: Radio Threatens to Reach into Country's Mailboxes," *Newsweek* 6, no. 21 (23 November 1935), 41-42; and "Facsimile Scramble," *Business Week* no. 435 (1 January 1938), 30-31.

7. "A Newspaper by Radio: The St. Louis P-D Pioneers with Facsimile Process," *Newsweek* 12, no. 25 (19 December 1938), 28-29; and "Newspapers via Air Waves," *Inland Printer* 103, no. 6 (September 1939), 88-89.

8. "Network Daily Paper," *Business Week* no. 495 (25 February 1939), 35-36; and "Radio Newspapers Found Popular; Future Unpredictable," *Bulletin of the American Society of Newspaper Editors* no. 188 (1 June 1939), 2.

9. "Chicago Tribune Radio Facsimile Makes Debut," *Inland Printer* 117, no. 4 (July 1946), 46; "Facsimile Forward," *Newsweek* 31, no. 9 (1 March 1948), 46-48; and Edgar H. Felix, "Miami Herald Transmits Facsimile Newspaper," *FM and Television* 7, no. 4 (April 1947), 36-39; and "First Fax," *Time* 51, no. 2 (12 January 1948), 60-61.

10. "Facsimile Goes Commercial," *Electronics* 21, no. 8 (August 1948), 97; and "Radio Facsimile in the Clear," *Business Week* no. 923 (10 May 1947), 42-44.

11. Ruth Brindze, "Next—The Radio Newspaper," *The Nation* 146, no. 6 (5 February 1938), 154-55.

12. G. Harris Danzberger, "Facsimile Newspapers Are As Certain As Wire Pictures," *The New York Press* 87, no. 6 (March 1940), 22-24.

13. Samuel Ostrolenk, "Home Facsimile Recording," *Electronics* 11, no. 1 (January 1938), 26-27, 60.

14. Franklyn K. Lauden, "Radio Facsimile May Print 'Newspapers of Tomorrow'," *Radio and Television News* 40, no. 2 (August 1948), 39, 148-49.

15. "Ultra-Fast Ultrafax," *Newsweek* 32, no. 18 (1 November 1948), 50.

16. George A. Brandenburg, "No Immediate Threat Seen in Facsimile," *Editor and Publisher* 82, no. 19 (30 April 1949), 100; and Jerry Walker, "N.Y. Times Starts Fax But Adds to Presses," *Editor and Publisher* 81, no. 9 (21 February 1948), 56.

17. Lee Hills and Timothy J. Sullivan, *Facsimile* (New York: McGraw-Hill, 1949).

18. A writer in *Editor and Publisher* noted the view of some facsimile supporters that "TV raided the pocketbooks of the advertisers and broadcasters and captured the popular fancy before the newspaper-of-the-air could be delivered to the parlor." See Jerry Walker, "History of Facsimile Experiment Recorded," *Editor and Publisher* 82, no. 45 (29 October 1949), 42. A more insightful view of the reason for television's victory would seem to lie in the words published by a former editor of *Editor and Publisher* in the mid-1930s. Skeptical of the efficiency of electronically delivered hardcopy news, Marlen Pew saw television as the more serious challenger to the daily newspaper. "I predict that [the facsimile newspaper] will be a flash in the pan," he asserted, noting that "if the facsimile newspaper

contains 32 pages and if 10 minutes are allowed for each page, our family will be kept sitting for a little more than five hours and I submit that would be the world's biggest bore, not to mention a pain in the neck, cramped legs and dazed eyes." See Marlen Pew, "Television and Tomorrow's Newspaper," *Pulp and Paper Magazine of Canada* 37, no. 7 (July 1936), 438-39.

19. An early reference to maps is in a two-page article in a 1938 issue of *Scientific American*. Separate halftone photos show a home receiver, a receiver with its cover removed, and a transmitter. All three photos show a facsimile page that includes a halftone. Maps are mentioned in only a single sentence: " 'Printing' of illustrated world events, bulletins with latest news flashes, photographs, market reports, weather maps, cartoons, recipes, and illustrated advertisements of all sorts, will thus be effected in homes while their occupants sleep, the machine being practically silent and entirely automatic in its operation." See "Home Newspaper by Radio," *Scientific American* 158, no. 6 (June 1938), 334-35. A short 1946 article in the *Inland Printer* showed a facsimile page with a promotional first-issue cartoon of a maternity ward, with baby "Facsimile" being presented to grandfather-publisher "Tribune" (complete with top hat!) by nurse WGN (the Chicago station owned by the *Tribune*). Only one sentence mentions maps: "Besides news and photographs, maps, graphs, comic strips, and crossword puzzles can be sent through the air." See "Chicago Tribune Radio Facsimile."

20. Hills and Sullivan, *Facsimile*, 16.

21. Ibid., 115.

22. Ibid., 28.

23. John V. L. Hogan, "Facsimile and Its Future Uses," *Annals of the American Academy of Politcal and Social Science* 213 (January 1941), 162-69 (ref. p. 162).

24. Ibid., 169.

25. "Improved Facsimile Systems Demonstrated for the Press," *Electrical Engineering* 65, no. 6 (June 1946), 292.

26. Milton Alden, "Will Newspapers Sell Their Presses?" *FM and Television* 6, no. 10 (October 1946), 32-33.

27. For a concise examination of the economic liabilities of the facsimile newspaper, see Mary A. Koehler, "Facsimile Newspapers: Foolishness or Foresight?" *Journalism Quarterly* 46, no. 1 (Spring 1969), 29-36.

28. In the late nineteenth and early twentieth centuries, a number of inventors in Europe and North America experimented with the electronic transmission of photoelectrically scanned images and their reproduction using a cathode ray beam. The major drawback to these early television systems was the camera, which in the 1920s was based on a mechanical scanner using mirrors and rotating perforated disks. The breakthrough that made television commercially practicable came in the early 1930s, when Vladimir Zworykin and Philo Farnsworth, working independently, developed the all-electronic scanner. See Elizabeth Antébi, *The Electronic Epoch* (New York: Van Nostrand Reinhold Company, 1982), 149-55; and D. G. Tucker, "Electrical Communication," in *A History of Technology*, ed. Trevor I. Williams, vol. 7, part 2, *The Twentieth Century, c. 1900 to c. 1950* (Oxford: Clarendon Press, 1978), 1220-80, esp. pp. 1257-62.

29. Leo Bogart, *The Age of Television*, 3rd ed. (New York: Frederick Ungar Publishing Co., 1972), 8-10.

30. Ibid., 50-51.

31. Columbia Broadcasting System News, *Television News Reporting* (New York: McGraw-Hill, 1958), 3-6.

32. Ibid., 104, 112, 174.

33. Ibid., 100.

34. Jack E. Kruger, "Comparing News by Radio, Television, and Newspapers," in *Radio and Television News* ed. Donald E. Brown and John Paul Jones (New York: Rinehart and Company, 1954), 13-25 (ref. p. 14).

35. I. E. Fang, *Television News* (New York: Hastings House, 1968), 45.

36. Ibid., 198-201.

37. For a concise explanation of chroma keying, see Richard Robinson, *The Video Primer*, 3rd ed. (New York: Perigee Books, 1983), 263-66, 306-7.

38. Patricia Suzanne Caldwell, "Television News Maps: An Examination of Their Utilization, Content, and Design" (Ph.D. dissertation, University of California at Los Angeles, 1979), 39-48. Also see Patricia S. Caldwell, "Television News Maps: The Effects of the Medium on the Map," in *Technical Papers*, American Congress on Surveying and Mapping, 41st annual meeting, Washington, D.C., 22-27 February 1981, 382-92.

39. European television has a slightly higher resolution, with 625 scan lines per frame. Britain has a 405-line standard for black-and-white television, and France a dual 819- and 625-line standard. See Robinson, *Video Primer*, 311-17.

40. Caldwell, "Television News Maps: An Examination," 105-24.

41. I. E. Fang, *Television News*, 2nd ed. (New York: Hastings House, 1972), 285.

42. Caldwell, "Television News Maps: An Examination" 180.

43. In 1975, 66 percent of households had a color television, and 97 percent had at least black-and-white. The average number of sets per household was 1.54, and a black-and-white receiver was still in use in many homes with color. In 1983, when the proportion of households with color sets had risen to 85 percent, 47 percent of all households still owned a black-and-white set. See U.S. Bureau of the Census, *Statistical Abstract of the United States: 1986*, 106th ed. (Washington, D.C., 1985), 545, 844.

44. To avoid a loss of detail caused by bright areas "blooming," or growing into surrounding parts of the picture, the brightest part of a televised image should not be more than 20 times brighter than the darkest part of the screen. This limitation is called the *contrast ratio*. For further discussion of brightness limitations, see Caldwell, "Television News Maps: An Examination," 52-57; and Herbert Zettl, *Television Production Handbook*, 3rd ed. (Belmont, Cal.: Wadsworth Publishing Co., 1976), 89-92, 334-36.

45. Full or partial color blindness mars the vision of about 8 percent of men and 1 percent of women. See "Color Blindness," *McGraw-Hill Encyclopedia of Science and Technology* (New York: McGraw-Hill, 1982), 3:411-12.

46. See, for example, Thomas D. Burrows and Donald N. Wood, Television Production: Disciplines and Techniques (Dubuque, Ia.: William C. Brown, 1982), 212.

47. Robinson, *Video Primer*, 290-94.

48. The picture is scanned 60 times each second but for alternating sets of odd and even scan lines only. That is, in one second there are 30 scans of even-numbered scan lines *interlaced* with 30 scans of odd-numbered scan lines. Thus, the entire "field" of the televised picture is scanned 60 times per second, whereas the complete "frame" of 525 scan lines is only scanned 30 times per second. There are two fields of picture information for every frame. Ibid., 215-20.

49. Some color receivers use the Trinitron color system, developed by American Ernest Lawrence but exploited commercially by Sony. The Trinitron system uses alternating thin vertical strips of color-emitting phosphors, instead of triads of color dots, and an *aperture grille* of thin vertical slits, instead of a shadow mask of tiny holes. Trinitron color is considered higher in quality than the triad system, for which RCA holds the patents. See Ibid., 302-3.

50. For discussion of various systems used to describe color for CRT displays, see David F. Rogers, *Procedural Elements for Computer Graphics* (New York: McGraw-Hill, 1985), 383-408.

51. For example, the leading general cartographic textbook in the United States, the fifth edition of Arthur Robinson's *Elements*, devotes less than two pages in its 30-page chapter on color to CRT color. See Arthur H. Robinson and others, *Elements of Cartography*, 5th ed. (New York: John Wiley and Sons, 1984), 162-91.

52. Caldwell, "Television News Maps: An Examination," 164-71.

53. Ben Blank and Mario R. Garcia, *Professional Video Graphic Design* (Englewood Cliffs, N.J.: Prentice-Hall, 1986).

54. Ibid., 99.

55. Balchin, "Map Media Watch: a Report," 339-43.

56. A public outcry followed publication of Vance Packard's *The Hidden Persuaders* in 1957, but despite widespread calls for legislation prohibiting insertion of frames with "subthreshold" messages, no laws were passed. See Wilson Bryan Key, *Subliminal Seduction: Ad Media's Manipulation of a Not So Innocent America* (Englewood Cliffs, N.J.: Prentice-Hall, 1973), 21-22; and Vance Packard, *The Hidden Persuaders* (New York: David McKay, 1957), 42-43. For a more recent examination of subliminal perception, see Bruce Bowers, "Subliminal Messages: Changes for the Better?" *Science News* 129, no. 10 (8 March 1986), 156-58.

57. Caldwell, "Television News Maps: An Examination," 27-28.

58. Caldwell observed that on an average day the *Los Angeles Times* rarely had more than one news map in addition to the daily weather map. In contrast, network-affiliated stations in the Los Angeles area commonly used two or three maps in a single broadcast. See Ibid., 23.

59. With the artwork produced in the 1970s by freelance artists, the set consists mostly of logos but includes state, region, and country maps. Harold G. Buell, Notes from an interview held at New York, 20 November 1984.

60. Blank and Garcia, *Professional Video Graphic Design*, 131-38.

61. Robert Watkins, "Computer Graphics in Broadcasting," *Computer Graphics World* 6, no. 6 (June 1983), 65-70. For a concise history and overview

of television graphics, see Stuart Silverstone, "Graphics Hit the Network News Beat," *Computer Graphics World* 10, no. 2 (February 1987), 35-38.

62. Ingeborg Hutzel, "Computer Graphics in Broadcasting: A Survey of the Field," *Computer Graphics World* 8, no. 4 (April 1985), 10-24, 122-26. For a discussion of electronic paint systems and other concepts, see Rodney Stock, "Introduction to Digital Computer Graphics for Video," *Society of Motion Picture and Television Engineers Journal* 90, no. 12 (December 1981), 1184-89; and Joseph L. Streich, "Increasing Graphics Productivity at WABC-TV," *Computer Pictures* 4, no. 4 (July/August 1986), 60-68.

63. "Dial-Line Modems Send Complex Graphics Quickly and Reliably to Weather Forecasters," *Communications News* 22, no. 12 (December 1985), 60-61; and "Weathercasters Shine News in Survey," *Television/Radio Age* 33, no. 5 (16 September 1985), 48, 91.

64. "Weather Graphics Now Seen in Markets Big and Small," *Broadcasting* 108, no. 16 (22 April 1985), 97.

65. As of Summer 1986, the American Meteorological Society had certified 420 television broadcasters since the inception of its Seal of Approval program in 1959. The AMA also certifies radio weathercasters. See "Seal of Approval Program for Radio and Television Weathercasting," *Bulletin of the American Meteorological Society* 67, no. 8 (August 1986), 1028-35. Also see Richard L. Tobin, "Weather or Not," *Saturday Review* 49, no. 46 (12 November 1966), 95-96.

66. Jack Shelley, "Weather News," in *Radio and Television News*, ed. Donald E. Brown and John Paul Jones (New York: Rinehart and Company, 1954), 151-68 (ref. p. 153).

67. Ibid., 160.

68. Fang, *Television News*, 204-5.

69. Fang, *Television News*, 2nd ed., 294-99.

70. "TV Radar Attracts Buyers," *Electronics* 31, no. 48 (28 November 1958), 17.

71. "KSTP-TV Uses Weather Radar and Laserfax," *Communications News* 15, no. 7 (July 1978), 30-31.

72. In 1985 several firms produced computer graphics paint systems designed primarily for television weather graphics, but Colorgraphics, which had a more general system, had captured about half the market for weather graphics systems. See Hutzel, "Computer Graphics in Broadcasting," 10, 16. Also see "Color Adds 'Pizzaz' to Weather Radar," *Design News* 36, no. 12 (23 June 1980), 13-14; John Hambleton, "Computer Mapping and the Television Weather Report," *Proceedings*, Auto-Carto Six, Ottawa, 16-21 October 1983, 1:469; and Gerald M. Heymsfield, Koushik K. Ghosh, and Lily C. Chen, "An Interactive System for Compositing Digital Radar and Satellite Data," *Journal of Climate and Applied Meteorology* 22, no 5 (May 1983), 705-13.

73. "Weather Graphics in Markets Big and Small."

74. Louis J. Battan, *Weather in Your Life* (San Francisco: W. H. Freeman and Company, 1983), 83-89.

75. Joseph A. Ryan, "Using the Home Computer As a Facsimile Receiver," *Weatherwise* 36, no. 6 (December 1983), 308-10. Also see National Oceanic and Atmospheric Administration, Environmental Data and Information Service, Na-

tional Climatic Center, *Environmental Satellite Data and Information*, Environmental Information Summaries C-16 (Asheville, N.C., 1980).

76. See, for example, Joel Olson, "Image Processing in Weather Forecasting," *Computer Graphics World* 6, no. 11 (November 1983), 41-54.

77. Glenn Garelik, "The Weather Peddlers," *Discover* 6, no. 4 (April 1985), 18-29.

78. "Dial-Line Modems Send Complex Graphics"; and "Weather Info a Breeze for Radio/TV Stations," *Communications News* 17, no. 4 (April 1980), 62.

79. Leonard Ray Teel, "The Weather Channel," *Weatherwise* 35, no. 4 (August 1982), 156-63.

80. In most writings *videotex* is a general term for a one- or two-way mass-market video message service provided by broadcast television, cable, telephone, or other electronic carrier. But some writers have used other terms to refer to specific variations. *Videotext*, with the final "t", has referred to telephone-delivered systems, whereas *teletext* has referred to systems using broadcast television. *Viewdata* has been a synonym, referring like *videotex* to any electronic delivery medium. See Richard H. Veith, *Television's Teletext* (New York: North-Holland, 1983), 1-2.

81. For a fuller discussion of frame numbers for pages in a tree structure, see Jan Gecsei, *The Architecture of Videotex Systems* (Englewood Cliffs, N.J.: Prentice-Hall, 1983), 168-72.

82. For a fuller discussion of the vertical blanking interval and its capacity for data transmission, see Ibid., 46-51; and Veith, *Television's Teletext*, 3-4.

83. See note 48 for a concise explanation of interlacing.

84. Colin McIntyre, "Teletext in Britain: The CEEFAX Story," in *Videotext: The Coming Revolution in Home/Office Information Systems*, ed. Efrem Sigel and others (White Plains, N.Y.: Knowledge Industry Publications, 1980), 23-55.

85. For further discussion of two-way cable TV systems see William Grant, *Cable Television* (Reston, Va.: Reston Publishing Co., 1983), 241-57; and Loy A. Singleton, *Telecommunications in the Information Age* (Cambridge, Mass.: Ballinger Publishing Co., 1983), 37-48.

86. Grant, *Cable Television*, 5-8.

87. See, for examples, James Martin, *The Wired Society* (Englewood Cliffs, N.J.: Prentice-Hall, 1978); John Naisbitt, *Megatrends: Ten New Directions Transforming Our Lives* (New York: Warner Books, 1982), esp. pp. 1-33; and Alvin Toffler, *The New Wave* (New York: William Morrow and Company, 1980), esp. pp. 171-83 and 210-23.

88. For discussion of videotex coding with picture description instructions (PDIs), see Veith, *Television's Teletext*, 88-95.

89. D. R. F. Taylor, "The Cartographic Potential of Telidon," *Cartographica* 19, nos. 3 and 4 (Autumn and Winter 1982), 18-30.

90. In the late 1970s and early 1980s, the United States substantially deregulated its telecommunications industries; the ultimate technological mix of hardware, software, and network media was uncertain, and legal issues, such as copyright, privacy, competition, and free entry to the marketplace dominated much of the discussion of videotex. See, for examples, Paul Hurly, "The Promises

and Perils of Videotex," *The Futurist* 19, no. 2 (April 1985), 7-13; and Richard M. Neustadt, "The Regulation of Electronic Publishing," *Federal Communications Law Journal* 33, no. 3 (Summer 1981), 331-417.

91. Brian Gaines, "Videotex—the Electronic Challenge," *Penrose Annual* 74 (1982), 47-56.

92. Minitel had 1.4 million subscribers in 1985, and looked forward to serving 8 million terminals by 1990. The subtitle of a 1986 *Newsweek* article reflected some Americans' marvel at Minitel's success: "France has seized the world lead in videotex." See Robert B. Cullen, Ruth Marshall, and Ken Pottinger, "The Minitel Revolution," *Newsweek* 107, no. 14 (7 April 1986), 75G. In 1986 a story in *Editor and Publisher* also used a subtitle to herald French success in marketing videotex: "Unlike in the United States, where videotex services are experiencing tough times, use of French videotex has been increasing each year." See George Nahon, "Videotex Proving to Be Successful in France," *Editor and Publisher* 119, no. 25 (21 June 1986), 40, 122, 124.

93. Joseph Roizen, "Videotext in Other Countries," in *Videotext: The Coming Revolution in Home/Office Information Retrieval*, ed. Efrem Sigel and others (White Plains, N.Y.: Knowledge Industry Publications, 1980), 23-55; and Veith, *Television's Teletext*, 20-22.

94. John C. Madden, *Videotex in Canada* (Ottawa: Canada, Ministry of Supply and Services, 1979), 19-28.

95. Veith, *Television's Teletext*, 117-19. Following its Canadian counterpart, the American National Standards Institute (ANSI) adopted NAPLPS as the United States videotex standard in 1983. See Jennifer Strothers and Mark Dietrich, "A Perspective on NAPLPS and Its Impact on the Development of Videotex in North America," *Videodisc and Optical Disk* 5, no. 1 (January-February 1985), 52-61.

96. See, for example, "Bank of Montreal's At-Home Pilot," *American Banker* 149, no. 86 (1 May 1984), 24; Canada, Department of Communications, *Telidon Trials + Services* (Ottawa: Ministry of Supply and Services, 1983), 41-43; and Richard Larratt, "Market Factors," in *The Telidon Book: Designing and Using Videotex Systems*, ed. David Godfrey and Ernest Chang (Reston, Va.: Reston Publishing Co., 1981), 7-80, esp. pp. 51-56.

97. See, for example, John Wicklein, *Electronic Nightmare* (New York: Viking Press, 1981), esp. pp. 100-155.

98. See Kathleen Criner and Jane Wilson, "An Uncertain Marketplace Takes Telecommunications Toll," *Presstime* 6, no. 12 (December 1984), 23-25 (ref. p. 25).

99. See, for example, Arnold Rosenfeld, "What the Hell Is Going On in New Technology?" *Bulletin of the American Society of Newspaper Editors* no. 616 (July/August 1978), 6-8. For a partial transcript of remarks on videotex by Rosenfeld and others at the 1979 meeting of the ASNE, see "Home Delivery by Electronics," *Problems of Journalism—Proceedings of the American Society of Newspaper Editors*, New York, 29 April - 2 May 1979, 26-30.

100. Smith, *Goodbye Gutenberg*.

101. "Half of Dailies Using or Studying New Technology," *Presstime* 5, no. 4 (April 1983), 16.

102. See, for example, David M. Reed, "Steps One Company Took to Start Cable Service," *Presstime* 4, no. 11 (November 1982), 16-17; and "Videotex-Cable," *APME Red Book* 36 (1983), 73-75.

103. Celeste Huenergard, "Milwaukee Dailies to Offer 24-hr. Electronic Newspaper," *Editor and Publisher* 115, no. 18 (1 May 1982), 63.

104. Tim Miller, "Times Mirror Videotex Under Way," *Editor and Publisher* 118, no. 10 (9 March 1985), 27-28.

105. C. David Rambo, "It's Still a 'Maybe' Market for New Technologies," *Presstime* 5, no. 1 (January 1983), 20-22.

106. For accounts of Viewtron's unsteady progress see Barbara J. Friedman, "It Was a Rocky First Year for Viewtron," *Presstime* 6, no. 12 (December 1984), 25; Tim Miller, "Videotex Market Shaken Up by PC Boom," *Editor and Publisher* 118, no. 10 (9 March 1985), 26-27; Efrem Sigel, "Videotext in the United States," in *Videotext: The Coming Revolution in Home/Office Information Retrieval*, ed. Efrem Sigel and others (White Plains, N.Y.: Knowledge Industry Publications, 1980), 87-111; Henry Urrows and Elizabeth Urrows, "Early Viewtron," *Videodisc and Optical Disk* 4, no. 4 (July-August 1984), 269-300; and "Viewtron Will Lower Its Rates," *Editor and Publisher* 118, no. 10 (9 March 1985), 26.

107. "Knight-Ridder 'Pulls Plug' on Its Videotex Operation," *Editor and Publisher* 119, no. 13 (29 March 1986), 14.

108. Kathleen Criner, "Newspapers at Mid-decade and Beyond: Telecommunications," *Presstime* 7, no. 1 (January 1985), 26.

109. For a detailed examination of the journalistic potential of European videotex—but with little attention to news maps—see David H. Weaver, *Videotex Journalism* (Hillsdale, N.J.: Lawrence Erlbaum Associates, 1983). For an account of newspaper involvement in PAV ventures, see "Public Access Videotex," *Editor and Publisher* 118, no. 12 (23 March 1985), 40-42.

110. C. David Rambo, "Newspaper Companies Go Back to Basics," *Presstime* 9, no. 1 (January 1987), 22-30; and Richard W. Stevenson, "Videotex Players Seek a Workable Formula," *New York Times*, 25 March 1986, D1 and D21.

111. See, for example, John L. Perry, "Technology Is Not Journalism," *Quill* 71, no. 4 (April 1983), 26-29.

112. "ANPA Urges Senate Panel to Retain Curbs on Phone Companies' Electronic Publishing," *Presstime* 8, no. 8 (August 1986), 41; Kathleen Criner and Jane Wilson, "New-Technology Players Jockey for Position," *Presstime* 6, no. 11 (November 1984), 23-25; and Kathleen Criner and Jane Wilson, "Telecommunications History Is Short But Stormy," *Presstime* 6, no. 10 (October 1984), 24-26.

Chapter 6: Concluding Remarks

1. For discussion of the limitations of MacDraw and its effects upon newspaper cartography, see Mark Monmonier, "A Geographer's View of Newspaper Maps and a Cartographer's Guide," *Design, The Journal of the Society of Newspaper Design* no. 26 (1987), 14-17.

2. George Garneau, "Getting Photos Faster," *Editor and Publisher* 120, no. 23 (6 June 1987), 72, 126-27.

3. "AP Will Also Be Delivering News Graphics Faster," *Editor and Publisher* 120, no. 23 (6 June 1987), 126.

4. Among other differences between the samples for 1985 and 1987, by July 1987 all three newspapers had adopted the new national weather map prepared for AP members by Accu-Weather, Inc., of State College, Pennsylvania. The Syracuse *Herald-Journal* also used the AP/Accu-Weather regional map for the Northeast, whereas the Utica *Observer Dispatch* had introduced a weather map of the state and local region produced by its own artist. Oddly, all three newspapers sometimes used Laserphoto versions of AP maps for which sharper, more aesthetically pleasing images might have have been obtained with the LaserWriter. On occasion these papers also used copies of foreign-area basemaps from Tribune Media Services.

5. The *Watertown Daily Times*, as Figure 6.3 indicates, produced new artwork for most nonlocal maps as well.

6. For examples of the niche-space concept applied to history, see Paul Colinaux, *The Fates of Nations: A Biological Theory of History* (New York: Simon and Schuster, 1980). This ecological concept is particularly useful in Colinaux's discussions of the effects of crowding on war and revolution and the effects of technology on the broadening of niche-spaces and the rise of a middle class.

7. For an insightful essay on the various roles of the map, see Wood and Fels, "Designs on Signs: Myth and Meaning in Maps," 54-103. Wood and Fels employ the term *extrasignificant* to describe exploitation of the entire map as a social or political sign.

8. For a seminal study of the effects of technology and legislation on the development of mass communications, see Harold A. Innis, *The Bias of Communication* (Toronto: University of Toronto Press, 1951).

9. A daily newspaper in large city might spend $10 million on a new press. Depreciated over 12 to 15 years, the press has a much longer actual life. In the mid-1980s many newspapers were making substantial investments in new offset or flexographic printing systems. See, for example, Paul Kruglinski, "Buying Equipment," *Presstime* 7, no. 2 (February 1985), 21-27.

10. For discussion of newspapers' use of zoned editions to promote circulation in suburban and peripheral areas, see Mary A. Anderson, "Doing Battle in the Suburbs: Locals vs. Metros, Locals vs. Locals," *Presstime* 9, no. 6 (June 1987), 8-12; Boyd L. Miller, "More Dailies Zoning for Suburban Readers," *Journalism Quarterly* 42, no. 3 (Summer 1965), 460-62; and Arnie Rosenberg, "Zoned Editions," *Editor and Publisher* 118, no. 17 (27 April 1985), 17, 20.

11. For a humanistic interpretation of the effects upon culture of typography and the linear print media, see Marshall McLuhan, *The Gutenberg Galaxy* (Toronto: University of Toronto Press, 1962).

12. In a perceptive essay on the integration of letters and graphic symbols on a common printing surface, David Woodward asserts that maps are "in the border zone between graphic and typographic printing." See David Woodward,

"Maps, Music, and the Printer: Graphic or Typographic?" *Printing History* 8, no. 2 (1986), 3-14 (ref. p. 3). To extend Woodward's proposition, electronic printing technology has promoted journalistic cartography by shifting newspapers into this zone, which has become the center of image transfer technology rather than an intersection of disparate technologies.

13. For an empirical examination of map use by American and British newspapers, see Kristina Ferris, "A Cross-National Comparison of Journalistic Cartography in British and American Daily Newspapers" (M.A. thesis, Syracuse University, 1987). Ferris examined twenty newspapers in each country for a six-day period in May 1986. She noted that 70 percent of the 142 nonweather maps found in the American newspapers were locator maps, in contrast to only 25 percent of the 48 maps found in the British newspapers (pp. 46-49). Ferris attributed the lower rate of map use in England to fewer staff artists and less ease in obtaining maps from a central producer.

14. In his masterful monograph on the design and use of statistical graphics, Edward Tufte compared 15 American and foreign news publications for the "sophistication" of their statistical graphics. The five American newspapers and magazines ranked in the middle or toward the bottom. However, Tufte did not include maps and time-series graphics among "relational graphics," based on two or more statistical variables, and thus indirectly he chided the American media for its news maps, most of which are locator maps. See Tufte, *Visual Display*, 83-84. Yet, in the sense that the graphic plane constitutes two or three dimensions, all maps—even simple locator maps—are inherently relational. Tufte might have addressed the question of comparative sophistication more effectively by examining maps separately and by considering for each publication the relative frequency of relational graphics as well as their proportion.

15. See E. D. Hirsch, Jr., *Cultural Literacy: What Every American Needs to Know* (Boston: Houghton Mifflin, 1987), 152-215. Hirsch compiled his list with the assistance of Joseph Kett, a historian, and James Trefil, a physicist. His thesis is that a basic factual vocabulary is indispensable for efficient communication within a culture.

16. Even Eric Hirsch, who asserted the need for factual knowledge about significant places, apparently has little regard for map use. Said Hirsch, "By stressing the essential role of content in reading, this book should have punctured the myth that reading and writing are like bike riding or map reading, skills that require only a narrow range of specific information plus some practice." Ibid., 144.

17. Among the better general books on the effective use of data graphics are Jacques Bertin, *Semiology of Graphics: Diagrams, Networks, Maps*, trans. William J. Berg (Madison, Wis.: University of Wisconsin Press, 1983); William S. Cleveland, *The Elements of Graphing Data* (Montery, Cal.: Wadsworth, 1985); and Tufte, *Visual Display*.

18. For an essay on how authors, editors, and publishers might promote the integration of maps and text, see Mark Monmonier, "Map-Text Coordination in Geographic Writing," *The Professional Geographer* 33, no. 4 (November 1981), 406-12.

19. Digital photography, based upon the still video camera, might even replace the photojournalist with an observant reporter backed up by image processing technology and a good picture editor. For discussion of developments in electronic still photography, see George Garneau, "Electronic Photos for Newspapers," *Editor and Publisher* 119, no. 28 (12 July 1986), 30-32.

Bibliography

Abramson, Howard S. *National Geographic: Behind America's Lens on the World.* New York: Crown, 1987.

Albert, Jeffrey L. "Gannett's Macintosh Menu." *Deadline Mac* 1, no. 4 (July 1986): 2.

Alden, Milton. "Will Newspapers Sell Their Presses?" *FM and Television* 6, no. 10 (October 1946): 32-33.

Alexander, Louis. *Beyond the Facts: A Guide to the Art of Feature Writing.* 2nd ed. Houston: Gulf Publishing Company, 1982.

Allen, John E. *Newspaper Makeup.* New York: Harper and Brothers, 1936.

Anderson, Douglas A., and Anderson, Claudia J. "Weather Coverage in Dailies." *Journalism Quarterly* 63, no. 2 (Summer 1986): 382-85.

Anderson, Mary A. "Doing Battle in the Suburbs: Locals vs. Metros, Locals vs. Locals." *Presstime* 9, no. 6 (June 1987): 8-12.

Anderson, Steven. Interview with author. Arlington, Va., 24 October 1984.

"ANPA Urges Senate Panel to Retain Curbs on Phone Companies' Electronic Publishing." *Presstime* 8, no. 8 (August 1986): 41.

Antébi, Elizabeth. *The Electronic Epoch.* New York: Van Nostrand Reinhold Company, 1982.

Anthony, Kate Newton. "Base Maps, Files and Simple Rules Guide USA Today's Graphics." *Design, The Journal of the Society of Newspaper Design* no. 18 (Winter 1985): 9.

"(AP)." *Fortune* 15, no. 2 (February 1937): 88-93, 148-62.

"AP Announces Three Major Service Enhancements." *Editor and Publisher* 118, no. 19 (11 May 1985): 20, 48.

"AP Enters the Commercial Market." *Editor and Publisher* 118, no. 16 (20 April 1985): 31.

"A.P. Explains New Telephoto Service." *Editor and Publisher* 66, no. 50 (28 April 1934): 6, 111.

AP Photo and Graphics Committee. "Continuing Study Report." *APME Red Book* 31 (1978): 168-71.

"AP Photofax Device Hits at Moisture." *Editor and Publisher* 89, no. 40 (29 September 1956): 63.

"AP to Expand Satellite System." *Editor and Publisher* 116, no. 52 (24 December 1983): 34.

"AP Will Also Be Delivering News Graphics Faster." *Editor and Publisher* 120, no. 23 (6 June 1987): 126.

"A.P. Wirephotos Flash Across Nation." *Editor and Publisher* 67, no. 34 (5 January 1935): 7.

APME Red Book. New York: Associated Press Managing Editors Association, annually.

"Apple Computer Reaches for Its Macintosh." *Economist* 290, no. 7326 (28 January 1984): 68-69.

Arnold, Edmund C. *Designing the Total Newspaper*. New York: Harper and Row, 1981.

———. *Functional Newspaper Design*. New York: Harper and Row, 1956.

———. *Ink on Paper 2: A Handbook for the Graphic Arts*. New York: Harper and Row, 1972.

———. "Map Out Future Use of Expository Art." *Publisher's Auxiliary* 121, no. 8 (22 April 1985): 5.

———. *Modern Newspaper Design*. New York: Harper and Row, 1969.

Association of American Geographers. Committee on Geography and International Studies. *Geography and International Knowledge*. Washington, D.C., 1982.

Bagdikian, Ben H. *The Information Machines: Their Impact on Men and the Media*. New York: Harper and Row, 1971.

———. *The Media Monopoly*. Boston: Beacon Press, 1983.

Bakeless, John. *Magazine Making*. New York: Viking Press, 1931.

Baker, Alan M.; Falundi, Rowan; and Green, David R. "An Evaluation of the SAS/GRAPH Software for Computer Cartography." *Professional Geographer* 37, no. 2 (March 1985): 204-14.

Baker, T. Thorne. *The Telegraphic Transmission of Photographs*. London: Constable and Company, 1910.

———. "The Telegraphy of Photographs, Wireless and by Wire." *Nature* 84, no. 2129 (18 August 1910): 220-26.

———. *Wireless Pictures and Television*. New York: D. Van Nostrand, 1926.

Balchin, W. G. V. "Graphicacy." *American Cartographer* 3, no. 1 (April 1976): 33-38.

———. "Media Map Watch." *Geographical Magazine* 57, no. 8 (August 1985): 408-9.

———. "Media Map Watch: A Report." *Geography* 70, no. 309, pt. 4 (October 1985): 339-43.

"Bank of Montreal's At-Home Pilot." *American Banker* 149, no. 86 (1 May 1984): 24.

Barnes, Andrew. "AP's Graphics." *APME Red Book* 33 (1980): 242-44.

Barnouw, Erik. *A Tower in Babel*. New York: Oxford University Press, 1966.

Bastian, George C. *Editing the Day's News*. New York: Macmillan, 1924.

Battan, Louis J. *Weather in Your Life*. San Francisco: W. H. Freeman and Company, 1983.

Beaudat, Daniel, and Bonin, Serge. *Cartes et Figures de la Terre*. Paris: Centre George Pompidou, 1980.

Beechey, St. Vincent. "Photography Applied to Engraving on Wood." *The Art Journal*, n.s. 6 (1854): 244.

Belloc, Alexis. *La Telegraphie Historique*. Paris: Librarie de Firmin-Didot et Cie, 1889.

Berchtold, William E. "More Fodder for Photomaniacs." *North American Review* 239, no. 1 (January 1935): 19-30.

Berger, Meyer. *The Story of the New York Times, 1851-1951*. New York: Simon and Schuster, 1951.

Bernstein, Theodore M. "Maps: An Important Reader Tool." *Bulletin of the American Society of Newspaper Editors* no. 483 (1 February 1965): 1-3.

Berry, W. Turner. "Printing and Related Trades." In *A History of Technology*. Vol. 5, *The Late Nineteenth Century*, 683-715. Oxford: Clarendon Press, 1958.

Bertin, Jacques. *Semiology of Graphics: Diagrams, Networks, Maps*. Trans. William J. Berg. Madison, Wis.: University of Wisconsin Press, 1983.

The Best of Newspaper Design. 6th ed. Washington, D.C.: Society of Newspaper Design, 1985.

Betcher, Robert J. "How to Produce Art and Graphics." *Design, The Journal of the Society of Newspaper Design* no. 3 (September 1980): 19-20.

Blakemore, M. J., and Harley, J. B. *Concepts in the History of Cartography: A Review and Perspective*. Cartographica monograph no. 26. Toronto: University of Toronto Press, 1980.

Blank, Ben, and Garcia, Mario R. *Professional Video Graphic Design*. Englewood Cliffs, N.J.: Prentice-Hall, 1986.

Blum, Andre. *The Origins of Printing and Engraving*. New York: Charles Scribner's Sons, 1940.

Bogart, Leo. *The Age of Television*. 3rd ed. New York: Frederick Ungar Publishing Company, 1972.

Booth, Alan. "The Recall of News Items." *Public Opinion Quarterly* 34, no. 4 (Winter 1970-71): 604-10.

Bowers, Bruce. "Subliminal Messages: Changes for the Better?" *Science News* 129, no. 10 (8 March 1986): 156-58.

Bradley, Barry. "You Can Be a Weather Writer—and a Newsroom Star." *Bulletin of the American Society of Newspaper Editors* no. 664 (November 1983): 18-19.

Bradley, Bob. "Paging the Eightes." *Penrose Annual* 72 (1980): 77-84.

Brandenburg, George A. "No Immediate Threat Seen in Facsimile." *Editor and Publisher* 82, no. 19 (30 April 1949): 100.

Brindze, Ruth. "Next—The Radio Newspaper." *The Nation* 146, no. 6 (5 February 1938): 154-55.

Brooker-Gross, Susan R. "The Changing Concept of Place in the News." In *Geography, the Media, and Popular Culture*, edited by Jacqueline Burgess and John R. Gold, 63-85. London: Croom Helm, 1985.

Brown, Francis. *Raymond of the Times*. New York: W. W. Norton and Company, 1951.

Brown, Richard. "Nigel Holmes: Time's Graphic Statistician." *Penrose Annual* 73 (1981): 9-24.

Buehl, Harold G. Interview with author. New York, 20 November 1984.

Burke, Thomas J. M., and Lehman, Maxwell, eds. *Communication Technologies and Information Flow*. New York: Pergamon Press, 1981.

Burlingame, Roger. *Engines of Democracy: Inventions and Society in Mature America.* New York: Charles Scribner's Sons, 1940.

Burrows, Thomas D., and Wood, Donald N. *Television Production: Disciplines and Techniques.* Dubuque, Ia.: William C. Brown, 1982.

Bush, Chilton R. *News Research for Better Newspapers.* New York: American Newspaper Publishers Foundation, 1969.

Butler, Jacalyn Klein, and Kent, Kurt E. M. "Potential Impact of Videotext on Newspapers." *Newspaper Research Journal* 5, no. 1 (Fall 1983): 3-12.

Byxbee, O. F. *Establishing a Newspaper.* Chicago: Inland Printer Company, 1901.

Caldwell, Patricia Suzanne. "Television News Maps: An Examination of Their Utilization, Content, and Design." Ph.D. diss., University of California at Los Angeles, 1979.

———. "Television News Maps: The Effects of the Medium on the Map." In *Technical Papers*, American Congress on Survey and Mapping, 41st annual meeting, Washington, D.C., 22- 27 February 1981, 382-92.

Calvert, Edgar B. "Development of the Daily Weather Map." *U.S. Weather Bureau Bulletin* no. 24 (1899): 144-50.

Canada. Department of Communications. *Telidon Trials + Services.* Ottawa: Ministry of Supply and Services, 1983.

"Canada's Own." *Newsweek* 64, no. 8 (24 August 1964): 51-53.

Canadian Advertising Rates and Data 58, no. 10 (October 1985): 133-38.

Canale, Edward. "Finding an Artist." *Design, The Journal of the Society of Newspaper Design* no. 17 (Fall 1984): 30-31.

Canham, Erwin D. *Commitment to Freedom: The Story of the Christian Science Monitor.* Boston: Houghton Mifflin, 1958.

Carlson, C. Lennart. *The First Magazine: A History of The Gentleman's Magazine.* Providence, R.I.: Brown University, 1938.

Carlson, Oliver. *The Man Who Made News.* New York: Duell, Sloan and Pearce, 1942.

Carrell, Robert. Interview with author. Toronto, 7 November 1984.

Carver, Lloyd. "AP Buys Two Transponders from Western Union." *Editor and Publisher* 117, no. 28 (14 July 1984): 11.

Carwell, Rod. "Open Cleaner Style on AP Graphics Gets Rave Notices." In *Report of the APME Photo and Graphics Committee*, Toronto, 1981, 15-16.

Catalano, Dominic. Interview with author. Syracuse, N.Y., 14 September 1984.

Chalmers, Floyd S. *A Gentleman of the Press.* Garden City, N.Y.: Doubleday and Company, 1969.

Chapman, Brian. "Electronic Scanning." *Penrose Annual* 68 (1975): 130-36.

Chapman, Sherwood. Interview with author. Cortland, N.Y., 3 January 1985.

Chappell, Gary Bradford. "Newsweekly Magazine Maps: Coverage of the Falkland Islands Crisis in *Newsweek, Time*, and *U.S. News and World Report.*" M.S. thesis, University of Wisconsin-Madison, 1983.

"Chicago Tribune Radio Facsimile Makes Debut." *Inland Printer* 117, no. 4 (July 1946): 46.

Clair, Colin. *A History of European Printing.* London: Academic Press, 1976.

Clark, Ruth. "Relating to Readers in the '80s." In *ASNE-84*, 107-21. Washington, D.C.: American Society of Newspaper Editors, 1985.
Clarke, Arthur C. *Voices Across the Sea.* Rev. ed. New York: Harper and Row, 1974.
Clemow, Bice. "Picture Services Rushing into Field of Telephotograph Transmission." *Editor and Publisher* 69, no. 9 (29 February 1936): 3-4.
Cleveland, William S. *The Elements of Graphing Data.* Monterey, Cal.: Wadsworth, 1985.
Colebrook, Frank. "Newspaper Half-Tones." *Penrose Annual* 11 (1905): 93-96.
Colinaux, Paul. *The Fates of Nations: A Biological Theory of History.* New York: Simon and Schuster, 1980.
"Color Adds 'Pizzaz' to Weather Radar." *Design News* 36, no. 12 (23 June 1980): 13-14.
"Color Blindness." *McGraw-Hill Encyclopedia of Science and Technology.* Vol. 3, 411-12. New York: McGraw-Hill, 1982.
Colton, Arthur F. "Telephone Newspaper—A New Marvel." *Technical World Magazine* 16, no. 2 (February 1912): 666-69.
Columbia Broadcasting System News. *Television News Reporting.* New York: McGraw-Hill, 1958.
"Coming Attraction: Designer Graphics by Wire." *Design, The Journal of the Society of Newspaper Design* no. 15 (Spring 1984): 7.
Comparato, Frank. *Chronicles of Genius and Folly: R. Hoe and Company and the Printing Press As a Service to Democracy.* Culver City, Cal.: Labyrinthos, 1979.
"Convention 'First' Color via Wirephoto." *Editor and Publisher* 89, no. 34 (18 August 1956): 12.
Cook, David. Interview with author. Washington, D.C., 12 February 1985.
Cooper, Kent. *Kent Cooper and the Associated Press: An Autobiography.* New York: Random House, 1959.
Costigan, Daniel M. *Electronic Delivery of Documents and Graphics.* New York: Van Nostrand Reinhold Company, 1978.
———. *FAX: The Principles and Practice of Facsimile Communication.* Philadelphia: Chilton Book Company, 1971.
Couture, Ron. "Designing Graphics with a Computer." *Design, The Journal of the Society of Newspaper Design* no. 12 (Summer 1983): 28-29.
———. Interview with author. New York, 1 October 1984.
Crandall, George S. "Now, As to the Matter of Newspaper Pictures." *Bulletin of the American Society of Newspaper Editors* no. 86 (16 November 1934): 3-4.
Criner, Kathleen. "Newspapers at Mid-decade and Beyond: Telecommunications." *Presstime* 7, no. 1 (January 1985): 26.
Criner, Kathleen, and Johnson-Hall, Martha. "Videotex: Threat or Opportunity." *Special Libraries* 71, no. 9 (September 1980): 379-85.
Criner, Kathleen, and Wilson, Jane. "An Uncertain Marketplace Takes Telecommunications Toll." *Presstime* 6, no. 12 (December 1984): 23-25.
———. "New-Technology Players Jockey for Position." *Presstime* 6, no. 11 (November 1984): 23-25.

———. "Telecommunications History Is Short But Stormy." *Presstime* 6, no. 10 (October 1984): 24-26.

Cuff, David J., and Matson, Mark T. *Thematic Maps: Their Design and Production.* New York: Methuen, 1982.

Cullen, Robert B.; Marshall, Ruth; and Pottinger, Ken. "The Minitel Revolution." *Newsweek* 107, no. 14 (7 April 1986): 75G.

Curtis, Richard A. "Doing Graphics for Your Readers." *Proceedings of the American Society of Newspaper Editors*, 1983 convention, 291-302.

———. "The Evolution of USA Today." *Design, The Journal of the Society of Newspaper Design* no. 9 (Fall 1982): 5-12.

Cutlip, Scott M. "Content Flow of AP News—From Trunk to TTS to Reader." *Journalism Quarterly* 31, no. 4 (Fall 1954): 434-46.

D'Agostino, Dick. Interview with author. Baltimore, 13 February 1985.

D'Agostino, Dick, and Dresser, Michael. *The Baltimore Sun Typographic Design Stylebook.* Baltimore: The Sun, 1984.

Dahl, Folke. *A Bibliography of English Corantos and Periodical Newsbooks, 1620-1642.* London: The Bibliographical Society, 1952.

Dalgin, Ben. *Advertising Production: A Manual on the Mechanics of Newspaper Printing Techniques.* New York: McGraw-Hill, 1946.

Danzberger, G. Harris. "Facsimile Newspapers Are As Certain As Wire Pictures." *The New York Press* 87, no. 6 (March 1940): 22-24.

Davis, Dwight B. "Business Turns to In-House Publishing." *High Technology* 6, no. 4 (April 1986): 18-26.

Dean, John Ward. "The Gerrymander." *The New England Historical and Geneological Register* 46, no. 184 (October 1892): 374-83.

De Feria, Tony. Interview with author. Baltimore, 13 February 1985.

Desmond, Robert W. *The Press and World Affairs.* New York: D. Appleton-Century Company, 1937.

"Dial-Line Modems Send Complex Graphics Quickly and Reliably to Weather Forecasters." *Communications News* 22, no. 12 (December 1985): 60-61.

Dietterich, Mark. Interview with author. Watertown, N.Y., 7 August 1985.

Dill, William A. *Growth of Newspapers in the United States.* Lawrence, Kans.: Department of Journalism, University of Kansas, 1928.

Dinsdale, A. "Commercial Picture Transmission." *Wireless World* 20, no. 17 (27 April 1924): 510-16.

Dizard, Wilson P., Jr. *The Coming Information Age: An Overview of Technology, Economics, and Politics.* 2nd ed. New York: Longman, 1985.

Driver, David. Interview with author. London, 1 March 1985.

"Dry 8 x 10 Glossies Delivered by Wire." *Editor and Publisher* 97, no. 47 (21 November 1964): 48.

The Economist, 1843-1943: A Centenary Volume. London: Oxford University Press, 1943.

"An Electric Picture Telegraph." *Engineering News* 47, no. 18 (1 May 1902): 354-55.

Elson, Robert T. *Time Inc.: The Intimate History of a Publishing Enterprise, 1923-1941.* Edited by Duncan Norton-Taylor. New York: Atheneum, 1968.

Eltman, Frank. Interview with author. Oneida, N.Y., 6 December 1984.

"Enter the Business Mac—Apple in the Office." *Modern Office Technology* 30, no. 3 (March 1985): 34-41.

Erwin, Ray. "Facsimile Photo Machines Used by 300 Newspapers." *Editor and Publisher* 89, no. 30 (21 July 1956): 11, 50.

———. "Fairchild Improves Its Scan-A-Graver." *Editor and Publisher* 88, no. 24 (4 June 1955): 10, 69.

———. "N. Y. Times Developing New 'Fax' Machines." *Editor and Publisher* 83, no. 25 (17 June 1950): 18.

Evans, Harold. *Editing and Design*. Vol. 4, *Pictures on a Page: Photo-Journalism, Graphics and Picture-Editing*. London: Heinemann, 1978.

———. *Good Times, Bad Times*. New York: Atheneum, 1981.

"Facsimile." *APME Red Book* 7 (1954): 100-3.

"Facsimile Forward." *Newsweek* 31, no. 9 (1 March 1948): 46-48.

"Facsimile Goes Commercial." *Electronics* 21, no. 8 (August 1948): 97.

"FACSIMILE: Radio Threatens to Reach into Country's Mailboxes." *Newsweek* 6, no. 21 (23 November 1935): 41-42.

"Facsimile Scramble." *Business Week* no. 435 (1 January 1938): 30-31.

Fahrney, Ralph Ray. *Horace Greeley and the Tribune in the Civil War*. Cedar Rapids, Ia.: Torch Press, 1936.

Fang, I. E. *Television News*. New York: Hastings House, 1968.

———. *Television News*. 2nd ed. New York: Hastings House, 1972.

Felix, Edgar H. "Miami Herald Transmits Facsimile Newspaper." *FM and Television* 7, no. 4 (April 1947): 36-39.

Fenby, Jonathan. *The International News Services*. New York: Schocken Books, 1986.

Ferris, Kristina. "A Cross-National Comparison of Journalistic Cartography in British and American Daily Newspapers." M.A. thesis, Syracuse University, 1987.

Fidler, Roger. "Knight-Ridder Network On-Line." *Deadline Mac* 1, no. 4 (July 1986): 2.

Finberg, Howard I. "In the Beginning, Graphics Editors Were As Popular As OPEC." *Design, The Journal of the Society of Newspaper Design* no. 17 (Fall 1984): 4-6.

———. "Mapmaker, Mapmaker, Make Me a Map." *Design, The Journal of the Society of Newspaper Design* no. 15 (Spring 1984): 9-11

"First Fax." *Time* 51, no. 2 (12 January 1948): 60-61.

Fischer, Marty. Interview with author. Chicago, 14 May 1985.

Fitzgerald, Mark. "Flexo Is Passing the Test." *Editor and Publisher* 177, no. 45 (10 November 1984): 28-29.

———. "Freelance Graphics." *Editor and Publisher* 118, no. 45 (9 November 1985): 18.

———. "Photo Scanners Move into Pagination." *Editor and Publisher* 119, no. 27 (5 July 1986): 26-29.

Fitz-Gerald, W. G. "A Telephone Newspaper." *Scientific American* 96, no. 25 (22 June 1907): 507.

Floherty, John J. *Your Daily Paper*. Philadelphia: J. B. Lippincott, 1938.

Forbes, Joan. Interview with author. Boston, 19 November 1984.

Ford, Sheridan. "Illustrated Daily Journalism." *Inland Printer* 13 (July 1894): 346-47.

Friedman, Barbara J. "It Was a Rocky First Year for Viewtron." *Presstime* 6, no. 12 (December 1984): 25.

Friedrichs, Günter, and Schaff, Adam, eds. *Microelectronics and Society: For Better or For Worse.* Oxford: Pergamon Press, 1982.

Furno, Richard. Interview with author. Washington, D.C., 24 June 1985.

———. Telephone conversation with author. 31 March 1987.

GAIA Ltd. staff, and Myers, Norman. *GAIA: An Atlas of Planet Management.* New York: Doubleday, 1984.

Gaines, Brian. "Videotex—the Electronic Challenge." *Penrose Annual* 74 (1982): 47-56.

Gamble, William. "Photographic Processes of To-Day." *Penrose Annual* 1 (1895): 5-11.

"GANSAT: Another Different Voice Takes Shape." *Gannetteer* (June 1981): 4-7.

Garcia, Mario R. *Contemporary Newspaper Design: A Structural Approach.* Englewood Cliffs, N.J.: Prentice-Hall, 1981.

———. *Contemporary Newspaper Design: A Structural Approach.* 2nd ed. Englewood Cliffs, N.J.: Prentice-Hall, 1987.

———. Interview with author. Syracuse, N.Y., 10 December 1984.

Garcia, Mario R., and Price, Lynn. "The Front Page: Flexing to Fit the Big News." *Washington Journalism Review* 6, no. 2 (March 1984): 40-45.

Garelik, Glenn. "The Weather Peddlers." *Discover* 6, no. 4 (April 1985): 18-29.

Garneau, George. "Another Computer-to-Plate Test in the Works." *Editor and Publisher* 118, no. 12 (23 March 1985): 45.

———. "Color Quality Control: How It's Done at USA Today." *Editor and Publisher* 118, no. 3 (19 January 1985): 11, 22-33.

———. "Computer-to-Plate Project to End." *Editor and Publisher* 118, no. 12 (23 March 1985): 32-33.

———. "Electronic Photos for Newspapers." *Editor and Publisher* 119, no. 28 (12 July 1986): 30-32.

———. "Getting Photos Faster." *Editor and Publisher* 120, no. 23 (6 June 1987): 72, 126-27.

———. "Improving Wire Service Graphics." *Editor and Publisher* 120, no. 12 (21 March 1987): 44.

———. "Using Personal Computers to Put Out a Newspaper." *Editor and Publisher* 119, no. 11 (15 March 1986): 32.

———. "Weather Graphics Via PCs." *Editor and Publisher* 119, no. 40 (4 October 1986): 62-64.

Garratt, G. R. M. "Telegraphy." In *A History of Technology*, edited by Charles Singer and others. Vol. 4, *The Industrial Revolution*, 644-62. Oxford: Clarendon Press, 1958.

Gauthier, Magella-J., and Chamberland, Claude. "Communication graphique et magazines: les cas de Newsweek et de Time." *Revue de Carto-Quebec* 5, no. 1 (1984): 11, 21.

Gecsei, Jan. *The Architecture of Videotex Systems.* Englewood Cliffs, N.J.: Prentice-Hall, 1983.

Genovese, Margaret. "UPI: Asking for a Year." *Presstime* 8, no. 12 (December 1986): 6-9.

Geraci, Philip C. "Comparison of Graphic Design and Illustration Use in Three Washington, D.C., Newspapers." *Newspaper Research Journal* 5, no. 2 (Winter 1983): 29-39.

Gerola, Humberto, and Gomory, Ralph E. "Computers in Science and Technology." *Science* 225, no. 4657 (6 July 1984): 11-18.

Gidal, Tim Nachum. *Modern Photojournalism: Origin and Evolution.* Vol. 1 of *Photography: Men and Movements.* New York: Macmillan, 1973.

Gilmartin, Patricia. "The Design of Journalistic Maps: Purposes, Parameters, and Prospects." *Cartographica* 22, no. 4 (Winter 1985): 1-18.

———. "The Recall of Journalistic Maps and Other Graphics." Paper presented at the annual meeting of the Canadian Cartographic Association, July 1986, Burnaby, British Columbia.

Girard, Peter Henry. "A Comparison of National Atlas Programs in the United States and Canada." M.A. thesis, Syracuse University, 1986.

Given, John L. *Making a Newspaper.* New York: Henry Holt and Company, 1907.

Gloede, Bill. "UPI: Goodbye AT&T, Hello Federal?" *Editor and Publisher.* 116, no. 30 (23 July 1983): 11, 36.

Godfrey, David, and Parkhill, Douglas, eds. *Gutenberg Two: The New Electronics and Social Change.* 3rd ed., rev. Toronto: Press Porcépic Ltd., 1982.

Goltz, Gene. "Weather Pages." *Presstime* 9, no. 3 (March 1987): 16-19.

Goodacre, Clive. "Response 300: On-line Retouching." *Penrose Annual* 72 (1980): 97-100.

Goodman, Danny. "The Laser's Edge." *Macworld* 2, no. 2 (February 1985): 70-79.

Gramling, Oliver. *AP: The Story of the News.* New York: Farrar and Rinehart, 1940.

Grant, William. *Cable Television.* Reston, Va.: Reston Publishing Company, 1983.

Gray, David B. "Information Please." *Design, The Journal of the Society of Newspaper Design* no. 15 (Spring 1984): 3-8.

———. "Pagination Puzzle Still Missing Pieces." *Design, The Journal of the Society of Newspaper Design* no. 18 (Winter 1985): 7-8.

"The Gray Telautograph." *Engineering News* 29, no. 12 (23 March 1893): 271-73.

Gress, Edmund G. *The American Handbook of Printing.* 3rd ed. New York: Oswald Publishing Company, 1913.

Griffith, Elmer C. *The Rise and Development of the Gerrymander.* Chicago: Scott, Foresman and Company, 1907.

Grimshaw, Robert. "Korn's Photographic Fac-simile Telegraph." *Scientific American* 96, no. 7 (16 February 1907): 148.

Grimwade, John. Interview with author. London, 1 March 1985.

Gross, Lynne Schafer. *The New Television Technologies.* Dubuque, Ia.: William C. Brown, 1983.

Guzelimian, Vahé. *Becoming a MacArtist*. Greensboro, N.C.: Compute! Publications, 1985.

Hagedoorn, Leo. "The First Century of Photo-Engraving." *Penrose Annual* 19 (1913-1914): 189-97.

"Half of Dailies Using or Studying New Technology." *Presstime* 5, no. 4 (April 1983): 16.

Hambleton, John. "Computer Mapping and the Television Weather Report." *Proceedings*, Auto-Carto Six, Ottawa, 16-21 October 1983, 1: 469.

Hamilton, Frederick W. *A Brief History of Printing in America*. Chicago: Committee on Education, United Typothetae of America, 1918.

Harlow, Alvin F. *Old Post Bags*. New York: D. Appleton and Company, 1928.

———. *Old Wires and New Waves: The History of the Telegraph, Telephone, and Wireless*. New York: D. Appleton-Century Company, 1936.

Harrington, Harry Franklin, and Watson, Elmo Scott. *Modern Feature Writing*. New York: Harper and Brothers, 1935.

Harrington, Mark W. "History of the Weather Map." *U. S. Weather Bureau Bulletin* no. 11, pt. 2 (1894): 327-35.

Harrison, C. A. "The Fairchild Photo-Electric Engraver." *Penrose Annual* 45 (1951): 104-5.

Harrison, Richard Edes. Interview with author. New York, 6 June 1985.

Hassing, O., and Nielson, J. Oskar. "The Hassing Electro Optical Engraving Machine." *Penrose Annual* 40 (1983): 108-11.

Healey, Gerald B. "Photo Transmission Hit by Production Managers." *Editor and Publisher* 110, no. 37 (10 September 1977): 26, 50.

Heiskell, Henry L. "The Commercial Weather Map of the United States Weather Bureau." In *Yearbook of Agriculture: 1912*, U.S. Department of Agriculture, 537-39.

Herbert, David E. "Digitizing and Storing Graphics in the AP Electronic Darkroom." *Editor and Publisher* 115, no. 10 (6 March 1982): 26-29.

Hess, Stephen, and Kaplan, Milton. *The Ungentlemanly Art: A History of American Political Cartoons*. New York: Macmillan, 1968.

Heymsfield, Gerald M.; Ghosh, Koushik K.; and Chen, Lily C. "An Interactive System for Compositing Digital Radar and Satellite Data." *Journal of Climate and Applied Meteorology* 22, no. 5 (May 1983): 705-13.

Hides, Michael. "Fleet Street's New Face." *IPI Report* 15, no. 2 (June 1966): 6.

Hills, Lee, and Sullivan, Timothy J. *Facsimile*. New York: McGraw-Hill, 1949.

Hind, Arthur M. *An Introduction to the History of the Woodcut*. 2 vols. Boston: Houghton Mifflin, 1935. Reprint. New York: Dover Publications, 1963.

Hirsch, E. D., Jr. *Cultural Literacy: What Every American Needs to Know*. Boston: Houghton Mifflin, 1987.

Hislop, W. B. "Photo-engraving: A Survey of Six Decades." *Penrose Annual* 56 (1962): 112-15.

———. "The Work of Frederick E. Ives: An Appreciation." *Penrose Annual* 40 (1933): 105-7.

"History of AP Assessment Formula." *Editor and Publisher* 117, no. 12 (24 March 1984): 12.

Hobson, Harold; Knightley, Phillip; and Russell, Leonard. *The Pearl of Days: An Intimate Memoir of the Sunday Times, 1822-1972*. London: Hamilton, 1972.

Hodge, Carl. "APN's Pictorial Maps a Showcase for News." *Editor and Publisher* 81, no. 16 (10 April 1948): 50.

Hodgson, Pat. *The War Illustrators*. New York: Macmillan, 1977.

Hogan, John V. L. "Facsimile and Its Future Uses." *Annals of the American Academy of Political and Social Science* 213 (January 1941): 162-69

Hohenberg, John. *The Professional Journalist: A Guide to the Principles and Practices of the News Media*. 2nd ed. New York: Holt, Rinehart and Winston, 1969.

Holbein, Thomas J. "Trends in Newspaper Editing and Design." *Design, Journal of the Society for Newspaper Design* no. 16 (Summer 1984): 16-17.

Holmes, Nigel. *Designer's Guide to Creating Maps and Charts*. New York: Watson-Guptill Publications, 1984.

Holmes, Nigel, and DeNeve, Rose. *Designing Pictorial Symbols*. New York: Watson-Guptill Publications, 1985.

"Home Delivery by Electronics," *Problems of Journalism—Proceedings of the American Society of Newspaper Editors*, New York, 29 April - 2 May 1979, 26-30.

"Home Newspaper by Radio." *Scientific American* 158, no. 6 (June 1938): 334-35.

Horgan, Stephen Henry. *Horgan's Half-tone and Photomechanical Processes*. Chicago: Inland Printer Company, 1913.

———. "Journalism's Greatest Alliance." In *Achievement in Photo-engraving and Letterpress Printing*, edited by Louis Flader, 61-62. Chicago: American Photo-Engravers Association, 1927.

———. "The World's First Illustrated Newspaper." *Penrose Annual* 35 (1933): 23-24.

Horton, Brian. Interview with author. New York, 20 November 1984.

"How We Get Our News." *Harper's New Monthly Magazine* 34, no. 102 (March 1867): 511-22.

Huenergard, Celeste. "Milwaukee Dailies to Offer 24-hr. Electronic Newspaper." *Editor and Publisher* 115, no. 18 (1 May 1982), 63.

Hughes, Patrick. *A Century of Weather Service: A History of the Birth and Growth of the National Weather Service, 1870-1970*. New York: Gordon and Breach, 1970.

Hurley, T. F. W. "Wood-cuts and Photo-engravings." *The Month* 68, no. 308 (February 1890): 191-97.

Hurly, Paul. "The Promises and Perils of Videotex." *The Futurist* 19, no. 2 (April 1985): 7-13.

Hutt, Allen. *The Changing Newspaper: Typographic Trends in Britain and America, 1622-1972*. London: Gordon Fraser, 1973.

Hutzel, Ingeborg. "Computer Graphics in Broadcasting: A Survey of the Field." *Computer Graphics World* 8, no. 4 (April 1985): 10-24, 122-26.

Hyde, Grant Milnor. *Journalistic Writing*. 2nd ed. New York: D. Appleton-Century Company, 1929.

Hynds, Ernest C. *American Newspapers in the 1970s.* New York: Hastings House, 1975.

———. *American Newspapers in the 1980s.* New York: Hastings House, 1980.

"Improved Facsimile Systems Demonstrated for the Press." *Electrical Engineering* 65, no. 6 (June 1946): 292.

IMS/Ayer Directory of Publications. Fort Washington, Pa.: IMS Press, 1985.

Innis, Harold A. *The Bias of Communication.* Toronto: University of Toronto Press, 1951.

Jackson, John, with Chatto, W. A. *A Treatise on Wood Engraving, Historical and Practical.* 2nd ed. London: Henry G. Bown, 1861.

Jackson, Mason. *The Pictorial Press: Its Origin and Progress.* London: Hurst and Blackett, 1885.

Jaffe, Erwin. *Halftone Photography for Offset Lithography.* New York: Lithographic Technical Foundation, 1960.

Jareaux, Robin. Interview with author. Boston, 19 November 1984.

Jay, Margaret. "The *Economist* Formula: Elitism, Anonymity, 'Successful Coziness'." *Washington Journalism Review* 7, no. 4 (April 1985): 52-54.

Johnson, John, Jr. Interview with author. Watertown, N.Y., 7 August 1985.

Jones, Clarence R. *Facsimile.* New York: Murray Hill Books, 1949.

Jones, Jenk, Jr., and others. "AP Laserphoto: How it Works." In *Report of the APME Photo and Graphics Committee*, ca. 1978.

Junod, Joseph V. Interview with author. Ithaca, N.Y., 10 January 1985.

Kahn, Robert S. "Magazine Photography Begins: An Editorial Negative." *Journalism Quarterly* 42, no. 1 (Winter 1965): 53-59.

Kalish, Stanley E., and Edom, Clifton C. *Picture Editing.* New York: Rinehart and Company, 1951.

Kay, F. George. *Royal Mail: The Story of the Posts in England from the Time of Edward IVth to the Present Day.* London: Rockliff, 1951.

Keegan, Michael. Interview with author. Washington, D.C., 12 February 1985.

Kesterton, W. H. *A History of Journalism in Canada.* Toronto: McClelland and Stewart, 1967.

Key, Wilson Bryan. *Subliminal Seduction: Ad Media's Manipulation of a Not So Innocent America.* Englewood Cliffs, N.J.: Prentice-Hall, 1973.

Kidron, Michael, and Segal, Ronald. *The State of the World Atlas.* New York: Simon and Schuster, 1981.

Kindel, Stephen. "Applesauce." *Forbes* 133, no. 4 (13 February 1984): 39-41.

Klein, Stanley. "Coping with the Doldrums." *S. Klein Computer Graphics Review* 2, no. 1 (Fall 1986): 11-12.

"Knight-Ridder Launches Computer Graphics Network." *Editor and Publisher* 118, no. 36 (7 September 1985): 30.

"Knight-Ridder 'Pulls Plug' on Its Videotex Operation." *Editor and Publisher* 119, no. 13 (29 March 1986): 14.

Kobre, Sidney. *The Yellow Press and Gilded Age Journalism.* Tallahassee, Fla.: Florida State University, 1964.

Koehler, Mary A. "Facsimile Newspapers: Foolishness or Foresight?" *Journalism Quarterly* 46, no. 1 (Spring 1969): 29-36.

Koehler, S. R. "The Photo-Mechanical Processes." *Technological Quarterly and Proceedings of the Society of Arts* 5, no. 3 (October 1892): 164-204.
Kopec, Richard J. "Geography: No 'Where' in North Carolina." Department of Geography, University of North Carolina, 1984. Photocopy.
"Korn's New Telephotographic System," *Scientific American Supplement* 66, no. 1696 (4 July 1908): 14-16.
Kruger, Jack E. "Comparing News by Radio, Television, and Newspapers." In *Radio and Television News*, edited by Donald E. Brown and John Paul Jones, 13-25. New York: Rinehart and Company, 1954.
Kruglinski, Paul. "Buying Equipment." *Presstime* 7, no. 2 (February 1985): 21-27.
———. "Color in Small Dailies." *Presstime* 7, no. 11 (November 1985): 10-11.
———. "Pagination." *Presstime* 7, no. 3 (March 1985): 30-31.
"KSTP-TV Uses Weather Radar and Laserfax." *Communications News* 15, no. 7 (July 1978): 30-31.
Kubler, George A. *A New History of Stereotyping*. New York, 1941.
Lacy, Paul. "USA TODAY Artists Try Their Hand at New Computer Graphics System." *Gannetteer* (March 1985): 16-17.
Lankford, Charles. Interview with author. Baltimore, 13 February 1985.
Larratt, Richard. "Market Factors." In *The Telidon Book: Designing and Using Videotex Systems*, edited by David Godfrey and Ernest Chang, 7-80. Reston, Va.: Reston Publishing Company, 1981.
Lauden, Franklyn K. "Radio Facsimile May Print Newspapers of Tomorrow." *Radio and Television News* 40, no. 2 (August 1948): 39, 148-49.
Lebergott, Stanley. "Wage Trends, 1800-1900." In *Trends in the American Economy in the Nineteenth Century*, Conference on Research in Income and Wealth, 449-99. Princeton, N.J.: Princeton University Press, 1960.
Leckrone, Walter. "Those New Plastic Cuts." *Bulletin of the American Society of Newspaper Editors* no. 342 (1 April 1952): 14-15.
Lee, Alfred McClung. *The Daily Newspaper in America: The Evolution of a Social Instrument*. New York: Macmillan, 1937.
Leimer, Judith A. "The Influences of the Technical Constraints and Personnel Limitations on the Quantity of Maps in American Newspapers." M.S. thesis, University of Wisconsin-Madison, 1982.
Lescarboura, Austin C. "Sending Photographs Over Wires." *Scientific American* 123, no. 19 (6 November 1920): 474, 483-84.
Levy, Louis Edward. "Development and Recent Advances of the Techno-Graphic Arts." *Journal of the Franklin Institute* 180, no. 4 (October 1915): 387-408.
Library of Congress. Geography and Map Division. *Bibliography of Cartography*. 5 vols. Boston: G. K. Hall and Company, 1973. *Bibliography of Cartography*. First supplement, 2 vols. Boston: G. K. Hall and Company, 1980.
Litofsky, Jerry. Interview with author. New York, 20 November 1984.
Lockwood, Robert. Interview with author. New Tripoli, Pa., 3 June 1985.
———. Telephone conversation with author. 18 November 1986.
"London-based Firm Acquires Telecom Interest." *Editor and Publisher* 118, no. 8 (23 February 1985): 34.

"Looking at USA Today." *APME Red Book* 36 (1983): 46-56.

Lu, Cary. "Laser Printers Zap the Price Barrier." *High Technology* 4, no. 9 (September 1984): 52-57.

Lyman, Peter. *Canada's Video Revolution: Pay-TV, Home Video and Beyond.* Toronto: James Lorimer and Company, 1983.

McCain, Nina. "Newspapers vs. Computers: The Press Hedges Its Bets." *Technology Review* 83, no. 2 (November/December 1980): 72-73.

McClellan, Stephen T. *The Coming Computer Industry Shakeout.* New York: John Wiley and Sons, 1984.

McClelland, Kathy. "Sprucing Up an Old Gray Lady." *Design, Journal of the Society of Newspaper Design* no. 17 (Fall 1984): 23-26.

McClure, S. S. "Newspaper 'Syndicates'." *The Critic* 8, no. 186 (23 July 1887): 42-43.

McIntyre, Colin. "Teletext in Britain: The CEEFAX Story." In *Videotext: The Coming Revolution in Home/Office Information Systems*, edited by Efrem Sigel and others, 23-55. White Plains, N.Y.: Knowledge Industry Publications, 1980.

MacKean, Sidney H. "How a Modern News Service Operates." *The Publisher's Guide* 21, no. 6 (June 1914): 24-28.

McLuhan, Marshall. *The Gutenberg Galaxy.* Toronto: University of Toronto Press, 1962.

McMahon, Joseph E. Interview with author. Rome, N.Y., 6 December 1984.

McNay, Michael. Interview with author. London, 27 February 1985.

MacNeil, Neil. "The Presentation of the News." In *The Newspaper: Its Making and Its Meaning*, edited by the staff of the New York Times, 125-45. New York: Charles Scribner's Sons, 1945.

Madden, John C. *Videotex in Canada.* Ottawa: Canada, Ministry of Supply and Services, 1979.

"Making a Half-tone Engraving." *Scientific American*, n.s. 83, no. 10 (8 September 1900): 145, 153-54.

"A Marriage Made in Heaven?" *Editor and Publisher* 117, no. 9 (3 March 1984): 42.

Marriott, William. "The Earliest Telegraphic Daily Meteorological Reports and Weather Maps." *Quarterly Journal of the Royal Meteorological Society* 29, no. 126 (April 1903): 123-31.

Marsh, Jack. Interview with author. Utica, N.Y., 19 December 1984.

Marshall, C. F. Dendy. *The British Post Office from Its Beginnings to the End of 1925.* London: Oxford University Press, 1926.

Martin, James. *The Wired Society.* Englewood Cliffs, N.J.: Prentice-Hall, 1978.

Matthews, Albert. "The Snake Devices, 1754-1776, and the Constitutional Courant, 1765." *Publications of the Colonial Society of Massachusetts* 11 (1906-7): 409-53.

Matthews, George T., ed. *News and Rumor in Renaissance Europe.* New York: G. P. Putnam's Sons, 1959.

Maverick, Augustus. *Henry J. Raymond and the New York Press for Thirty Years.* Hartford, Conn.: A. S. Hals and Company, 1870.

Merrill, John C. *The Elite Press: Great Newspapers of the World.* New York: Pitman, 1969.

Meyer, Herbert W. *A History of Electricity and Magnetism.* Cambridge, Mass.: M.I.T. Press, 1971.

Meyer, Philip. *Precision Journalism.* Bloomington, Ind.: Indiana University Press, 1973.

Miller, Boyd L. "More Dailies Zoning for Suburban Readers." *Journalism Quarterly* 42, no. 3 (Summer 1965): 460-62.

Miller, Edward D. "Newspaper Design: Visual and Visionary." *Bulletin of the American Society of Newspaper Editors* no. 622 (March 1979): 3-5.

Miller, Edward D.; Lockwood, Robert; and Lindenmuth, Jeff. "Allentown: Designing for Each Day's News." *Newspaper Design Notebook* 1, no. 2 (March/April 1979): 1, 13-15.

Miller, Tim. "Times Mirror Videotex Under Way." *Editor and Publisher* 118, no. 10 (9 March 1985): 27-28.

———. "Videotex Market Shaken Up by PC Boom." *Editor and Publisher* 118, no. 10 (9 March 1985): 26-27.

Mintz, Louis. "Graphics Networks." *Presstime* 8, no. 9 (September 1986): 12-14.

Mitchell, Curtis. *Cavalcade of Broadcasting.* New York: Rutledge Books, 1970.

"Modern Newspaper Illustration." *The Newspaper Maker* 12, no. 314 (28 March 1901): 5.

Moen, Daryl R. *Newspaper Layout and Design.* Ames, Ia.: Iowa State University Press, 1984.

Monmonier, Mark. "A Geographer's View of Newspaper Maps and a Cartographer's Guide." *Design, The Journal of the Society of Newspaper Design* no. 26 (1987), 14-17.

———. "The Geography of Change in the Newspaper Industry of the Northeast United States, 1940-1980." *Proceedings of the Pennsylvania Academy of Science* 60, no. 1 (1986): 55-59.

———. "The Hopeless Pursuit of Purification in Cartographic Communication: A Comparison of Graphic-Arts and Perceptual Distortions of Graytone Symbols." *Cartographica* 17, no. 1 (1980): 24-39.

———. "Map-Text Coordination in Geographic Writing." *The Professional Geographer* 33, no. 4 (November 1981): 406-12.

———. "The Rise of Map Use By Elite Newspapers in England, Canada, and the United States." *Imago Mundi* 38 (1986): 46-60.

———. *Technological Transition in Cartography.* Madison, Wis.: University of Wisconsin Press, 1985.

Monmonier, Mark, and Pipps, Val. "Weather Maps and Newspaper Design: Response to *USA Today*?" *Newspaper Research Journal* 8, no. 4 (Summer 1987): 31-42.

Moran, James. *Printing Presses: History and Development from the Fifteenth Century to Modern Times.* Berkeley and Los Angeles: University of California Press, 1973.

Moran, Tom, and Foster, Ed. "IBM Models Offer Diverse Capabilities." *Info World* 9, no. 14 (6 April 1987): 3, 80.

Mott, Frank Luther. *The News In America.* Cambridge, Mass.: Harvard University Press, 1962.

Mullins, Marcy Eckroth, and Albert, Buzzy. "USA TODAY: How to Draw Maps." *Deadline Mac* 1, no. 2 (May 1986): 1-2.

Munsell, Joel. *Chronology of the Origin and Progress of Paper-Making.* 5th ed. Albany, N.Y.: Munsell, 1876.

Murrell, William. *A History of American Graphic Humor. Vol. 1, 1747-1865.* New York: Whitney Museum of American Art, 1933.

Nahon, George. "Videotex Proving to Be Successful in France." *Editor and Publisher* 119, no. 25 (21 June 1986): 40, 122, 124.

Naisbitt, John. *Megatrends: Ten New Directions Transforming Our Lives.* New York: Warner Books, 1982.

National Council for Geographic Education and the Association of American Geographers. Joint Committee on Geographic Education. *Guidelines for Geographic Education: Elementary and Secondary Schools.* Washington, D.C., 1984.

National Oceanic and Atmospheric Administration, Environmental Data and Information Service, National Climatic Center. *Environmental Satellite Data and Information.* Environmental Information Summaries C-16. Asheville, N.C., 1980.

Natkiel, Richard. *Decisive Battles of the Twentieth Century: Land-Sea-Air.* London: Sidgwick and Jackson, 1976.

———. Interview with author. London, 26 February 1985.

"Network Daily Paper." *Business Week* no. 495 (25 February 1939): 35-36.

Neustadt, Richard M. *The Birth of Electronic Publishing: Legal and Economic Issues in Telephone, Cable and Over-the-Air Teletext and Videotext.* White Plains, N.Y.: Knowledge Industry Publications, 1982.

———. "The Regulation of Electronic Publishing." *Federal Communications Law Journal* 33, no. 3 (Summer 1981): 331-417.

Newman, Peter C. "Each Venture Is a New Beginning, a Raid on the Inarticulate." *Maclean's* 91, no. 19 (18 September 1978): 2.

———. "Maclean's at 75: Still Keeping Shared Dreams Alive." *Maclean's* 93, no. 39 (29 September 1980): 3.

"News Pictures by Wire." *Electronics* 10, no. 11 (November 1937): 12-17, 82-83.

"A Newspaper by Radio: The St. Louis P-D Pioneers with Facsimile Process." *Newsweek* 12, no. 25 (19 December 1938): 28-29.

"Newspaper Designers, Editors Give Birth to a New Organization." *Newspaper Design Notebook* 1, no. 2 (March/April 1979): 10-11.

"Newspaper Illustrations." *Penrose Annual* 3 (1897): 17-32.

"Newspaper Printing." *Penrose Annual* 40 (1938): 166-67.

"Newspapers via Air Waves." *Inland Printer* 103, no. 6 (September 1939): 88-89.

"Newspictures Ignore Profit and Loss." *Fortune* 11, no. 3 (September 1930): 60-67, 107-8.

Nielson, Erfert. "Art to Go." *Macworld* 3, no. 12 (December 1986): 130-37.

Norman, Jim. "How USA Today Does It." *Design, Journal of the Society of Newspaper Design* no. 12 (Summer 1983): 4-6.

North, Anthony. "No, But I Saw the Pictures." *New Outlook* 163, no. 6 (June 1934): 17-21.

North, S. N. D. *History and Present Condition of the Newspaper and Periodical Press of the United States.* U. S. Department of the Interior, Census Office, 10th Census, 1880. Washington: Government Printing Office, 1884.

O'Brien, Frank M. *The Story of the Sun.* New York: D. Appleton and Company, 1928. Reprint. New York: Greenwood Press, 1968.

Olson, Joel. "Image Processing in Weather Forecasting." *Computer Graphics World* 6, no. 11 (November 1983): 41-54.

Olson, Kenneth E. *Typography and Mechanics of the Newspaper.* New York: D. Appleton and Company, 1930.

Ostendorf, Bill. "Graphics on a Budget." *Design, The Journal of the Society of Newspaper Design* no. 16 (Summer 1984): 26.

———. "Who Are We?" *Design, The Journal of the Society of Newspaper Design* no. 18 (Winter 1985): 12-13.

Ostrolenk, Samuel. "Home Facsimile Recording." *Electronics* 11, no. 1 (January 1938): 26-27, 60.

Packard, Vance. *The Hidden Persuaders.* New York: David McKay, 1957.

Park, Robert E. "The Natural History of the Newspaper." In *The City*, edited by Robert E. Park, Ernest W. Burgess, and Roderick D. McKenzie, 80-98. Chicago: University of Chicago Press, 1925.

Parker, Laurie. "Meeting the Needs of Changing Times." *Design, Journal of the Society of Newspaper Design* no. 8 (December 1981): 5-6.

Parton, James. *Caricature and Other Comic Art.* New York: Harper and Brothers, 1877.

Pearson, Karen Severud. "Mechanization and the Area Symbol: Techniques in the 19th Century Geographical Journals." *Cartographica* 20, no. 4 (1983): 1-34.

Perry, John L. "Technology Is Not Journalism." *Quill* 71, no. 4 (April 1983): 26-29.

Perry, John W. "A.P. Members Approve Telephoto Service After Heated Debate in Convention." *Editor and Publisher* 66, no. 50 (28 April 1934): 5, 110.

———. "Wirephoto Found Flexible and Speedy; Operates with Uncanny Precision." *Editor and Publisher* 67, no. 35 (12 January 1935): III, XI.

"Personal Computers: And the Winner is IBM." *Business Week* no. 2810 (3 October 1983): 76-95.

Perugi, Deborah. Interview with author. Boston, 19 November 1984.

Peterson, Theodore. *Magazines in the Twentieth Century.* Urbana, Ill.: University of Illinois Press, 1964.

Petterssen, Sverre. *Introduction to Meteorology.* 3rd ed. New York: McGraw-Hill, 1969.

Pew, Marlen. "Television and Tomorrow's Newspaper." *Pulp and Paper Magazine of Canada* 37, no. 7 (July 1936): 438-39.

"Photography Applied to Engraving on Wood." *The Art Journal*, n.s. 6 (1854): 224.

"Photos Edited Like News Copy by Use of Fax." *Editor and Publisher* 87, no. 12 (13 March 1954): 56.

Picketts, Jack. Interview with author. Toronto, 7 November 1984.

"Pictures by Wire." *The Newspaper Maker* 9, no. 214 (27 April 1899): 5.

Plews, F. W. "The Electrical Transmission of Colour Photographs for Newspaper Printing." *Penrose Annual* 40 (1938): 166-67.

Popkin, Roy. *The Environmental Science Services Administration.* New York: Frederick A. Praeger, 1967.

Porter, William C. "Eventually—Pagination Saves Time." *Editor and Publisher* 119, no. 25 (21 June 1986): 32, 123.

Pred, Allan R. "Urban Systems Development and the Long-Distance Flow of Information through Preelectronic U.S. Newspapers." *Economic Geography* 47, no. 4 (October 1971): 498-524.

Press, Charles. *The Political Cartoon.* East Brunswick, N.J.: Associated University Presses, 1981.

"Public Access Videotex." *Editor and Publisher* 118, no. 12 (23 March 1985): 40-42.

Pugsley, Peter. "A New High Speed Digital Colour Scanner." *Penrose Annual* 69 (1976): 79-86.

"Radio Facsimile in the Clear." *Business Week* no. 923 (10 May 1947): 42-44.

"Radio Newspapers Found Popular; Future Unpredictable." *Bulletin of the American Society of Newspaper Editors* no. 188 (1 June 1939): 2.

Raggett, Michael. "Broadcast Telesoftware." *Computer Graphics World* 6, no. 9 (September 1983): 49-52.

Rambo, C. David. "It's Still a 'Maybe' Market for New Technologies." *Presstime* 5, no. 1 (January 1983): 20-22.

——. "Newspaper Companies Go Back to Basics." *Presstime* 9, no. 1 (January 1987): 22-30.

Read, James. Interview with author. Auburn, N.Y., 30 April 1985.

Reed, David M. "Steps One Company Took to Start Cable Service." *Presstime* 4, no. 11 (November 1982): 16-17.

Rees, Nancy. Interview with author. Chicago, 14 May 1985.

Reynolds, F. W. "A New Telautograph System." *Electrical Engineering* 55, no. 9 (September 1936): 996-1007.

Rhind, David. "Cartography: Metropolitan Canada in Maps." *Geographical Magazine* 49, no. 10 (July 1977): 664-65.

Ristow, Walter W. "Journalistic Cartography." *Surveying and Mapping* 17, no. 4 (October 1957): 369-90.

Ritzenberg, Phillip. "Backshop Blues Got You Down?" *Design, The Journal of the Society of Newspaper Design* no. 12 (Summer 1983): 30-31.

———. "The Coming Effects of Technology." *Design, The Journal of the Society of Newspaper Design* no. 3 (September 1980): 8-9.

Robinson, Arthur H. *Early Thematic Mapping in the History of Cartography.* Chicago: University of Chicago Press, 1982.

———. "Mapmaking and Map Printing: The Evolution of a Working Relationship." In *Five Centuries of Map Printing*, edited by David Woodward, 1-23. Chicago: University of Chicago Press, 1975.

Robinson, Arthur H., and others. *Elements of Cartography.* 5th ed. New York: John Wiley and Sons, 1984.

Robinson, John P. "Daily News Habits of the American Public." *ANPA News Research Report* no. 15 (22 September 1978): 5.
Robinson, Phillip. "The Macintosh Plus." *Byte* 11, no. 6 (June 1986): 85-90.
Robinson, Richard. *The Video Primer*. 3rd ed. New York: Perigee Books, 1983.
Rogers, David F. *Procedural Elements for Computer Graphics*. New York: McGraw-Hill, 1985.
Roizen, Joseph. "Videotext in Other Countries." In *Videotext: The Coming Revolution in Home/Office Information Retrieval*, edited by Efrem Sigel and others, 23-55. White Plains, N.Y.: Knowledge Industry Publications, 1980.
Rosenberg, Arnie. "Zoned Editions." *Editor and Publisher* 118, no. 17 (27 April 1985): 17, 20.
Rosenfeld, Arnold. "What the Hell Is Going On in New Technology?" *Bulletin of the American Society of Newspaper Editors* no. 616 (July/August 1978): 6-8.
Rosewater, Victor. *History of Cooperative News-Gathering in the United States*. New York: D. Appleton and Company, 1930.
Rosner, Roy D. *Distributed Telecommunications Networks Via Satellites and Packet Switching*. Belmont, Cal.: Wadsworth, 1982.
Ross, W. Curtis. "More Pictures at Less Cost for this Small Town Paper." *Bulletin of the American Society of Newspaper Editors* no. 370 (1 November 1954): 10.
"The Rotogravure Quick-Printing Process and Its Possibilities." *Scientific American Supplement* 75, no. 1946 (19 April 1913): 248-49.
Ryan, Joseph A. "Using the Home Computer As a Facsimile Receiver." *Weatherwise* 36, no. 6 (December 1983): 308-10.
Sabbatini, Andrew. (Comments on setting up a map department.) *APME Red Book* 31 (1978): 168-169.
———. Interview with author. New York, 1 October 1984.
Scheele, Carl H. *A Short History of the Mail Service*. Washington, D.C.: Smithsonian Institution Press, 1970.
Schlender, Brenton R. "Apple to Unveil Two Macintosh Models to Step Up Role in Business Computers." *Wall Street Journal*, 2 March 1987, 4.
Schraubstadter, Carl, Jr. "Newspaper Illustration. No. 2." *Inland Printer* 5, no. 3 (December 1887): 173-74.
Schuneman, R. Smith. "Art or Photography: A Question for Newspaper Editors of the 1890s." *Journalism Quarterly* 42, no. 1 (Winter 1965): 43-52.
———. *Photographic Communication*. New York: Hastings House, 1972.
Schwarzlose, Richard A. "Early Telegraphic News Dispatches: Forerunner of the AP." *Journalism Quarterly* 51, no. 4 (Winter 1974): 596-601.
———. "Harbor News Association: The Formal Origin of the AP." *Journalism Quarterly* 45, no. 2 (Summer 1968): 253-60.
Scopin, Joseph. Interview with author. Washington, D.C., 12 February 1985.
Scott, R. H. "Weather Charts in Newspapers." *Journal of the Society of Arts* 23, no. 1183 (23 July 1875): 776-82.
"Seal of Approval Program for Radio and Television Weathercasting." *Bulletin of the American Meteorological Society* 67, no. 8 (August 1986): 1028-35.

Segalowitz, Sid J. *Two Sides of the Brain: Brain Lateralization Explored.* Englewood Cliffs, N.J.: Prentice-Hall, 1983.

Settel, Irving. *A Pictorial History of Radio.* New York: Grosset and Dunlap, 1967.

Sexton, William C. "The Explosion in Graphics." *Bulletin of the American Society of Newspaper Editors* no. 547 (January 1971): 16-22.

Shapiro, Howard S. "Giving a Graphic Example: The Increased Use of Charts and Maps." *Nieman Reports* 36, no. 1 (Spring 1982): 4-7.

Sharzer, Shirley. Interview with author. Toronto, 7 November 1985.

Shelley, Jack. "Weather News." In *Radio and Television News,* edited by Donald E. Brown and John Paul Jones, 151-68. New York: Rinehart and Company, 1954.

Shuman, Edwin L. *Practical Journalism.* New York: D. Appleton and Company, 1903.

Sieghart, Paul, ed. *Microchips with Everything: The Consequences of Information Technology.* London: Comedia Publishing Group, 1982.

Sigel, Efrem. "Videotext in the United States." In *Videotext: The Coming Revolution in Home/Office Information Retrieval,* edited by Efrem Sigel and others, 87-111. White Plains, N.Y.: Knowledge Industry Publications, 1980.

Silverstone, Stuart. "Graphics Hits the Network News Beat." *Computer Graphics World* 10, no. 2 (February 1987): 35-38.

———. "Newspapers Turn to Mac's Graphics." *Computer Graphics World* 10, no. 5 (May 1987): 81-86.

———. "Newsroom Graphics." *Macworld* 4, no. 2 (February 1987): 130-35.

Silverstone, Stuart, and Webb, Craig L. "Many Newspapers, Small and Large, Turn to 'The Mac'." *Presstime* 8, no. 4 (April 1986): 23-26.

Simon, Herbert A. "Designing Organizations for an Information Rich World." In *Models of Bounded Rationality,* 171-84. Cambridge, Mass.: M.I.T. Press, 1982.

Singleton, Loy A. *Telecommunications in the Information Age.* Cambridge, Mass.: Ballinger Publishing Company, 1983.

Smillie, Thomas W. "Photographing on Wood for Engraving." *Smithsonian Miscellaneous Collections* 47 (1905): 497-99.

Smith, Anthony. *Goodbye Gutenberg: The Newspaper Revolution of the 1980s.* New York: Oxford University Press, 1980.

———. *The Newspaper: An International History.* London: Thames and Hudson, 1979.

Smith, R. H. "All the 'Firsts' of the *Illustrated London News.*" *Penrose Annual* 67 (1974): 101-12.

Society of Newspaper Design 1986 Membership Directory. Reston, Va., 1986.

"Society of Newspaper Designers on the Drawing Board." *Newspaper Design Notebook* 1, no. 1 (January/February 1979): 3.

Solarzano, Lucia. "Why Johnny Can't Read Maps, Either." *U.S. News and World Report* 98, no. 11 (25 March 1985): 50.

Sorrells, John H. *The Working Press: Memos from the Editor about the Front and Other Pages.* New York: Ronald Press, 1930.

Springer, Sally P., and Deutsch, Georg. *Left Brain, Right Brain.* San Francisco: W. H. Freeman, 1981.

Stapler, Harry. "Attitudes on Design." *Design, The Journal of the Society of Newspaper Design* no. 3 (September 1980): 17-18.

Steinberg, S. H. *Five Hundred Years of Printing.* 2nd ed., rev. Harmondsworth, Middlesex: Penguin Books, 1966.

Stephenson, Richard W. "Maps for the General Public: Commercial Cartography of the American Civil War." Paper presented at the Eleventh International Conference on the History of Cartography, Ottawa, 8-12 July 1985.

Stevenson, Richard W. "Videotex Players Seek a Workable Formula." *New York Times*, 25 March 1986, D1, D21.

Stewart, Duncan. Interview with author. London, 1 March 1985.

Stewart, E. R. "Telegraphing Pictures to the 'Daily Mirror'." *Penrose Annual* 13 (1907-8): 147-48.

Still, Alfred. *Communications Through the Ages: From Sign Language to Television.* New York: McGraw-Hill, 1946.

Stock, Rodney. "Introduction to Digital Computer Graphics for Video." *Society of Motion Picture and Television Engineers Journal* 90, no. 12 (December 1981): 1184-89.

Stoddard, Henry Luther. *Horace Greeley: Printer, Editor, Crusader.* New York: G. P. Putnam's Sons, 1946.

Stone, Gerald C.; Schweitzer, John C.; and Weaver, David H. "Adoption of Modern Newspaper Design." *Journalism Quarterly* 55, no. 4 (Winter 1978): 761-66.

Stone, M. David. "Has IBM Finally Done It?" *Science Digest* 93, no. 2 (February 1985): 78-79.

Stonecipher, Harry W.; Nicholls, Edward C.; and Anderson, Douglas A. *Electronic Age News Editing.* Chicago: Nelson Hall, 1981.

Strauss, Victor. *The Printing Industry.* Washington, D.C.: Printing Industries of America; New York: R. R. Bowker Company, 1967.

Streich, Joseph L. "Increasing Graphics Productivity at WABC-TV." *Computer Pictures* 4, no. 4 (July/August 1986): 60-68.

Strothers, Jennifer, and Dietrich, Mark. "A Perspective on NAPLPS and Its Impact on the Development of Videotex in North America." *Videodisc and Optical Disk* 5, no. 1 (January-February 1985): 52-61.

Sutton, Albert A. *Design and Makeup of the Newspaper.* New York: Prentice-Hall, 1948.

Swanberg, W. A. *Pulitzer.* New York: Charles Scribner's Sons, 1967.

Swanson, Rod. Telephone conversation with author, 31 March 1987.

Taft, William H. *American Magazines for the 1980s.* New York: Hastings House, 1982.

Talese, Gay. *The Kingdom and the Power.* New York: World Publishing Company, 1969.

Talman, C. F. "The Elements of Forecasting." *New York Times*, 12 August 1934, section 8, p. 2.

Tatarian, Roger. "UPI's Woes: A Historical Perspective." *Editor and Publisher* 118, no. 36 (7 September 1985): 52, 42, 31.

Taylor, Craig. Interview with author. Ottawa, 5 November 1984.

Taylor, D. R. F. "The Cartographic Potential of Telidon." *Cartographica* 19, nos. 3 and 4 (Autumn and Winter 1982): 18-30.

Taylor, Edward F. "Telling the Weather Story." *Weatherwise* 36, no. 2 (April 1983): 52-59.

Tebbel, John. *The American Magazine: A Compact History*. New York: Hawthorne Books, 1969.

"Technical Progress." *APME Red Book* 11 (1958): 40-44.

Teel, Leonard Ray. "The Weather Channel." *Weatherwise* 35, no. 4 (August 1982): 156-63.

"The Telautograph." *Nature* 64, no. 1648 (30 May 1901): 107-9.

"Telephoning Our Press Photographs." *Scientific American* 131, no. 2 (April 1924): 87, 139.

Teskey, Frank. Interview with author. Toronto, 7 November 1985.

Thackray, Arnold, and Merton, Robert K. "On Discipline Building: the Paradoxes of George Sarton." *Isis* 63, no. 219 (December 1972): 473-95.

Thomas, Isaiah. *The History of Printing in America*. 1810. Reprint. New York: Weathervane Books, 1975.

Thompson, Robert Luther. *Wiring a Continent: The History of the Telegraph Industry in the United States, 1832-1866*. Princeton, N.J.: Princeton University Press, 1947.

Thrower, Norman J. W. "Edmund Halley As a Thematic Geo-Cartographer." *Annals of the Association of American Geographers* 59, no. 4 (December 1969): 652-76.

The Times (London). *"The Thunderer" in the Making, 1785-1841*. Vol. 1 of *The History of The Times*. London: Office of The Times, 1935.

———. *The Tradition is Established, 1841-1885*. Vol. 2 of *The History of The Times*. London: Office of The Times, 1939.

"The 'Times' Weather Chart." *Nature* 11, no. 285 (15 April 1875): 473-74.

Tobin, Nancy. *Understanding Changes in the Growth and Shape of Newspaper Art Departments*. Reston, Va.: Society of Newspaper Design, 1985.

Tobin, Richard L. "Weather or Not." *Saturday Review* 49, no. 46 (12 November 1966): 95-96.

Toffler, Alvin. *The New Wave*. New York: William Morrow and Company, 1980.

Toong, Hoo-Min D., and Gupta, Amar. "The Computer Age Gets Personal." *Technology Review* 86, no. 1 (January 1983): 26-27.

"The Transmission of Photographs Over Telephone Wires." *Scientific American* 106, no. 22 (1 June 1912): 493-94.

Trent, William Peterfield, and others. *The Cambridge History of American Literature*. New York: G. P. Putnam's Sons, 1921.

"Troy Member Sees Need for Morgue of Maps." *Bulletin of the American Society of Newspaper Editors* no. 133 (1 November 1936): 2.

Tucker, D. G. "Electrical Communication." In *A History of Technology*, edited by Trevor I. Williams. Vol. 7, *The Twentieth Century, c. 1900 to c. 1950*, part 2, 1220-80. Oxford: Clarendon Press, 1978.

Tufte, Edward. *The Visual Display of Quantitative Information.* Cheshire, Conn.: Graphics Press, 1983.

Turner, Eugene. "Review of *MacChoro* Version 1.2." *American Cartographer* 14, no. 1 (January 1987): 69-71.

"TV Radar Attracts Buyers." *Electronics* 31, no. 48 (28 November 1958): 17.

"TV Tests Reflect Pervasive Geographic Ignorance." *Update*, National Geographic Society, Educational Media Division, no. 7 (Spring 1987): 2.

"2 Wire Services Comment on Solar Noise Interference to Satellites." *Editor and Publisher* 115, no. 15 (10 April 1982): 28.

Ulrich's International Periodicals Directory. 24th ed. New York: R. R. Bowker, 1985.

"Ultra-Fast Ultrafax." *Newsweek* 32, no. 18 (1 November 1948): 50.

Ungaro, Joseph M. "The Electronic Newsroom." *Bulletin of the American Society of Newspaper Editors* no. 568 (April 1973): 9-11.

"Unifax II Ranks As One of Best New Products." *Editor and Publisher* 108, no. 40 (4 October 1975): 13.

United Nations Educational, Scientific and Cultural Organization (UNESCO). *News Agencies: Their Structure and Operation.* New York: Greenwood Press, 1969.

"UPI Adds New 'Special Sections Packages'." *Editor and Publisher* 117, no. 42 (20 October 1984): 26.

"UPI Chairman Seeks Newspapers' Understanding." *Editor and Publisher* 118, no. 19 (11 May 1985): 34.

"UPI Distributes Accu-Weather." *Editor and Publisher* 118, no. 3 (19 January 1985): 9.

"UPI Expands Graphics Dept." *Editor and Publisher* 117, no. 33 (18 August 1984): 10, 35.

"UPI Photo Transmission All Automatic." *Editor and Publisher* 104, no. 6 (6 February 1971): 13.

"UPI Wants to Void Guild Contract." *Editor and Publisher* 118, no. 30 (27 July 1985): 14.

Urrows, Henry, and Urrows, Elizabeth. "Early Viewtron." *Videodisc and Optical Disk* 4, no. 4 (July-August 1984): 269-300.

U.S. Bureau of the Census. *Historical Statistics of the United States, Colonial Times to 1970*. Washington, D.C.: U.S. Government Printing Office, 1975.

———. *Statistical Abstract of the United States: 1986.* 106th ed. Washington, D.C., 1985.

———. *Statistical Abstract of the United States: 1987.* 107th ed. Washington, D.C., 1986.

U.S. Weather Bureau. *Report of the Chief of the Weather Bureau, 1911-1912.* Washington, D.C.: Government Printing Office, 1913.

Veith, Richard H. *Television's Teletext.* New York: North-Holland, 1983.

"Videotex-Cable." *APME Red Book* 36 (1983): 73-75.

"Viewtron Will Lower Its Rates." *Editor and Publisher* 118, no. 10 (9 March 1985): 26.

Walker, Jerry. "History of Facsimile Experiment Recorded." *Editor and Publisher* 82, no. 45 (29 October 1949): 42.

———. "New York Times Starts Fax But Adds to Presses." *Editor and Publisher* 81, no. 9 (21 February 1948): 56.

Walker, Tom. "Pagination Breakthrough May Finally Be At Hand." *Presstime* 4, no. 12 (December 1982): 38-40.

Ward, Robert De Courcy. "The Newspaper Weather Maps of the United States." *American Meteorological Journal* 11, no. 3 (July 1894): 96-107.

"Washington Post Sets Modest Circulation Goal for Weekly." *Editor and Publisher* 116, no. 18 (30 April 1983): 45.

Watkins, Robert. "Computer Graphics in Broadcasting." *Computer Graphics World* 6, no. 6 (June 1983): 65-70.

Watson, Elmo Scott. *History of Auxiliary Newspaper Service in the United States.* Champaign, Ill.: Illini Publishing Company, 1923.

———. *A History of Newspaper Syndicates in the United States, 1865-1935.* Chicago, 1936.

"Weather Graphics Now Seen in Markets Big and Small." *Broadcasting* 108, no. 16 (22 April 1985): 97.

"Weather Info a Breeze for Radio/TV Stations." *Communications News* 17, no. 4 (April 1980): 62.

"Weathercasters Shine News in Survey." *Television/Radio Age* 33, no. 5 (16 September 1985): 48, 91.

Weaver, David H. *Videotex Journalism.* Hillsdale, N.J.: Lawrence Erlbaum Associates, 1983.

Webb, Craig L. "Syndicated and Computer Graphics." *Presstime* 7, no. 8 (August 1985): 26-28.

Weeks, Lyman Horace. *A History of Paper-Manufacturing in the United States, 1690-1916.* New York: Lockwood Trade Journal Company, 1916.

Weitenkampf, Frank. *Political Caricature in the United States.* New York: The New York Public Library, 1953.

Wendt, Lloyd. *The Wall Street Journal.* Chicago: Rand McNally, 1982.

Westley, Bruce H. *News Editing.* 3rd ed. Boston: Houghton Mifflin, 1980.

Whipple, Leon. *How to Understand Current Events: A Guide to an Appraisal of the News.* New York: Harper and Brothers, 1941.

Wicklein, John. *Electronic Nightmare.* New York: Viking Press, 1981.

Willens, Doris. "Reuters Celebrates Its 100th Anniversary, London, July 11." *Editor and Publisher* 84, no. 24 (9 July 1951): 9, 75-76.

Williams, Trevor I., ed. *The Twentieth Century, c. 1900 to c. 1950.* Vol. 7, part 2 of *A History of Technology.* Oxford: Clarendon Press, 1978.

Williamson, Lenora. "Management and Staff Must Care." *Editor and Publisher* 119, no. 23 (7 June 1986): 18-19, 41.

Willings Press Guide, 1985. 111th ed. West Sussex: Thomas Skinner Directories, 1985.

Wilson, Harold S. *Magazine and the Muckrakers.* Princeton, N.J.: Princeton University Press, 1970.

Wilson, Harold W. "Electronics and Plastics in Photo-Journalism." *Journalism Quarterly* 29, no. 3 (Summer 1952): 316-19.

"Wirephoto Glossies from New Receiver." *Editor and Publisher* 98, no. 14 (3 April 1965): 67.

Wong, Ray. "The Clarion Ledger." *Design, The Journal of the Society of Newspaper Design* no. 7 (September 1981): 13-14.

Wood, Denis, and Fels, John. "Designs on Signs: Myth and Meaning in Maps." *Cartographica* 23, no. 3 (Autumn 1986): 54-103.

Wood, James Playsted. *Magazines in the United States.* New York: Ronald Press, 1971.

Woodward, David. "The Decline of Commercial Wood-Engraving in Nineteenth-century America." *Journal of the Printing Historical Society* no. 10 (1974-75): 56-83.

———. "Maps, Music, and the Printer: Graphic or Typographic?" *Printing History* 8, no. 2 (1986): 3-14.

———. "The Study of the History of Cartography." *American Cartographer* 1, no. 2 (October 1974): 101-15.

———. "The Woodcut Map." In *Five Centuries of Map Printing*, edited by David Woodward, 25-50. Chicago: University of Chicago Press, 1975.

———, ed. *Art and Cartography: Six Historical Essays.* Chicago: University of Chicago Press, 1987.

Wright, Ian. Interview with author. London, 27 February 1985.

Wye, Brad. Interview with author. Washington, D.C., 12 February 1985.

Yasaki, Ed. "Big Mac Attack." *Datamation* 30, no. 2 (February 1984): 61-64.

Young, James C. "Is the Radio Newspaper Next?" *Radio Broadcast* 7, no. 5 (September 1925): 575-80.

Zettl, Herbert. *Television Production Handbook.* 3rd ed. Belmont, Cal.: Wadsworth Publishing Company.

Index